Acid Rain and Environmental Degradation

NEW HORIZONS IN ENVIRONMENTAL ECONOMICS

General Editor: Wallace E. Oates, *Professor of Economics, University of Maryland*

This important new series is designed to make a significant contribution to the development of the principles and practices of environmental economics. It will include both theoretical and empirical work. International in scope, it will address issues of current and future concern in both East and West and in developed and developing countries.

The main purpose of the series is to create a forum for the publication of high quality work and to show how economic analysis can make a contribution to understanding and resolving the environmental problems confronting the world in the late 20th century.

Titles in the series include:

Principles of Environmental and Resource Economics
A Guide for Students and Decision-Makers
Edited by Henk Folmer, H. Landis Gabel and Hans Opschoor

The Contingent Valuation of Environmental Resources
Methodological Issues and Research Needs
Edited by David J. Bjornstad and James R. Kahn

Environmental Policy with Political and Economic Integration
The European Union and the United States
Edited by John B. Braden, Henk Folmer and Thomas S. Ulen

Energy, Environment and the Economy
Asian Perspectives
Paul R. Kleindorfer, Howard C. Kunreuther and David S. Hong

Estimating Economic Values for Nature
Methods for Non-Market Valuation
V. Kerry Smith

Models of Sustainable Development
Edited by Sylvie Faucheux, David Pearce and John Proops

Contingent Valuation and Endangered Species
Methodological Issues and Applications
Kristin M. Jakobsson and Andrew K. Dragun

Acid Rain and Environmental Degradation
The Economics of Emission Trading
Ger Klaassen

The Economics of Pollution Control in the Asia Pacific
Robert Mendelsohn and Daigee Shaw

Economic Policy for the Environment and Natural Resources
Techniques for the Management and Control of Pollution
Edited by Anastasios Xepapadeas

Welfare Measurement, Sustainability and Green National Accounting
A Growth Theoretical Approach
Thomas Aronsson, Per-Olov Johansson and Karl-Gustaf Löfgren

Acid Rain and Environmental Degradation
The Economics of Emission Trading

Ger Klaassen
International Institute for Applied Systems Analysis,
Laxenburg, Austria

NEW HORIZONS IN ENVIRONMENTAL ECONOMICS

Edward Elgar
Cheltenham, UK • Brookfield, US

IIASA
International Institute for Applied Systems Analysis
Laxenburg, Austria

© International Institute for Applied Systems Analysis 1996

All rights reserved. No part of this publication may be reproduced, stored in a retrieval system or transmitted in any form or by any means, electronic, mechanical or photocopying, recording, or otherwise without the prior permission of the publisher.

Published by
Edward Elgar Publishing Limited
8 Lansdown Place
Cheltenham
Glos GL50 2HU
UK

Edward Elgar Publishing Company
Old Post Road
Brookfield
Vermont 05036
US

Typeset by
Publications Department
IIASA
A-2361 Laxenburg
Austria

A catalogue record for this book
is available from the British Library

Library of Congress Cataloguing in Publication Data
Klaassen. Ger, 1958–
  Acid rain and environmental degradation : the economics of
emission trading / Ger Klaassen.
  (New horizons in environmental economics)
  Includes bibliographical references and index.
  1. Air—Pollution—Economic aspects. 2. Acid rain—Economic
aspects. 3. Emissions trading. I. Title. II. Series.
  HC79.A4K58   1996
  333.9'2—dc20   96–16620
                                                            CIP

ISBN 1 85898 489 0

Printed and bound in Great Britain by
Hartnolls Limited, Bodmin, Cornwall

The International Institute for Applied Systems Analysis

is an interdisciplinary, nongovernmental research institution founded in 1972 by leading scientific organizations in 12 countries. Situated near Vienna, in the center of Europe, IIASA has been for more than two decades producing valuable scientific research on economic, technological, and environmental issues.

IIASA was one of the first international institutes to systematically study global issues of environment, technology, and development. IIASA's Governing Council states that the Institute's goal is: *to conduct international and interdisciplinary scientific studies to provide timely and relevant information and options, addressing critical issues of global environmental, economic, and social change, for the benefit of the public, the scientific community, and national and international institutions.* Research is organized around three central themes:

- Global Environmental Change;
- Global Economic and Technological Change;
- Systems Methods for the Analysis of Global Issues.

The Institute now has national member organizations in the following countries:

Austria
The Austrian Academy of Sciences

Bulgaria
The National Committee for Applied Systems Analysis and Management

Canada
The Canadian Committee for IIASA

Czech Republic
The Czech Committee for IIASA

Finland
The Finnish Committee for IIASA

Germany
The Association for the Advancement of IIASA

Hungary
The Hungarian Committee for Applied Systems Analysis

Italy
The Italian Committee for IIASA

Japan
The Japan Committee for IIASA

Kazakstan
The National Academy of Sciences

Netherlands
The Netherlands Organization for Scientific Research (NWO)

Poland
The Polish Academy of Sciences

Russian Federation
The Russian Academy of Sciences

Slovak Republic
The Slovak Committee for IIASA

Sweden
The Swedish Council for Planning and Coordination of Research (FRN)

Ukraine
The Ukrainian Academy of Sciences

United States of America
The American Academy of Arts and Sciences

Contents

List of Figures	xi
List of Tables	xiii
Foreword *Tomasz Zylicz*	xv
Acknowledgements	xvii

1 Introduction — 1
  1.1 Background — 1
  1.2 Objectives — 5
  1.3 Approach and Structure of the Study — 6

Part I Theory, Models, and Practice — 9

2 Economic Instruments in Theory — 11
  2.1 Introduction — 11
  2.2 Efficiency and Instrument Choice — 12
  2.3 The Cost-Efficiency Concept — 20
  2.4 Emission Charges — 22
  2.5 Regulation — 26
  2.6 Tradable Emission Permits — 27
  2.7 Administrative Practicability — 31
  2.8 Dynamic Efficiency — 33
  2.9 Political Acceptability — 37
  2.10 Conclusions — 41

3 Instruments in Theory when Location Matters — 44
  3.1 Introduction — 44
  3.2 Cost-Efficiency Conditions — 45
  3.3 Ambient or Deposition Charges — 46

	3.4	Regulation and Location	47
	3.5	Emission Trading	47
	3.6	Administrative Practicability, Dynamic Efficiency, and Political Acceptability	60
	3.7	Concluding Comments	61
4	**Economic Instruments in an International Context: Theory**		**64**
	4.1	Introduction	64
	4.2	Conceptual Framework	65
	4.3	Emission Charges	75
	4.4	Tradable Emission Permits	80
	4.5	Concluding Observations	96
5	**Empirical Simulation Models and Emission Trading**		**98**
	5.1	Introduction	98
	5.2	Total Emission Levels as the Objective	99
	5.3	Air Quality Standards as Objectives	106
	5.4	Administrative Practicability, Innovation, and Distributional Issues	121
	5.5	Conclusions	127
6	**Emission Trading for Air Pollution in Practice**		**130**
	6.1	Introduction	130
	6.2	The EPA's Emission Trading Policy	131
	6.3	Lead Trading	138
	6.4	Chlorofluorocarbons and Halons	141
	6.5	Sulfur Trading under the 1990 Clean Air Act Amendments	145
	6.6	Power Plants Quota in Denmark	157
	6.7	Sector Covenants in the Netherlands	160
	6.8	Offsets in Germany	164
	6.9	Transfers under the Montreal Protocol	166
	6.10	Conclusions	170
7	**Theory, Models, and Practice: Concluding Observations**		**175**
	7.1	Introduction	175
	7.2	Cost Efficiency of Emission Trading	175
	7.3	Relevant Conditions	178
	7.4	Administrative Practicability, Innovation, and Political Acceptability	179

Contents ix

Part II Application to Sulfur Emissions in Europe   183

**8   The Institutional Framework for Controlling
     Sulfur Emissions in Europe**   185
   8.1   Introduction   185
   8.2   Acidification in Europe   186
   8.3   The United Nations Economic Commission for Europe   188
   8.4   The European Community   200
   8.5   National Legislation and Policies   204
   8.6   Concluding Remarks   208

9   Trading with Exchange Rates   210
   9.1   Introduction   210
   9.2   The Concept of Exchange Rate Trading   211
   9.3   An Exchange Rate Equal to the Ratio of the
         Marginal Costs in the Optimum   214
   9.4   Conclusions   237

10 Sulfur Emission Trading in Europe: A Model Simulation   239
   10.1   Introduction   239
   10.2   Optimization   240
   10.3   Simulation Method for Joint Implementation   242
   10.4   Data on Costs and Atmospheric Transport   247
   10.5   Simulation Results   249
   10.6   Conclusions and Discussion   271

11 An Institutional Design for Joint Implementation   273
   11.1   Introduction   273
   11.2   Background and Objectives   275
   11.3   The Potential Costs and Benefits of Different Schemes   276
   11.4   Rules and Conditions   282
   11.5   Joint Implementation and Domestic Policy   290
   11.6   Organizing International Transactions   292
   11.7   Monitoring and Enforcement   296
   11.8   Conclusions   300

12 Conclusions   302
   12.1   Focus of the Study   302
   12.2   Findings of Theory, Empirical Models, and Practice   302
   12.3   Application to Sulfur Emissions in Europe   308

Appendix: List of Symbols   **312**

References   **314**

Index   **331**

List of Figures

2.1	Coasian bargaining.	15
2.2	Uncertainty about costs and welfare losses of instruments.	18
2.3	Emission charges and emission trading.	24
2.4	Innovation with no response from regulator.	34
2.5	Innovation with response from regulator.	35
3.1	Deposition permits.	51
3.2	Three trading rules.	53
4.1	Cooperative and non-cooperative Nash equilibriums.	67
4.2	Joint implementation of emission reduction.	72
4.3	Joint implementation of deposition targets.	73
4.4	Countries trade emission permits.	83
4.5	Firms trade emission permits.	84
4.6	Countries trade deposition permits.	89
4.7	Deposition permits with firms trading.	92
4.8	Trading rules on deposition.	94
5.1	Cost savings and percentage emission reduction.	101
5.2	Cost savings of ambient permits in relation to ambient standards.	108
5.3	Cost savings of trading rules.	113
5.4	Multiple-zone trading with full information.	120
5.5	Multiple-zone trading with imperfect information.	120
6.1	Lead banking and lead standards.	140
6.2	The spot market for SO_2 allowances in 1993.	147
6.3	Sulfur emissions and banking.	152
6.4	Sulfur dioxide quota for power plants in Denmark.	158
8.1	The Convention on Long-range Transboundary Air Pollution.	189
8.2	Critical sulfur deposition values.	195
8.3	EC emission standards for SO_2 for solid and liquid fuels.	203
9.1	Exchange rate trading with two sources and one receptor.	216

9.2	Exchange rate trading with two sources and two binding receptors.	220
9.3	Violation with two receptors and two constraints.	222
9.4	Exchange rate trading with three countries and one receptor.	224
9.5	Exchange rate trading with three countries and two binding receptors.	230
9.6	Constant-deposition trading.	236
10.1	Flow diagram of joint implementation.	243
10.2	Cost function.	245
10.3	Cost function combined with regulation.	249
10.4	Deposition goals below Second Sulfur Protocol levels or five-percentile critical loads.	251
10.5	Demand and supply on Norway's market.	265
10.6	Demand for and supply of Bulgarian emission reductions.	266
10.7	Constant-deposition rates and sequence.	268
11.1	Environmental impact assessment.	289
11.2	A clearinghouse.	295

List of Tables

2.1	Evaluation of instruments.	42
2.2	Conditions affecting cost efficiency and environmental effectiveness.	42
3.1	Features of instruments when source location is a factor.	62
4.1	Overview of equilibrium conditions for marginal costs.	74
5.1	Potential cost savings of emission trading.	100
5.2	Cost savings of ambient permits.	107
5.3	Environmental impacts of ambient discharge permits.	109
5.4	Costs of single-zone emission trading to meet ambient standards.	116
5.5	Administrative costs of sulfur trading in USA (1993–2010).	122
5.6	Financial burden and permit expenditures when permits are sold.	125
6.1	Trading options and emission limits.	132
6.2	Trades and cost savings under the EPA's emission trading policy.	135
6.3	Expected cost savings of sulfur trading (1995–2010).	149
6.4	Market prices and traded volumes.	150
6.5	International transfers of ozone-depleting substances.	169
6.6	Evaluation of trading schemes.	171
7.1	Cost efficiency of different emission trading schemes.	177
7.2	Conditions influencing the cost efficiency of tradable emission permits.	177
8.1	Development of SO_2 emissions in Europe.	191
8.2	National ceilings in the Second Sulfur Protocol.	196
8.3	Emission standards in the Second Sulfur Protocol.	199
8.4	EC emission ceilings for SO_2 for existing LCPs.	204
8.5	National emission standards.	205
8.6	National fuel standards.	206
9.1	Two countries and one receptor.	218
9.2	Three countries and one receptor.	228

9.3	Trades with three countries and one receptor.	229
9.4	Data for the case of three countries and two receptors.	233
9.5	Trades with three countries and two receptors.	234
9.6	Trade with imperfect information about costs.	237
10.1	Constant-deposition rates.	253
10.2	Annual costs for the year 2000.	257
10.3	Emissions (kton SO_2/year).	258
10.4	Ecosystem protection.	259
10.5	Distributional impacts.	260
10.6	Joint implementation with cost-minimum exchange rates.	262
10.7	Potential buyers and sellers with constant-deposition rates.	262
10.8	Emissions and ecosystem protection without non-signatories and with emission and fuel standards.	263
10.9	Summary of emission trading schemes.	272
11.1	Joint implementation rules.	277
11.2	Summary of model results.	279
11.3	Transaction costs.	281
11.4	Example of an emission trading agreement (emissions in kton SO_2).	285
11.5	Banking of sulfur emissions by former CSFR.	287
11.6	Joint implementation and domestic policy.	291

Foreword

Many European economists have jealously regarded the growing American experience with tradable permits. Extremely complex legal systems in force in European countries often make it difficult to depart from administrative routines and let markets determine the allocation of abatement efforts where appropriate. It turns out rather unexpectedly that what is next to impossible on the national scale may become a reality on the continental scale. Some protocols to international conventions provide for a flexibility in complying with emission reduction targets that many national regulations lack. This book by Ger Klaassen comes in handy right at the moment when environmental policy makers in Europe are seriously considering emission trading under the Geneva Convention on Long-range Transboundary Air Pollution.

At last we have a succinct yet comprehensive survey of an important research and policy field. Dr. Klaassen makes a meaningful and original contribution to the discussion of tradable permits against a background of environmental policy instruments in general, and the possible application of these permits in the European acid rain context. These two contributions correspond to two distinct parts of the book. In the first part he gives a state-of-the-art survey of both theory and actual applications of tradable permits. In the second part he examines the European acid rain issue and discusses how it could be addressed by means of tradable permits. Both parts are important, and both are very well written.

The book will appeal to a wide audience. It will be read by environmental economists as well as by policy makers and "Euro-crats." Its mathematics is not very demanding (although it may discourage some lazy readers), and it is accessible for college students. All optimization arguments are consistently based on the Kuhn-Tucker theorem and thus the reader does not need to have extensive training in advanced microeconomics. Nevertheless, the book is also addressed to specialists, many of whom will find it stimulating and innovative. I am pleased to see Ger Klaassen's work becoming a vital element in the debate between researchers and practitioners on how to save our continent's natural heritage.

Tomasz Zylicz
Warsaw University
and Ministry of Environment

Acknowledgements

Wien, Wahlheimat der Genies is the title of one of the current bestsellers in Austria. The title best applies to Andries Nentjes for his decision to spend several months in Vienna during the last years to act as a sparring partner for the author of this book. Without his careful and critical, but always stimulating, support this book would not yet be finished. I would like to thank Hans Opschoor for being patient with me and for carefully reviewing the manuscript at several stages. I also appreciate the comments from Tomasz Zylicz, who acted as an independent reviewer.

Furthermore, I would especially like to mention Finn Førsund for his ideas and suggestions; Markus Amann for his critical comments and his skillful translation of my trading model design into computer language (Fortran); Sergey Shibayev for modeling the random trading routine, which proved to be a random process in itself; Janusz Cofala for his comments on earlier drafts of Chapter 6 and for reminding me that better is the enemy of good; and Clas-Otto Wene for translating Danish documents. I thank Wolfgang Schöpp for our discussions on how we would ever manage to finish our books.

I also owe thanks to my colleagues from the Task Force on Economic Aspects of Abatement Strategies, especially David Pearce and Johan Sliggers, for acting as a discussion forum for yet another part of this book disguised as a background paper.

Special thanks go to Margret Gottsleben, who translated my handwriting into text readable to everyone and provided assistance well beyond the call of duty. I appreciate the graphics support of Peter Dörfner and Zbigniew Klimont and the efforts of the International Institute for Applied Systems Analysis (IIASA) library, which expeditiously dealt with my continual requests for new materials. I must also record my thanks to Ellen Bergschneider for final English editing of the text.

This work was initiated through the financial support of the Norwegian Ministry of the Environment, it took off due to the Dutch National Member Organization of IIASA, which financed Andries Nentjes' stay at IIASA, and it was completed because my home institute, IIASA, enabled me to focus on economic instruments as an integral part of the Transboundary Air Pollution Project. I also acknowledge the financial support of the Free University in Amsterdam for my participation in a research project on ecologically sustainable economic development at the Institute for Environmental Studies and the Department of Spatial Economics. This project enabled me to do fundamental research and to start working on this volume.

Last but not least, the emotional support of my wife Brigitte and our sons Daniel and Sebastian needs to be mentioned for all the time "der Ger" had to work on his book instead of spending time with them.

Ger Klaassen, Vienna

Chapter 1

Introduction

1.1 Background

1.1.1 Policy context

One important example of environmental disruption is the damage that acid rain currently does to the forests, lakes, groundwater, and soils of Europe. Acid rain mainly results from the discharge of sulfur and nitrogen oxides into the air. These oxides remain in the air long enough to be transported across national boundaries but are ultimately deposited in wet (rain or snow) or dry form on the earth's surface or vegetation. During transportation these pollutants can be transformed into sulfuric and nitric acids (Persson, 1982). The impact of acidification depends on the level of acid deposition and the natural sensitivity of soils and water bodies. On the one hand, current levels of sulfur and nitrogen deposition in Europe clearly exceed critical loads (Nilsson and Grennfelt, 1988; Hettelingh *et al.*, 1991), the levels of deposition below which no significant damage to sensitive ecosystems is expected to occur according to present knowledge. Simulations with RAINS (the Regional Acidification INformation and Simulation model), developed at the International Institute for Applied Systems Analysis (IIASA) in Laxenburg, Austria, show that with the control policies planned in 1993 large parts of Europe will be exposed to levels of acid deposition that exceed the critical loads. This situation might be considered unsustainable from an ecological point of view (Amann *et al.*, 1993). On the other hand, the costs of controlling acidifying emissions increase exponentially with increasing emission reductions (Alcamo *et al.*, 1990; Amann *et al.*, 1993). Sulfur emissions, which mainly result from the burning of fossil fuels, are a major contributor to acidification in Europe. Studies indicate that the emission reductions achieved with current national and international legislation could have been achieved at significantly lower costs with coordinated differentiated policies instead of uniform cutbacks (Cesar and Klaassen, 1990; Klaassen *et al.*, 1994; Klaassen, 1993b).

Europe-wide negotiations in this field take place within the framework of the UN/ECE (United Nations Economic Commission for Europe) Convention on Long-range Transboundary Air Pollution. In 1985, a protocol was signed requiring all signatories to reduce sulfur emissions by 30 percent of their 1980 levels, starting in 1993. Recent negotiations focused on a new sulfur protocol. This Second Sulfur Protocol, signed in 1994, combines effect- and source-oriented measures. To reduce the negative impacts of acidification, the Protocol takes the critical loads for sulfur deposition as long-term objectives. Because these loads differ from location to location, reductions in sulfur deposition must differ as well; reductions must be higher in sensitive regions (such as Scandinavia) than in less susceptible areas (southern Europe, for example). Second, the emissions of some countries have more impact on sensitive areas than do those of others because of the atmospheric transport of sulfur. Therefore, the reductions must be higher in the latter countries to be equally effective. Finally, in some countries pollution control costs are lower than in other countries, which suggests the need for differentiated control for reasons of efficiency. Critical loads, atmospheric transport, and pollution control costs can be combined in integrated assessment models, such as the RAINS model (Alcamo *et al.*, 1990). These models were used in the negotiations on a new sulfur protocol to design an accord that to a certain degree met the critical loads for sulfur while minimizing total European control costs (Amann *et al.*, 1993). This effect-oriented approach resulted in an agreement on differentiated national emission reductions among countries, with some countries making significant reductions (up to 80 percent) and others undertaking smaller reductions. As a second tier of the Protocol, countries agreed to comply with emission standards for new large combustion installations and to reduce the sulfur content of some fuels.

Interestingly, and here we arrive at the topic of this study, the Protocol permits countries to trade their national emission reduction commitments, or in the terminology of the Protocol, "The Parties to the Protocol may ... decide whether two or more Parties may jointly implement the obligations set out in Annex II" (UN/ECE, 1994a, p. 6). The obligations referred to here are the national emission ceilings. In other words, one party can decrease its emission reductions if another party increases its reductions compared with the national emission reductions agreed on in the Protocol. As of March 1995, the parties to the Protocol had not reached an agreement on the rules to guide the joint implementation or trading of emission reduction commitments. This study sets out to elaborate possible rules and designs for joint implementation.

Introduction

The main reason for parties to be interested in joint implementation is that it may improve the cost efficiency of the Protocol. In spite of the fact that the Protocol was designed in a cost-efficient way, there might still be room for increasing cost efficiency for two reasons. First, the Protocol that has been signed does not correspond exactly to the modeled cost-minimum solution. Second, costs are uncertain and thus might differ from the model; trading would create incentives for countries to discover the actual costs and allow them to adapt their policies to reduce costs.

1.1.2 Economic instruments and environmental policy

The potential cost savings of economic or market-based instruments such as emission trading are well known in the literature. However, a clear-cut definition of economic instruments as alternatives or supplements to regulatory instruments is difficult to give. Common elements of economic instruments are the existence of financial stimuli, the possibility of voluntary action, the involvement of government authorities, and the intention to maintain or improve environmental quality (Opschoor and Vos, 1989). This might involve the following instruments: charges, subsidies, deposit-refund systems, market creation (such as tradable emission permits), and financial enforcement incentives.

According to economic theory a (Pigovian) tax equal to the marginal social damage leads to efficiency; welfare or net benefits are maximized. Emission trading would be equally efficient. With uncertainty about costs and benefits, the expected welfare impacts of charges and tradable permits will differ (Weitzman, 1974; Adar and Griffin, 1976). Due to deficiencies in estimating benefits, efficiency in the sense of maximizing net benefits is not usually a practical criterion.

Economic instruments are also more cost efficient in static terms than are emission standards; predetermined environmental standards are reached at lower costs than with regulatory instruments (Baumol and Oates, 1988; Tietenberg, 1985). The term static here means an unchanged environmental goal and given technology and location (Bohm and Russell, 1985). Furthermore, it is maintained that these instruments affect cost efficiency in a dynamic sense by promoting innovation (Wiersma, 1989; Nentjes and Wiersma, 1988).

Given the difficulties involved in finding optimal levels of pollution through cost-benefit calculation, some authors have proposed a more modest criterion for evaluating instruments, that of cost efficiency (see, for example, Baumol and Oates, 1988). Exogenously set environmental quality goals can be achieved at the lowest costs. It has been shown that economic cost efficiency (static or dynamic) alone is not a sufficient measure of the performance or applicability of economic instruments. The following dimensions are relevant as well (Bohm and Russell,

1985; Opschoor and Vos, 1989; Baumol and Oates, 1979; De Clerq, 1983; OECD, 1991; Sociaal Economische Raad, 1990):

- Environmental effectiveness (the extent to which the environmental objective is attained);
- Administrative practicability (information intensity, ease of enforcement, and level of administrative costs);
- Political feasibility (possible opposition, distribution of costs and benefits, compatibility with the existing institutional framework).

Whether or not the performance of economic instruments is better than that of command-and-control (CAC) types of regulations also depends on the circumstances under which these instruments are applied. The following appear to be among the relevant factors:

- The type of pollutant (uniformly mixed or not, assimilative or persistent, synergistic) (Tietenberg, 1985; Griffin, 1987; Zylicz, 1994);
- The source type (small/big; point sources/non-point sources) (Tietenberg, 1980);
- The number of receptor areas (Bohm and Russell, 1985);
- Uncertainty of cost functions (Bohm and Russell, 1985);
- Market imperfections (imperfect competition, transaction costs, public good characteristics) (Hahn, 1984; Adar and Griffin, 1976; Tietenberg, 1990).

These elements influence the relative success of tradable emission permits and effluent charges in terms of both efficiency and environmental effectiveness (Baumol and Oates, 1988).

In addition to this, there appears to be a significant difference between the results of economic theory and modeling exercises on the one hand and the cost efficiency of economic incentives in practice on the other hand (Tietenberg, 1989, 1990; Opschoor and Vos, 1989; Hahn and Hester, 1989b; Hahn and Stavins, 1992; Klaassen and Nentjes, 1995). This indicates the need for an analysis of the possible causes of this divergence. These causes might include the use of highly stylized benchmarks for economic instruments in theory and model studies, or the disregarding of market imperfections and the institutional context in which conflicts on property rights are resolved in practice, thus resulting in "imperfect" economic instruments (Opschoor and van der Straaten, 1992; Kling, 1988; Swaney, 1992; Hahn, 1990; Opschoor and Turner, 1994).

Finally, when discussing cost efficiency for transboundary pollution some additional elements are relevant. In a transboundary context no central agency exists to enforce an optimal solution (Baumol and Oates, 1988). Because some resources (e.g., air, water) have public good characteristics, countries may find it beneficial to act as a free rider. They may refuse to sign an agreement and a full cooperative solution might not be agreed on (Mäler, 1989, 1990; Barrett, 1990). This not only affects the attainment of an optimal solution (in terms of maximizing net benefits), but also might affect the cost efficiency of economic instruments (Barrett, 1990, 1991, 1992; Nentjes, 1994; Crocker, 1966; Bohm, 1990; OECD, 1976).

1.2 Objectives

Against the background described above, this study has the following objectives:

- To examine the general questions of whether, under which circumstances (conditions), and how (institutional design) the application of economic instruments – more specifically, emission trading – would be more cost efficient than regulatory instruments in theory, in empirical simulation models, and in practice, for non-uniformly dispersed pollutants, especially in a transboundary context;
- To analyze the more specific question of whether, under which conditions, and how emission trading can be implemented to improve the cost efficiency of current policies for controlling sulfur dioxide emissions in Europe, in particular the Second Sulfur Protocol.

The emphasis of the study is on static cost efficiency, the attainment of a goal at minimum costs. Environmental effectiveness will be dealt with explicitly. Administrative practicability is relevant here, because information requirements and costs to the environmental authorities are considered. Political acceptability is relevant insofar as the distribution of costs and benefits and the institutional context affect the design and the cost efficiency of the instruments under consideration. Dynamic efficiency considerations will only be mentioned in the first part of the study for the sake of completeness.

Furthermore, the study focuses on different emission trading schemes rather than on effluent charges, mainly because the trading of agreed-on national emission ceilings fits into the current Protocol much better than does a system of emission charges, as agreements have taken the form of quantitative ceilings. Trying to achieve these levels with a system of country-specific charges would be complicated, although not impossible in theory. However, the properties of emission charges will be dealt with in the theoretical part of this study (Chapters 2 to 4).

New elements in the study are as follows:

- The combination of theory, simulation models, and practice to arrive at more realistic conclusions and to enable more realistic modeling exercises;

- The international application of emission trading schemes for non-uniformly dispersed, transboundary pollutants;
- The analysis of a new trading scheme based on fixed exchange rates and simulated as a bilateral sequential process in combination with more realistic estimates of the potential cost savings.

1.3 Approach and Structure of the Study

This study consists of two parts. The more general Part I (Chapters 2 to 7) is a survey of economic instruments – particularly emission trading – in theory, in empirical simulation models, and in practice. Part II (Chapters 8 to 11) is more specific and examines the cost efficiency of trading emission reduction commitments for sulfur dioxide (SO_2) in Europe.

Chapter 2 is an overview of economic theory on environmental policy and on economic instruments. The focus is on the conditions under which economic instruments promote cost efficiency in a domestic context. This deals with conditions for cost efficiency and possible market failures (imperfect competition, transaction costs, free riding/public good characteristics). Efficiency, administrative practicability, and political acceptability are treated on the basis of major studies and reviews (see De Clerq, 1983; Cropper and Oates, 1990; Bohm and Russell, 1985; Kling, 1988; Bressers and Klok, 1988; Nentjes, 1990a). After a brief introduction to the economic approach (externalities, property rights, public good/common property), the chapter examines the efficiency of economic instruments (uncertainty, benefit estimation problems), cost efficiency and environmental effectiveness (the static concept, emission trading schemes, emission charges, regulation), as well as administrative practicability, dynamic efficiency, and political acceptability.

Chapter 3 is devoted to the cost efficiency of economic instruments for dealing with non-uniformly dispersed pollution – that is, for those pollutants where location significantly influences environmental impact. After a description of the conditions for cost efficiency, the extent to which emission charges, regulation, and various emission trading schemes can be cost efficient is investigated.

Chapter 4 expands on Chapters 2 and 3, addressing the special features of environmental policy and economic instruments in an international context. The emphasis is on how an international context affects the cost efficiency of emission trading and charges. Relevant material is reviewed (such as Barrett, 1990, 1991; Bohm, 1990; Nentjes, 1994; Mäler, 1989, 1990; Rose and Stevens, 1993). Because most of this material focuses on global pollutants and not on regional pollutants such as SO_2, new elements will be added.

Introduction 7

Chapter 5 gives an overview of the results of empirical simulation studies with assessments of the performance of economic instruments. Simulation studies are similar to engineering studies and usually, but not always, use an optimization model for assessing the cost efficiency and environmental effectiveness of economic incentives (see, for example, Atkinson, 1994; Rentz *et al.*, 1990; Welsch, 1989; Wiersma, 1989). Both national and international studies are surveyed on the conditions or assumptions that influence cost efficiency and environmental effectiveness.

Chapter 6 contains a survey of the experience with emission trading for air pollution control in the USA, in Europe, and worldwide. Studies of actual experiences that analyze the performance of these instruments in practice on a national or an international level are discussed (e.g., Bohm, 1990; Hahn and Hester, 1989a; Opschoor and Vos, 1989; Opschoor, 1994; Tietenberg, 1985, 1989, 1990). This material is augmented by more detailed national material on emission trading in the USA and information on European experiences.

In Chapter 7 the findings of the previous chapters on theory, models, and practical application are compared in order to find where theory, models, and practice diverge and where they overlap.

The institutional context of the problem of acid rain and sulfur emissions in Europe are introduced in Chapter 8. The chapter consists of two main parts: a broad overview of the problem of acidification in Europe and the role of sulfur, and an updated overview of current policies and institutional frameworks with an emphasis on international aspects.

Chapter 9 contains an analysis of the performance of a new system of emission trading according to a fixed exchange rate and the conditions for cost efficiency in a theoretical setting. As a special case, the exchange rate is based on the ratios of the marginal costs in the cost minimum (thus accounting for the transfer coefficients of those receptors that exactly meet deposition constraints in the cost minimum). The cost efficiency and environmental impacts of exchange rate trading are assessed for the following circumstances: one or more receptors; two, three, or more countries trading; and perfect and imperfect information on costs. The section compares the differences between the cost-minimum formulation and bilateral sequential trading.

The trading of sulfur emission reduction commitments among countries in Europe is simulated in Chapter 10. First, the chapter describes the optimization or cost-minimization routine that is implemented in the RAINS model (Alcamo *et al.*, 1990). Second, the emission trading model that simulates trading with a fixed exchange or offset rate as a bilateral sequential process accounting for fixed transaction costs is described. The data on costs and atmospheric transport used are summarized as well. Finally, Chapter 10 applies the method developed in Chapter 9 to the recent negotiations on a new sulfur protocol in Europe. The performance of the new protocol with and without trading provisions is assessed. The cost efficiency, environmental effectiveness, and distribution of costs and physical benefits for 38 countries or regions are described. The sensitivity of the results for a number of assumptions (trading sequence, participants in the trading, combined trade and regulations, transaction costs, and uncertainty) are tested.

Chapter 11 provides a closer look at the institutional design of emission trading or joint implementation, embarking on questions concerning possible designs and their cost efficiency, rules for trading, organization of transactions, the relation between domestic and international policy, and monitoring and enforcement. Conclusions, especially on Part II of the study, are the subject of Chapter 12.

Part I

Theory, Models, and Practice

"If it is feasible to establish a market to implement a policy, no policy maker can afford to do without one. Unless I am very much mistaken, markets can be used to implement any anti-pollution policy." (Dales, 1968, p. 100)

Chapter 2

Economic Instruments in Theory

2.1 Introduction

The proposal to establish markets in pollution rights as a cost-efficient means of reducing water pollution was put forward by Dales (1968). He suggested the creation of a number of rights to discharge one unit of waste during a certain period and the subsequent auctioning of these rights. With a sufficiently high price, some firms would find it more profitable to treat their pollution than to buy rights. Therefore, demand for pollution rights would be reduced and a market equilibrium would be attained. Dales proposed the creation of markets in pollution rights as an alternative to emission charges. The advantage of markets is that the environmental authority, faced with uncertainty about costs, does not need a trial-and-error process and adaptation of charges to attain a desired level of discharges (see Baumol and Oates, 1971, 1988).

The goal of this chapter is to survey economic theory to examine whether, under which circumstances, and how emission trading and emission charges could be implemented to improve the cost efficiency of meeting environmental objectives. More specifically, the chapter aims to evaluate emission trading in comparison with regulatory approaches and charges with respect to the following criteria:

- Efficiency (maximizing net benefits);
- Static cost efficiency;
- Environmental effectiveness (attainment of environmental objectives);
- Administrative practicability (ease of monitoring and enforcement, information requirements, and administrative costs);
- Dynamic incentives (innovation, location shifts, price distortions);
- Political acceptability (i.e., distribution of costs/benefits and compatibility with the institutional framework).

The emphasis is on the application of economic incentives in a domestic context where the goal of environmental policy is a given level of emissions in a region. In other words, the focus is on so-called uniformly dispersed (or global) pollutants where the location does not influence the damage. Chapter 3 will focus on the cost efficiency of instruments when dealing with non-uniformly dispersed pollutants such as SO_2. Chapter 4 will discuss transboundary pollution for the case of non-uniformly dispersed pollutants.

This chapter has the following structure. In Section 2.2 the relation between economic instruments and efficiency is described. The conditions for cost efficiency are introduced in Section 2.3. In Section 2.4 the conditions under which emission charges can be cost efficient are examined, and in Section 2.5 regulatory instruments are analyzed. Section 2.6 explores the cost efficiency of tradable emission permits. In Section 2.7 the administrative practicability of the different instruments is investigated. Section 2.8 contains an analysis of dynamic efficiency, and Section 2.9 contains an analysis of the political acceptability of the instruments.

2.2 Efficiency and Instrument Choice

2.2.1 Efficiency and market failure

In economic theory pollution is regarded as an externality whose presence leads to an inefficient or sub-optimal allocation of scarce resources. Consequently, corrective devices are necessary to establish an optimal solution – an outcome that maximizes the net benefits or welfare of society. An externality is present whenever (Baumol and Oates, 1988, pp. 17–18)

- An individual's utility or profit level depends on real (technical/physical) variables whose values are chosen by others without attention to the effects on the individual's utility;
- The individual does not receive (or make) payment in compensation for the costs (benefits) imposed on her by others.

An example of a negative externality is when a power plant in Poland emits SO_2. This sulfur is transported in the air and comes down on a Swedish lake, acidifying the lake and leading to a reduction of the welfare level of Swedish fishermen. Special examples of externalities are when a natural resource (such as the air over Europe) is a common property resource. In cases where the potential users have free access to the resource, it can be used by any agent without restriction and without consideration of any negative external impact imposed on other agents. In some cases free access is due to the public good characteristics of the resource in question. A pure public good is characterized by the fact that, technically, nobody can be excluded from using the good (non-excludability) and there is no rivalry in use: the use by one individual does not subtract from any other individual's use (Siebert, 1987, pp. 62–64). Stated differently, private property rights, which include the right and the physical ability to exclude other users, cannot be established for pure public goods. In cases where exclusion is technically possible but not desirable (meritorious) one speaks of a merit good. Environmental economics also characterize pollution as a public "bad," resulting from waste discharged in the production of private goods.

The notion of an efficient or optimal level of pollution has its roots in welfare economics. An allocation of resources is Pareto-optimal when it is not possible (through further reallocations) to make one person better off without making someone else worse off.

In the presence of externalities, the market mechanism might lead to a suboptimal allocation of resources (Cropper and Oates, 1990). This can be illustrated using the following simple model of an economy. The utility (U) of a group of individuals is a function of the goods consumed (X) and the disutility from the level of pollution (Q) [condition (2.1)]. Pollution results from emissions (E) in the course of the production of X. Production is based on conventional inputs (I), such as labor, capital, natural resources, and waste discharges; it is negatively influenced by the level of pollution [condition (2.2)]. The level of pollution is a function of the emissions from all sources [condition (2.3)]. The partial derivatives have the following properties: $U_X > 0$, $U_Q < 0$, $X_I > 0$, $X_E > 0$, $X_Q < 0$, and $Q_E > 0$.

$$U = U(X, Q) \quad \text{Welfare,} \tag{2.1}$$
$$X = X(I, E, Q) \quad \text{Production,} \tag{2.2}$$
$$Q = Q(E) \quad \text{Pollution.} \tag{2.3}$$

Maximizing the welfare of a group of individuals subject to conditions (2.2) and (2.3), along with a further constraint on the availability of resources (I), leads to a set of first-order conditions. The following condition for a Pareto-efficient allocation is crucial in an externality context:

$$\frac{\partial U}{\partial X} \cdot \frac{\partial X}{\partial E} + \frac{\partial U}{\partial X} \cdot \frac{\partial X}{\partial Q} \cdot \frac{\partial Q}{\partial E} + \frac{\partial U}{\partial Q} \cdot \frac{\partial Q}{\partial E} = 0 \ . \tag{2.4}$$

After rearranging terms, this takes the form

$$\frac{\partial X}{\partial E} = \frac{\partial U}{\partial Q} \cdot \frac{\partial Q}{\partial E} \bigg/ \frac{\partial U}{\partial X} - \frac{\partial X}{\partial Q} \cdot \frac{\partial Q}{\partial E} \ . \tag{2.5}$$

Equation (2.5) states that production should be expanded only to the level where the marginal product of emitting one unit more ($\partial X / \partial E$) equals the marginal damage the additional emissions impose on the consumers' utility [the first term on the right-hand side of equation (2.5)] and on the producers (the second term). Stated differently, each agent should take pollution control measures to the point where marginal abatement costs equal the marginal benefits from reduced pollution (the sum of individuals and firms).

Firms operating in a competitive market with free access to environmental resources do not generate an optimal level of pollution. These firms continue to increase production (X) and pollution to the point where the private marginal return is zero: that is, until $\partial X/\partial E = 0$. Consequently, the level of pollution is too high, because the negative impacts of pollution on utility and production are ignored. In a competitive setting, producers maximize their producer surplus and expand production (and hence emissions) to the point where price equals marginal private costs. With negative externalities this is not efficient, because net benefits (or welfare) would be maximized at the point where price equals marginal private costs plus marginal external costs. In conclusion, in a competitive setting with free access to the environment (no restrictions on pollution), the level of production, and thus pollution, is too high and the prices of the polluting products are too low.

2.2.2 Coasian bargaining

One line of thinking, the property rights approach, regards the failure of the market in the case of externalities as stemming from a failure to define exclusive property rights (Coase, 1960). Property rights are the legally defined rights to use property in certain ways, to prevent others from exercising these rights (exclusivity), and to sell the property (transferability) (Dales, 1968). Two kinds of goods can be distinguished: private and public (Siebert, 1987). Private goods are owned by individuals, potential users can be excluded from their use, and there is competition or rivalry in use. A pure public good (such as a lighthouse) is consumed in equal amounts by all (Samuelson, 1954); excluding individuals from use is not possible and there is no rivalry in use (the use one individual makes does not reduce the use others can make). Common property is property shared with others. When no rules exist on the use of a common property resource, there is free or open access (Bromley, 1989); exclusive property rights are not defined and rivalry in use prevails.

Coase (1960) showed that a bargaining solution will result in a Pareto-optimal allocation of the environment if private, transferable property rights are defined, if transaction costs are negligible, and if individuals maximize their utility. Moreover, this allocation is independent of who has the initial property rights. *Figure 2.1* shows that if the polluter obtains the right to pollute, the level of pollution will be Q_m [maximizing net private benefits (MNPB): price minus marginal private

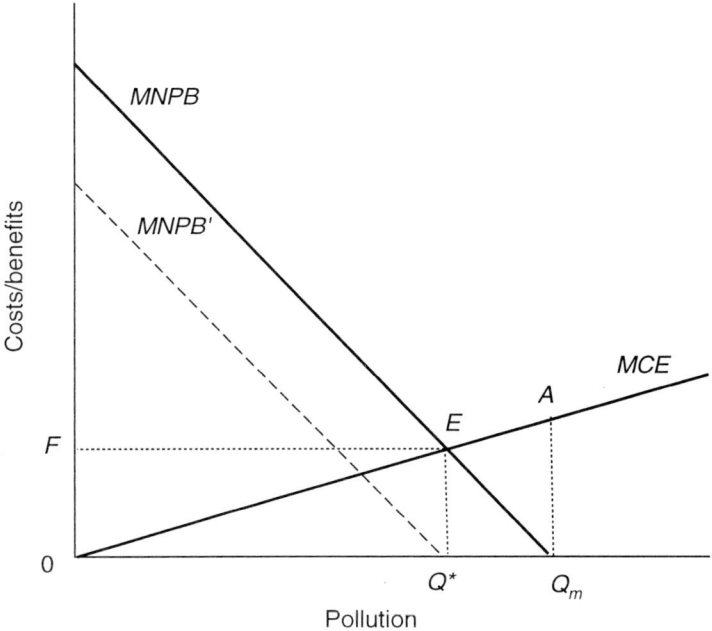

Figure 2.1. Coasian bargaining.

costs]. In this case the victim could pay the polluter to move to point Q^*, because the welfare of the victim would increase with the reduction in external costs (MCE) to equal area Q^*EAQ_m. This is sufficient to compensate the polluter, whose loss in private net benefits would be only the area EQ^*Q_m. The net gain in welfare would equal the area EAQ_m. Clearly, beyond point Q^* the reduced benefits for the polluter would exceed the reduced external costs for the victim. If the victim were to have the right not to be polluted, the starting point would be 0. In this case the polluter could buy the right to pollute from the victim. Again bargaining would take place up to point Q^*, where the difference between private benefits and external costs is maximized.

Serious criticism of the Coase theorem exists, however (see Baumol and Oates, 1988, p. 10; Pearce and Turner, 1990, pp. 70–83; Siebert, 1987, pp. 91–98). In the presence of transaction costs, the resulting allocation depends on the initial distribution of property rights. Transaction costs include costs of gathering parties together, organizing polluters, making the bargain, and so forth. This is most easily seen when transaction costs exceed the net benefits to be obtained from trade. In this case, no bargain would be struck and the level of pollution would depend on who initially obtained the property rights.

Another serious objection is that in most cases of pollution there is more than one polluter and pollutee (Baumol and Oates, 1988; Siebert, 1987). The problem is that in its role as a receptacle of waste, the environmental resource (be it air, water, or soil) is a free-access, common property resource; non-exclusive property rights predominate. In its role as provider of environmental quality, the resource is a "public good." In this case, the free rider problem arises because pollutees cannot be excluded from the benefits (reduced pollution) that result from a bargain between one pollutee and one polluter to reduce pollution. If the pollutees have the initial rights, the polluter must bargain with all of them to increase pollution to an optimal level, because the solution must include benefits for all pollutees. This might increase transaction costs considerably. Furthermore, it might not be easy to identify all pollutees and polluters (liability for pollution damage might be unclear, or the relationship between ambient quality and waste discharge might be complex) and some pollutees (future generations) might not be present (Rose-Ackermann, 1973). Consequently, simply assigning property rights might not work in a large number of cases.

2.2.3 Uncertainty about costs and damage functions

An instrument that at least in theory leads to an optimal solution is a Pigovian tax, a levy on the polluter equal to the marginal social damage. In terms of *Figure 2.1*, the tax would have to be equal to length *OF* (at the optimal solution) per unit of waste emissions. In this case the marginal net private benefit function (*MNPB*) of the firm facing perfect competition would become $MNPB'$ (*Figure 2.1*), and the firm would restrict pollution to the level Q^*. To attain such an optimal allocation, compensation of victims (except through lump-sum transfers) is not permitted. This is because if victims were to be fully compensated they would no longer have an incentive to undertake efficient levels of defensive measures (Cropper and Oates, 1990).

The classic arguments for optimal charges (Kneese, 1964; Kneese and Bower, 1968) hinge on a number of restrictive assumptions about the nature of the damage function. More specifically, it is assumed that the relationship between damage and emissions (or discharges) is linear and that the location of the source is irrelevant. In such a case, the public authority could determine the optimal charge level without knowledge of the cost functions (see Bohm and Russell, 1985), because the marginal damage would be constant and independent of the level of emissions. However, in the case of a nonlinear damage function the optimal charge cannot be

Economic Instruments in Theory

determined without knowledge of the cost functions (Bohm and Russell, 1985, pp. 408–409).

If the cost and damage functions are both known, different instruments can attain the optimum, including emission charges (per unit of emission), standards (quantitative controls on the emission level generated by each source), and a system of transferable emission rights allowing sources to buy and sell rights. Instead of setting the Pigovian tax at the optimal level, the environmental agency could issue the optimal number of emission permits whose total equals the optimal emission level and then either allow sources to bid for them at an auction or allocate them initially for free. In brief, both quantity and price approaches would yield the same result in terms of efficiency in a setting of perfect information where perfect enforcement and no differences in administrative costs between the different instruments are assumed.

If, however, the cost and damage functions are uncertain, none of the policy instruments continues to have the same impact on welfare. The welfare losses depend on the relative steepness of cost and damage functions. As Adar and Griffin (1976) have shown, uncertainty about the marginal damage functions does not influence the welfare losses of the different instruments. This is because standards, tradable permits, and charges result in the same level of emissions if the marginal cost function is known.

If the marginal cost function is not known with certainty, welfare losses will differ. In *Figure 2.2*, MC is the marginal control cost function perceived by the environmental agency and MDF is the marginal damage function. With the perceived cost and damage functions, the environmental agency could either set the optimal tax level at P_s or fix the total number of emission rights at E_s. With perfect information the impact on welfare would be the same. However, if the true marginal costs (MC') were to differ from the perceived costs (MC) (that is, if they were to be higher than perceived), the welfare impacts of taxes and tradable permits would differ. With these true costs (MC') the real optimum would be point C and not point E. Given the emission tax P_s (based on the wrong costs), polluters would now reduce emissions to point E_t, because they would equate the true marginal costs with the tax level. At this point pollution control costs (area CBE_tE^*) would be lower than in the true optimum, but damage (area CAE_tE^*) would be higher. The net loss of welfare (damage plus costs) would be equal to the area ABC. With tradable permits, total emissions would remain at E_s but the permit price (and thus marginal costs) would increase to P_d. In this case, pollution control costs would be higher (area CDE_sE^*) than in the true optimum, damage would be lower (area ECE^*E_s), and the total welfare loss would be equal to the area DCE. In this particular case, the welfare loss of tradable permits would exceed that of emission charges. Whether or not welfare losses under an emission charge exceed those

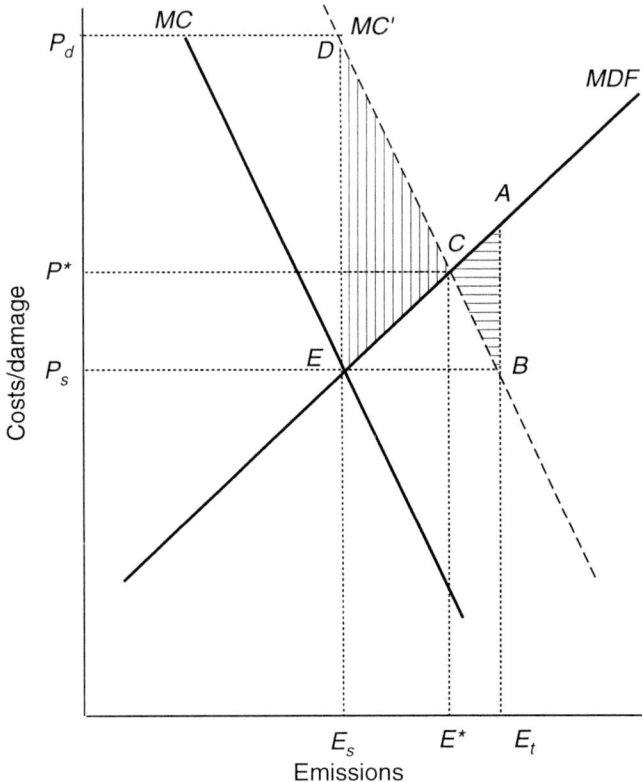

Figure 2.2. Uncertainty about costs and welfare losses of instruments.

under a tradable permit system depends on the steepness (elasticity) of both the cost and damage functions. If the damage function is relatively steep (inelastic), a small change in emissions results in considerable welfare losses and quantitative controls are preferable. If the cost function is relatively steep (inelastic), emission charges are the appropriate instrument to minimize expected welfare losses because they allow control of the marginal costs (Weitzman, 1974).

2.2.4 Limitations of designing optimal policies

A number of serious problems limit the applicability of the above model, which maximizes net benefits. These problems concern both practical application aspects and fundamental assumptions. Because these problems pertain to the determination of the optimal level of pollution rather than the choice of the instruments, they are only briefly summarized here (see De Clerq, 1983, for a more extensive treatment). The information requirements for the environmental agency that wants to set the optimal tax or emission level are enormous, because the tax rate should be set at the level of the marginal damage in the optimal solution. This not only requires information on the existing marginal damage due to pollution, but also requires knowledge of the marginal damage in the optimum solution. Despite the increased willingness to carry out benefit studies in the recent past (see, for example, Barde and Pearce, 1991), one tends to agree with Baumol and Oates (1988, p. 160) that "it is hard to be sanguine about the availability of a comprehensive body of statistics reporting the marginal net damage of the various externality generating activities." Recent overviews of the damage done by acid rain in Europe (UN/ECE, 1994b) confirm that we are far from knowing marginal damage functions, although progress undeniably has been made. Consequently, a basis for determining optimal emission levels by way of cost-benefit calculations simply does not exist.

This already suggests that there are serious limitations to using an approach that attempts to determine the level of pollution that maximizes net benefits. But there are also more fundamental concerns as well. The choice of an optimal level of pollution might have consequences for the distribution of welfare if compensation of welfare losses does not take place. A choice is then necessary between optimality and equity (De Clerq, 1983, p. 74). This is the case for intra- as well as intertemporal distribution of welfare (Tietenberg, 1992a, p. 37). Finally, optimal levels of pollution may exceed the assimilative capacity ("the critical loads," so to speak) of the natural environment, which may irreversibly destroy sensitive ecosystems. This irreversibility and the fact that natural resources may not have substitutes and may perform as yet undiscovered life-support functions are additional arguments for a prudent, risk-averse approach emphasizing levels of pollution that are lower than "optimal" (Klaassen and Opschoor, 1991; Pearce, 1989).

In view of the above discussion, the practical possibilities, as well as the usefulness, of determining optimal pollution levels and choosing instruments to meet these levels appear to be limited. An approach that starts from either ecological standards on the assimilative capacity of the environment or from politically determined environmental standards and selects the instruments to meet such standards at the lowest cost to society, accounting for distributional implications and uncertainty, might be more fruitful. This is the issue to which attention is turned in the next sections of this chapter. In Part II of this study it will be shown that this does not solve the problem of how these ecologically correct and politically acceptable levels should be set.

2.3 The Cost-Efficiency Concept

2.3.1 Introduction

The prime objective of environmental policy is the improvement or conservation of environmental quality. However, the achievement of acceptable levels of emissions, depositions, or ambient concentrations is often used as a more practical, operational environmental objective, particularly when it comes to discussing instruments for environmental policy. Uniformly dispersed pollutants, such as chlorofluorocarbons or carbon dioxide, have the same global impact on concentration levels irrespective of where they were emitted. In this case, controlling the total level of emissions is sufficient for controlling concentrations. With non-uniformly dispersed pollutants, such as SO_2 or ammonia, deposition or concentration levels are affected not only by the amounts emitted but also by the location of the sources. This spatial aspect complicates the design of environmental policy and will be analyzed in Chapter 3. In this chapter the focus is on uniformly dispersed pollutants.

2.3.2 Controlling emission levels

If the objective of environmental policy can be changed to controlling the total amount of emissions, the necessary conditions for a cost-efficient allocation of emission reductions can be derived using the following conceptual framework (Tietenberg, 1985). Cost efficiency is interpreted in a static sense; the environmental goal, the available technology, and the location of the source are fixed (Bohm and Russell, 1985). The framework uses the following variables: the total level of emissions (E), the uncontrolled emissions of source i ($\overline{E}_i$), the amount of emission reduction by source i (R_i), the background emissions (N), the desired level of emissions (E^0), and the function representing the minimum costs of reducing emissions for each source $[C_i(R_i)]$.

The total level of emissions is the sum of the emissions from all sources ($i = 1, \ldots, I$) plus the background emissions (for example, from other regions or from natural sources):

$$E = \sum_{i=1}^{I}(\overline{E}_i - R_i) + N \ . \tag{2.6}$$

Cost efficiency will be achieved if the allocation of emission reductions among the sources is chosen so that the costs of reaching the desired level of emissions

Economic Instruments in Theory

are minimized subject to the condition that the sum of emissions and background emissions is less than or equal to the desired level of emissions:

$$\min \sum_{i=1}^{I} C_i(R_i) \qquad (2.7)$$

subject to

$$\sum_{i=1}^{I} (\overline{E}_i - R_i) + N \leq E^0 . \qquad (2.8)$$

Furthermore, $\overline{E}_i \geq R_i \geq 0$. That is, emissions and reductions are nonnegative.

Finally, the cost function has the following properties: $C'_i(R_i) > 0$ and $C''_i(R_i) > 0$; that is, the marginal costs increase with further emission reductions.

The most important of the necessary and sufficient Kuhn–Tucker conditions for an optimum solution is the following:

$$R_i \left[\frac{dC_i(R_i)}{dR_i} - \lambda \right] = 0 . \qquad (2.9)$$

In equation (2.9), λ is the Lagrange multiplier, or shadow price, which reflects the change in the value of the objective function (the decrease in costs) when the constraint on the desired level of emission is relaxed (or increased) by one unit. The interpretation of equation (2.9) is as follows: in the optimum solution the marginal costs of each source are equal to λ or the source does not have to reduce its emissions (R_i is zero). The latter can be the case if the marginal costs of reducing the first unit of emissions for that specific source are rather high. The important conclusion is that for the cost-efficient solution the marginal costs per ton of emission control for every source must be equal.

2.3.3 Accumulation and multiple pollutants

If uniformly dispersed persistent pollution exceeds the assimilative capacity of the environment, pollutants (such as carbon dioxide and chlorofluorocarbons) accumulate over time. If environmental damage is caused by the stock of accumulated emissions over time and not by the flow of emissions at one specific point in time, it makes sense to define the environmental constraints as a restriction on accumulated emissions and not as a restriction on the flow (as in the previous section). In this case, a cost-efficient allocation is one that minimizes the present value of the control costs subject to the constraint that accumulated emissions do not exceed the desired level (E^0) at any point in time (t) during the planning horizon T:

$$\min \sum_{t=1}^{T} \sum_{i=1}^{I} \frac{C_i(R_i)}{(1+\tau)^{t-1}} \qquad (2.10)$$

subject to

$$N + \sum_{t=1}^{T}\sum_{i=1}^{I} (\overline{E}_{i,t} - R_{i,t}) \leq E_t^0 \quad \text{for every } t ,$$

where τ is the rate of interest used to translate future costs into present value. A basic condition for an optimal solution for a uniformly dispersed pollutant is

$$\frac{\partial C_i(R_{i,t})}{\partial R_{i,t}} = (1+\tau)\frac{\partial C_i(R_{i,t-1})}{\partial R_{i,t-1}} \quad i = 1, \ldots, I . \tag{2.11}$$

Hence, marginal abatement costs must increase over time by a certain percentage equal to the time preference τ (the interest rate) (Tietenberg, 1985) to arrive at a cost-efficient solution. If the pollutants not only accumulate but also are non-uniformly dispersed, similar but more complex conditions can be deducted (Griffin, 1987).

Policy makers are often faced with multiple pollutants drawing on the same assimilative capacity of the environment. These pollutants might interact in a linear or nonlinear (synergistic) way. The latter, for example, is the case with ground-level ozone concentrations, whose precursors are volatile organic compounds (VOCs) and nitrogen oxides. As long as multiple pollutants interact linearly there are no serious problems in deriving conditions for cost efficiency. A cost-efficient solution simply requires that the marginal control costs per ton of emissions abated for different pollutants reflect their respective impacts on pollutant concentrations (Endres, 1986; for an application see Amann and Klaassen, 1995). With nonlinear interactions the Kuhn–Tucker conditions may not be representative of the cost minimum, because the second-order quasi-convexity condition of the constraint function is violated.

Moreover, multiple local optima may occur (Endres, 1986; Zylicz, 1994). In this case, locating the global optimum requires full knowledge of cost functions.

2.4 Emission Charges

2.4.1 The perfect case

The basic economic argument in favor of emission charges is that they bring about the combination of control among polluters that minimizes the costs of waste disposal in a region (Kneese, 1964, p. 56) without requiring knowledge of the costs. Under the assumption that each source minimizes costs, it can be straightforwardly proven that imposition of a uniform charge (F) per unit of emission reaches the

Economic Instruments in Theory

cost-minimum solution. With a charge on its level of uncontrolled emissions each source choice is characterized by

$$\min C_i(R_i) + F(\overline{E}_i - R_i) \ . \tag{2.12}$$

The condition for an optimal solution is

$$C_i'(R_i) - F = 0 \ . \tag{2.13}$$

The minimum level of costs for the individual source is reached when the marginal costs of emission reduction equal the uniform charge. Because the charge is uniform, the marginal costs for each source are equal, which is exactly the condition for a cost-efficient solution. A more formal proof is supplied by Baumol and Oates (1971). A graphic illustration may underscore this. *Figure 2.3* presents the marginal costs for two sources (MC_1 and MC_2) as functions of the emissions after abatement for each individual source. In the reference case both sources reduce emissions to 10 units each, a total of 20 units. Clearly, in this case marginal costs differ and total costs are too high. One can easily see that a tax equal to US$800 per unit of emission would bring about the optimal allocation; both sources would reduce emissions to the point where the tax equals marginal costs. This implies that emissions after abatement would equal 15 units for MC_1 and 5 units for MC_2.

2.4.2 Uncertainty, growth, market imperfections, and discontinuities

Although the cost efficiency of emission charges is guaranteed, at least in a competitive market, environmental effectiveness is not. If pollution control costs are not fully known with certainty, a trial-and-error procedure is needed to reach the desired level of emissions (De Clerq, 1983). This might imply a certain delay in meeting the environmental objective or an overshooting of the emission goal if the charge set is too high. In the latter case costs might be higher than necessary, particularly if the fixed (and sunk) costs of pollution control are relatively important. This suggests that charges are more appropriate if meeting the emission goal immediately is less urgent than avoiding high costs. Charges have the advantage that they allow control of the marginal costs, although they do not allow control of the total costs. By comparison, however, none of the other instruments (emission trading or emission standards) give any control over costs. Charges may thus be preferred if marginal costs are expected to be high.

In the case of economic growth and inflation, the charge must be adapted to achieve the environmental objectives in a cost-efficient way. The same is true in the case of cost-reducing technical progress, which should result in a reduction of the charge level in order to avoid overshooting.

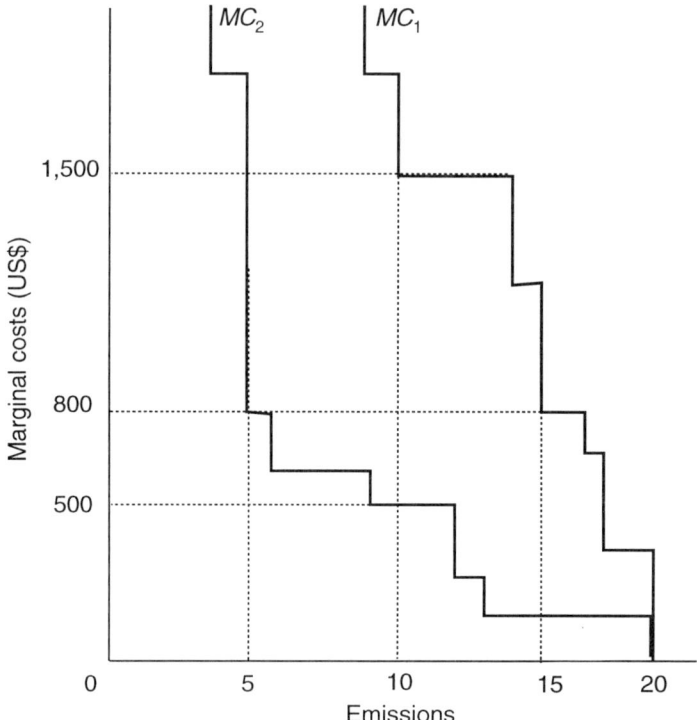

Figure 2.3. Emission charges and emission trading.

Market imperfections might also influence the results of a charge. In particular, collusive behavior might occur: a small number of polluters might form a cartel and overinvest in pollution control in order to achieve a reduction in emission charges and total costs (pollution control plus charges) in the future (Rose-Ackermann, 1973). This, of course, is only profitable if the additional costs are expected to be outweighed by a subsequent decrease in charge levels and associated payments.

Furthermore, cost functions are not always continuous, because there may be technological indivisibilities. This means that an emission charge set at a certain level of marginal costs might not determine a unique treatment level (Rose-Ackermann, 1973). In *Figure 2.3*, this could occur if the charge were set at a level of 700. In this case, source 2 (MC_2) would be indifferent to reducing emissions at any level between 5 and 10 units. Although cost efficiency is assured, the environmental impact would be uncertain and more or less purification than desired could occur.

However, it should be noted that although this may be a serious problem at the level of a single firm or a small number of firms, such indivisibilities can be expected to be less relevant at the level of the aggregate economy.

Imperfect enforcement of the emission charge would, in general, affect neither its environmental effectiveness nor its cost efficiency. Harford (1978, p. 41) finds that even with imperfect enforcement a cost-minimizing firm will always equate marginal costs with the emission charge level. If enforcement is imperfect and if charge payments are based on reported emissions, a polluting firm might find it profitable to avoid charges by underreporting its emissions. Profits are the sum of net sales revenues minus charge payments for reported emissions minus the expected fine and the emission charge to be paid for any unreported emissions that are discovered. The expected fine is a function of the size of the violation (actual emissions minus reported emissions). The firm will choose output, actual emissions, and reported emissions so as to maximize profits. The first-order conditions imply that the marginal costs of actual emission reduction must equal the marginal increase in the expected penalty (fine plus emission charge). Moreover, the increase in the expected penalty per unit decrease in reported emissions should equal the per unit emission charge. Both conditions taken together imply that the marginal costs of emission reduction always equal the emission charge level (Harford, 1978, pp. 34–35). If the emission charge level is always higher than the expected marginal penalty, then marginal costs are no longer equal to the charge but are equal to the (lower) marginal penalty. In that case the firm would report zero emissions and increase actual emissions (see also De Clerq, 1983, p. 138). So, in general, imperfect enforcement does not have a negative impact on the cost efficiency or the environmental effectiveness of the emission charge. However, if the emission charge is always higher than the expected marginal penalty, the environmental goal will not be met.

In summary, the cost efficiency of emission charges might be impaired if costs are uncertain and the fixed costs of pollution control are important, or if collusive behavior or cost-saving technical progress occurs. Technological indivisibilities in pollution control and imperfect enforcement do not affect cost efficiency. With imperfect information about costs or discrete control options, or with economic growth or inflation, the environmental effectiveness of charges is not guaranteed. With imperfect enforcement the environmental effectiveness might be impaired if the emission charges always exceed the expected marginal penalty for noncompliance. Therefore, charges are more appropriate for those types of pollution for which keeping marginal costs low is more relevant than immediately meeting the environmental goal.

2.5 Regulation

Regulation consists of directives that require individual decision makers to set one or more input/output quantities, such as emissions per unit of input, at a specified level or to prohibit them from exceeding a certain level (Bohm and Russell, 1985). Regulation can take the following forms: performance standards, regulation of variables correlated with emissions, design standards, and bans on products or processes. Performance or effluent emission standards are effective means for meeting environmental objectives if monitoring of compliance and penalties for noncompliance are adequate. Their advantage is that no trial-and-error procedure or costly adjustments are needed. The best results are obtained if the emissions or closely correlated variables are regulated. Regulation is a faster and less expensive way to temporarily control sudden emission peaks or reduced assimilative capacity (De Clerq, 1983). This is because to control sudden peaks the reduction of production or the use of different inputs are usually the only means to cut emissions. In the case of an emission charge, knowledge of marginal costs would be required; if information on costs is wrong or uncertain, the polluter might decide to pay the charge, thus the desired environmental effects might not occur. If charges that do not vary over time are used, the charges might have to be very high to avoid all peaks (Baumol and Oates, 1988). In this case, regulation (e.g., by curtailing production) and charges to ensure a base level of emission reduction might form a cost-efficient combination. When monitoring is costly, impossible, or unreliable, design standards prescribing the use of a specific technology are obviously a potentially optimal regulation alternative. Design standards appear to be politically attractive and reliable, especially when there is no doubt as to what the best available technology is, although the environmental impact also depends on how the technology is used once installed.

Regulation in the form of performance standards or emission ceilings offers individual polluters the same options for reaching emission standards as are available with economic incentives unless the technology is prescribed and options to partly reduce production or change input variables influencing emission levels are ruled out. So regulation is not necessarily cost inefficient at the firm level. Allocative inefficiency is likely, however, when marginal costs for polluters differ and the regulating agency does not have sufficient information on the individual cost curves of the polluters (Førsund, 1992a; De Clerq, 1983).

Emission standards are also more rigid, because they do not automatically adapt to changes in costs, technical progress, and economic growth. If more polluters enter the region than expected, standards must be tightened to achieve the same global emission goal. Inflation does not impair their environmental effectiveness. Bans on products might be efficient if close substitutes are available at low costs. Collective facilities (such as communal water purification plants) might be preferable if their costs are lower than those for individual polluters.

With discontinuous control options, emission standards are more effective than a charge if the regulator knows the emission reductions achieved by each technology but does not have perfect knowledge of the associated costs. Cost efficiency, however, is even less guaranteed, because incorrect standards might have to be met by pollution control equipment operating at less than full capacity (De Clerq, 1983, p. 189).

With imperfect monitoring and enforcement the environmental effectiveness of regulations is not guaranteed. Tightening the standard, and thus increasing the marginal costs, will only result in lower emission levels if the expected penalty for noncompliance is also increased (Harford, 1978, p. 41). This is because the profit-maximizing firm will compare the cost of noncompliance with the cost of meeting the standard when deciding its actual emission levels.

To summarize, without sufficient information on marginal control costs, regulation is likely to be inefficient when marginal costs differ between polluters. This is especially true when control options are discontinuous. Regulations can be effective in meeting environmental goals if monitoring and enforcement are adequate and if regulations are adapted to economic growth.

2.6 Tradable Emission Permits

2.6.1 Perfect competition

The question is whether Dales (1968) was very much mistaken when he stated that markets could be used to implement any anti-pollution policy and, building on work by Crocker (1966), suggested the creation of a number of rights to discharge one unit of waste during a certain period with the option of auctioning these rights (Dales, 1968, p. 100). Dales' argumentation was clear. Some firms might find it more profitable to treat their waste and will therefore buy fewer rights than others. When the price is high enough to reduce demand, the market for rights will be at an equilibrium.

A trading system for emission permits could be designed in the following way. An emission permit is defined in terms of an allowable emission rate per year. A central environmental agency responsible for the overall pollution control policy of a region determines the total amount of issued permits (L) by taking the emission objective and subtracting the expected background emissions. The permits are then distributed to the sources, for example, according to emission levels in the past ("grandfathering") or through an auction.

It can be shown that under a number of restrictive conditions trading of such emission permits can achieve the least-cost solution. These conditions are that the permit market is competitive, that transaction costs are low, and that sources minimize their pollution control costs (i.e., the sum of expenses for abatement measures and the net revenues of trading permits).

The following additional variables are defined: the initial permits of source i (L_i^0), and the price of the permit (P). The goal of each individual source is to minimize costs. Costs consist of pollution control costs plus the cost of buying additional permits:

$$C_i(R_i) + P\left[(\overline{E}_i - R_i) - L_i^0\right] . \tag{2.14}$$

The cost minimum is achieved if marginal costs of emission reductions equal the price of the permit:

$$\frac{dC_i(R_i)}{dR_i} - P = 0 . \tag{2.15}$$

Under perfect market conditions there will be only one price, leading to equal marginal costs for all polluters. Note that this is exactly the condition for arriving at a cost minimum in the previous formulation. At the same time, the emission ceiling will be met if enforcement works properly, because no more permits are issued than are allowed for in the total emission objective. A more intuitive understanding of this cost minimum can also be given. *Figure 2.3* shows the marginal cost curves of two sources (MC_1 and MC_2) as a function of the amount of emissions. Drawn in this particular manner, the figure shows all possible combinations of emissions after the two sources are controlled that lead to a total emission reduction of 20 units. If the initial solution is a uniform reduction, each source will initially emit 10 units. With emission trading, source 1, facing higher marginal costs, will buy permits for 5 units and increase its emissions to 15 units. Source 2 will sell 5 units and thus reduce emissions to 5. The equilibrium permit price will be US$800. If the market works perfectly, the advantages are plentiful: the cost minimum is attained, the emission objective is reached, no centralized information on pollution control costs is required, and the permits can be distributed initially in any way that is politically acceptable.

The environmental agency would have to define and allocate property rights, set up rules for their transferability, and ensure their proper enforcement. The environmental agency would create the institutions for Coasian bargaining. Whether small emitters, such as passenger cars and private households, should be included depends on monitoring and enforcement costs; however, this doesn't seem very appropriate (Tietenberg, 1980). The advantage of emission trading is that the price

Economic Instruments in Theory

automatically adapts to economic growth, inflation, and technical progress (Nentjes, 1990a). Moreover, it reduces uncertainty about environmental effectiveness and possible adjustment costs.

2.6.2 Market and policy imperfections

Cost-efficiency properties depend on smoothly functioning primary and secondary permit markets. Market power, strategic behavior, high transaction costs, uncertainty about property rights, and imperfect enforcement may limit the workings of a permit market.

Market imperfections may occur if a firm with market power acts as a price setter (that is, administers the price rather than taking it as given) in order to minimize the sum of abatement costs and permits expenditures (Hahn, 1984). Hahn (1984) has shown that the deviation of abatement costs from the cost minimum is related to the extent to which the initial distribution of permits differs from the equilibrium distribution. In formula,

$$P = \frac{C'_1}{1 + \frac{1}{\eta}\left(\frac{L^0_1 - L^*_1}{L^0_2}\right)}, \quad (2.16)$$

where P is the permit price, C'_1 is the marginal cost of the price-setting firm 1, η is the elasticity of demand from the fringe (the other, price-taking sources), and L^0_1 and L^0_2 are the initial distribution of permits to the price setter and to the fringe, respectively. Equation (2.16) shows that if the initial distribution of firm 1 equals the equilibrium distribution (L^*_i), price equals marginal costs. If this is not the case, the extent to which prices (and hence abatement costs) differ from the optimal price depends on the marginal cost curve (the supply curve for firm 1), the price elasticity of the demand from the fringe, and the extent to which the initial permit distribution deviates from the optimal distribution.

Another type of strategic behavior occurs if firms use the permit market to drive up rivals' costs (exclusionary manipulation). Misiolek and Elder (1989) conclude that, surprisingly, this may not necessarily have a negative impact on cost efficiency compared with cost-minimizing manipulations, which always have a negative impact. Moreover, this can only occur if firms operating in the same industry also participate in the same permit market.

The effect that transaction costs can have appears to be more relevant. Transaction costs incurred by firms when trading may consist of search and information or bargaining and decision costs (Stavins, 1994; Fisher et al., 1994). With potential transaction costs (G) depending on the volume traded, each firm faces the following problem (Stavins, 1994):

$$\min C(R_i) + P\left[\overline{E}_i - R_i - L_i^0\right] + G\left(\overline{E}_i - R_i - L_i^0\right) , \qquad (2.17)$$

where $\overline{E}_i - R_i - L_i^0$ is the volume of emissions traded. The solution to this problem is the following:

$$R_i\left[C_i'(R_i) + G'(R_i) - P\right] = 0 . \qquad (2.18)$$

In contrast to the case without transaction costs, marginal pollution control costs will now be equal, not to the permit price, but to the permit price minus marginal transaction costs. The existence of transaction costs restricts the volume of permits traded. Like a direct tax, transaction costs drive a wedge between the price paid by the buyer and the price received by the seller (Nicholson, 1989, pp. 418–420). Stavins (1994) shows that in the presence of transaction costs, total expenditures on pollution control (apart from the transaction costs) will be higher than the cost-minimum solution (exclusive of transaction costs). The impact on the quantities traded depends on the form transaction costs take. If these costs consist of a lump sum per trade or if marginal costs decrease with the volume traded they could decrease trade or have no effect (for example, if the fixed transaction costs are below the expected cost savings of each trade). If marginal transaction costs are constant (for example, US$10 per ton traded), increasing, or dependent on the value of the trade, they will decrease traded volumes. Stavins goes on to find that the presence of transaction costs may also imply that the final equilibrium, and hence cost efficiency, is no longer independent of the initial distribution of permits. In conclusion, transaction costs reduce trading levels, increase abatement costs, and may cause the final equilibrium to depend on the initial permit allocation.

With imperfect enforcement, tradable emission permits may not lead to a cost-efficient solution (Malik, 1990) or to the meeting of the environmental goal (Keeler, 1991). Malik (1990, pp. 101–102) demonstrates that with imperfect compliance firms may no longer equate marginal profits, a function of pollution discharge, with the permit price. If firms maximize profits, they set the level of emissions where marginal profits equal the permit price plus the expected fine for noncompliance. In general, the potential cost savings of noncompliance, similar to transaction costs, drive a wedge between the permit price and marginal profits (or marginal pollution control costs if output is given). Malik (1990) finds this to be true unless the audit probability is independent of both the firm's level of emissions and its permit holdings or unless the audit probability increases with the size of the violations (the difference between actual emissions and permits). If these conditions do not hold, noncompliance is worthwhile and pollution levels are higher than desired. Moreover, emission trading is no longer cost efficient because noncompliance affects demand and supply and the permit price (Malik, 1990, p. 104). Keeler (1991) adds that the structure of noncompliance penalties is relevant. If the marginal penalty is constant (independent of the violation), tradable emission permits lead to less noncompliance than does regulation, because tradable permits lead to lower marginal abatement costs and thus lower the gains from noncompliance. With increasing marginal penalties (as a function of the violation), all firms will comply if the permit price is below the expected per unit violation penalty. If the permit price is higher, all firms will violate the emission limits. Although pollution will be higher than the goal, it is unclear whether the size of violation would be higher with trading or with regulation (Keeler, 1991, p. 187). With decreasing marginal penalties, firms that decide not to comply will pollute more than under regulation and will sell their unused permits to complying firms. Although more firms will comply than under regulation, total pollution levels might be higher under tradable emission permits than under regulation. Therefore, with imperfect enforcement, whether or not tradable permits meet the environmental goal depends on the structure of the penalty function. Pollution might be higher or lower than under regulation.

In conclusion, tradable emission permits might not be cost efficient with market power on the permit market, with positive transaction costs, or with imperfect enforcement. The cost efficiency loss depends on the demand and supply elasticities and the shapes of the transaction cost and penalty functions. With imperfect enforcement, emission trading is also no longer environmentally effective, but the size of the violation might be different from that under regulation.

2.7 Administrative Practicability

The administrative practicability of policy instruments depends on two elements: the information intensity and the ease of monitoring and enforcement. Information intensity means the required level of data availability and the modeling skills needed for an effective and efficient policy. Ease of monitoring and enforcement pertains to the difficulty of measuring and monitoring discharges and the action required to prod violators back into line. The administrative practicability is influenced by the type of pollutant being regulated (uniformly or non-uniformly mixed) (Bohm and Russell, 1985; De Clerq, 1983).

Regarding the information requirements for the uniform mixing case, it is clear that all instruments (charges, emission trading, regulation) require an emission inventory. In addition, implementing an emission charge would require some global information on control costs in order to improve the environmental effectiveness and to limit the trial-and-error procedure of setting the correct charge level. Information on costs is not necessary for regulation to be effective in terms of meeting the emission goal. Physical regulation would, however, require source-specific data on costs to be cost efficient, and some information on the expected costs would also be needed in the face of political reluctance to accept costly measures or unequally distributed costs. By contrast, a charge would not require detailed information on costs to be cost efficient: a uniform charge would simply equalize marginal costs. In principle, no information on costs is required for an emission trading system to be both environmentally effective and cost efficient.

Whether the administrative costs of implementation, monitoring, and enforcement are different with economic instruments than with regulation is a question open to debate. On the one hand, an emission charge faces higher costs because the charge revenues must be collected. On the other hand, regulation might require cumbersome legal procedures to ensure enforcement (Klink *et al.*, 1989, pp. 31–32). The drawback of emission trading is that the environmental agency must keep track of the trades that take place and must compare the emission rights possessed with actual emissions. Charges require more administration in the cases of inflation, economic growth, and cost changes in order to ensure that the environmental objectives are still met. The same is true for regulation in the case of economic growth. With emission trading the price automatically adjusts to these external changes without affecting environmental quality. An interesting question is whether enforcement and monitoring costs need to be different with charges, tradable permits, and regulation. Bohm and Russell (1985, p. 416) find that monitoring is no more difficult with an emission charge than with emission standards or marketable permits. Harford (1978) demonstrates that even under imperfect enforcement, with an emission charge a firm's actual level of emissions equals the per unit tax level and is independent of the shape of the penalty function for noncompliance. This is because the firm always minimizes costs to pollute at the level where marginal pollution control costs equal the tax (De Clerq, 1983, p. 138). A firm may underreport their emissions to evade tax payments. With direct regulation, a firm's emissions depend on the shape of the expected penalty function (Harford, 1978, p. 27). This suggests that with taxes, less enforcement (and fewer costs) would be needed than with direct regulation if the environmental objective is to be met. Kambhu (1990, p. S-72), however, suggests that when the regulator's enforcement power is limited, direct regulation in the form of technology standards could lead to higher compliance levels than would be achieved with charges. Keeler (1991) shows that whether or not tradable discharge permits allow higher noncompliance than regulation depends on the shape of the penalty function. In general, there seems to be little reason to believe that economic instruments require more or less enforcement and more administration than do regulatory instruments.

Economic Instruments in Theory   33

In conclusion, emission trading requires less information on costs to be cost efficient and environmentally effective than do regulations or charges. Administrative costs could be higher or lower depending on the cost of collecting charges and keeping track of trades, on the structure of the penalty function that applies for noncompliance, and on whether one wants to meet environmental goals with certainty and in a cost-efficient way.

2.8 Dynamic Efficiency

2.8.1 Introduction

The approach followed until now was mainly placed in the convenient framework of a static, partial equilibrium approach; the number of sources, their location, and the available pollution control technologies were assumed to be fixed. This section highlights some aspects of the relations between policy instruments and dynamic efficiency by considering their impacts on innovation, factor prices, and long-term adjustment (dynamic and general equilibrium impacts) and locational shifts.

2.8.2 Innovation

Regulation, effluent charges, tradable permits, and subsidies do not all have the same impact on innovation (Wenders, 1975; Downing and White, 1986; Milliman and Prince, 1989; Nentjes and Wiersma, 1988). Their impact depends on whether the regulatory agency reacts when innovation actually takes place.

Figure 2.4 depicts the impact of cost-reducing innovation, which shifts the marginal costs of the innovating firm downward (from MC_1 to MC_2) when the regulator does not respond to innovation. The target emission level is E_0, which in the initial situation with MC_1 in principle can be achieved with an emission charge T_0, a tradable permit system with E_0 permits and equilibrium price P_0, or regulation prescribing E_0 emissions. Under a charge regime, the cost-reducing innovation would reduce emissions further (from E_0 to E^*), yielding abatement cost savings and a reduction in charge payments (for unabated emissions) equal to the area $AA'E_{\max}$. A similar scenario applies to an emission trading scheme if the firm's innovation has no impact on the permit price P_0. Cost savings and permit expenditures would be reduced by the area $AA'E_{\max}$. In the absence of a price response, both economic instruments would result in the same savings in expenditures and the same impact on the emissions of the individual firm. Clearly, this is different under a regulatory regime: cost-saving innovation would reduce pollution control costs (area $ABE_{\max}$), but no further savings in tax or permit expenditures would be obtained. In brief, economic incentives would tend

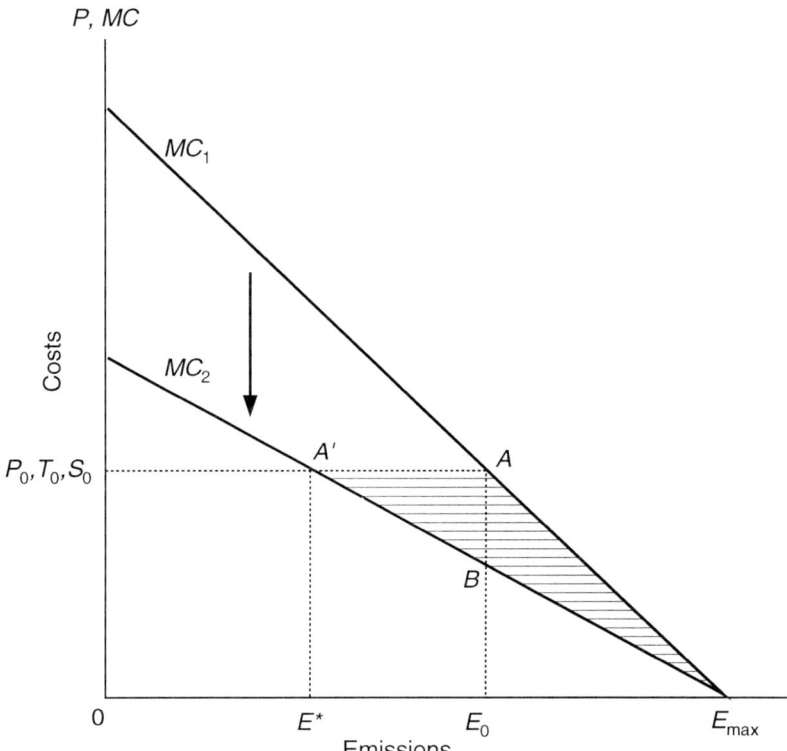

Figure 2.4. Innovation with no response from regulator.

to promote innovation more than regulation would, because a firm's expected expenditures would be higher under economic incentives than under regulation.

If the environmental agency (or the permit market) responds to the additional, unexpected reduction in emissions, the impact of charges and tradable permits differs (Figure 2.5). Such a response would be warranted if level E_0, for example, were to equal the assimilative capacity of the environment. If this were the case, the agency would want to lower the charge to T_1 and the price on the permit market would drop to P_1. Consequently, under a charge regime innovation would result in ultimate cost savings in pollution control cost equal to area ABE_{max} (pollution shifts from E^* to E_0) and in charge expenditures equal to area ABT_1T_0. With tradable permits, cost savings (ABE_{max}) and savings on permit expenditures (ABP_1P_0) would be similar to those with a charge. With regulation, cost savings still would only be equal to area ABE_{max}. Therefore, if the regulator responds to the

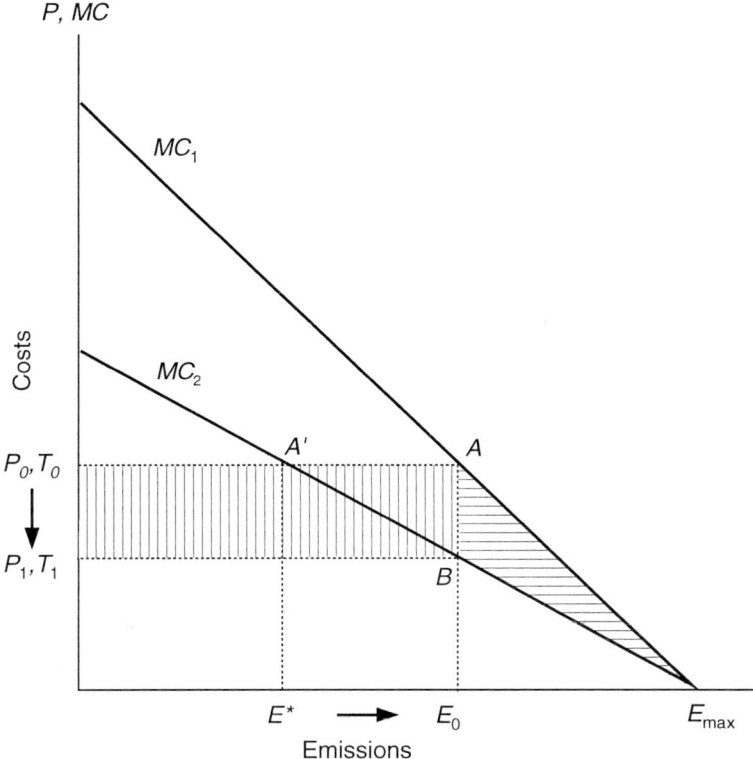

Figure 2.5. Innovation with response from regulator.

emission-reducing impact of innovation, the benefits of innovation, and therefore innovative push, would be higher with charges and emission trading than with regulation. It is important to note that with tradable emission permits, the market price increases with inflation or economic growth and, in contrast to regulation or charges, it automatically increases incentives to spur innovation in response to increased scarcity (Nentjes, 1990a). Similarly, cost-saving innovation will be reflected in a lower permit price.

These results are subject to some qualifications. Innovation under a regulatory regime might imply a risk of emission standards being tightened, because it results in a change of the definition of what can be considered the best available technology or the best practicable means. With regulation, the innovative push also depends on the type of regulation; it will be less with design or technology standards than with emission standards (Bohm and Russell, 1985). Nentjes and Wiersma (1988, p. 60) suggest that the argument in favor of the innovative impact of economic instruments relies on the assumption that polluters do their own environmental research and development. If innovation is undertaken by specialized firms, market incentives might create less clear-cut, more diffuse impacts on innovation (Wiersma, 1989). Innovation, then, depends on the demand for innovative pollution control, which is a function not only of the use of market incentives but also of the stringency of the environmental policy, because stringent standards increase the demand for cost-saving pollution control technology.

2.8.3 Long-term impacts on output, emissions, and prices

For ease of exposition, three different long-term impacts of environmental policy can be distinguished:

- Impacts on the long-run output and emissions within one particular industry;
- Impacts on output and factor prices in a multiple-sector context (general equilibrium approach);
- Impacts on the location of industry.

Regarding the long-term output levels of a competitive industry, emission charges, tradable permits, and regulation are likely to have different impacts on total industry output and emissions in the long run. This is because the instruments result in different incentives for firms' entry to or exit from the industry. A charge increases not only marginal costs of production but also average costs. This shifts the industry supply curve to the left and reduces total industry output and associated emissions. The same occurs with an emission permit market, because in a world of perfect knowledge this has the same impact. What is not clear from the outset is the difference between regulation on the one hand and charges/tradable permits on the other hand. Regulation is less cost efficient than charges and it raises short-term average costs more. However, with charges and permit trading, industry average costs are likely to be higher due to the price that must be paid for unabated emissions, which will lead to a reduction of output in that particular sector. In this context how the regulatory agency reacts is also important. If the policy target is a given maximum level of total emissions in industry, market prices will be lower, industry output will be higher, and emissions per unit of output will be lower under regulation than under charges or tradable permits.

The approach followed thus far is incomplete because it considers only one market. General equilibrium broadens the scope to more sectors and resource inputs (e.g., Førsund, 1985; Christiansen and Tietenberg, 1985; Peterson and Fisher, 1977). In a general equilibrium approach regulation changes the relative price of the polluting commodity, which affects sector structure and the allocation of resources.

According to Siebert (1987), one can set up a two-sector model where resources are used as input and pollutants resulting from commodity production can be abated. In such a context an emission tax, or any other instrument, affects environmental quality, sectoral output, and the allocation of resources. The tax reduces the use of resources in the pollution-intensive sector. Moreover, resource use declines in production and decreases in pollution abatement. The sector structure shifts in favor of the less pollution-intensive sector. Whether or not production in this sector increases depends on the strength of both substitution and income effects and on how emission permits are distributed. Environmental quality improves and the relative price of the pollution-intensive commodity increases so that demand declines. The price elasticity of demand for the pollution-intensive good influences the outcome. The eventual effects are not so straightforward, because there might be more than one factor of production, the structure of industry within a sector might change, production factors might not be fully employed, and curbing pollution might have positive effects on productivity. Furthermore, if one compares different instruments the results will also depend on the redistribution of charge revenues, on the initial distribution of tradable permits, and on whether measures are taken domestically or in an internationally coordinated way (Nentjes, 1990a). Moreover, and this is perhaps more important, the impact on environmental quality depends on how the different instruments adapt, if at all, to growth in production levels. Here, again, tradable permits have the advantage of automatic adjustment.

Finally, the choice of instruments might influence the location of industry in the long run (Bohm and Russell, 1985). Such locational shifts are more likely with instruments that are tailored to the location of the source: ambient charges, deposition permits or alternatives, and point-by-point regulation. Such systems tend to create a stronger incentive to locate away from "hot spots" (Dales, 1968) than do uniform or across-the-board regulations and charges.

2.9 Political Acceptability

No satisfactory framework exists that explains the direction of environmental regulation and the increasing role of economic instruments, although several attempts have been made (e.g., Bromley, 1989; Downing, 1984; Hahn, 1990; Kling, 1988; Opschoor and van der Straaten, 1992). Two elements seem to play a crucial role in explaining changes in regulation:

- Changes in property rights occur when the gains of internalizing externalities exceed the costs and environmental policy measures are taken.

- The form (the type or mixture of instruments) this change in property rights takes depends on the distribution of costs and benefits to all involved and the extent to which these actors are able to transform their interests.

Changes in the overall cost-benefit configuration appear to be a prerequisite for regulatory evolution. In other words, the benefits of making the change should exceed the costs, where costs also include transaction costs. The cost-benefit ratio may change due to economic, ecological, socio-cultural, or political changes. According to Bromley (1989), new economic and technological developments and changes in relative prices may foster changes in institutional arrangements because they may shift the limits of production possibilities so that net improvement is possible. Innovation might, for example, result in reduced pollution control costs and transaction costs, opening the way for changes in institutional arrangements. Environmental changes may result in increased benefits from reducing pollution. Socio-cultural shifts in collective attitudes about income share and consumption also induce changes. Similar arguments can be put forward for political changes. Kling (1988) states that regulatory evolution is likely to occur when new (potential) sources of costs and benefits are identified or recognized, or when changes in information availability and transaction cost barriers occur. This suggests that, because in principle they offer the possibility of reducing costs, the extended use of economic incentives creates an impetus for net beneficial regulatory changes.

Nentjes (1991) believes the available instruments might evolve. Technical or design standards are applied in the first, primitive stage, when environmental policies are in their inception and monitoring of emissions is out of reach. Performance standards belong to the next stage. Tradable permits become feasible at a higher stage of policy evolution, when national emission targets have been set, emissions per firm are registered and monitored, and enforcement is well developed.

Opschoor *et al.* (1994) stress the importance of the context in which instruments are applied and distinguish between external and internal elements. External elements include operative national and international economic, fiscal, and environmental policies. Internal elements are related to specific application and pertain to the characteristics of the sources of pollution (their number, the compliance costs, and the market scale) and of the polluting substance (the link between emissions and environmental impact and the ability to monitor it). These characteristics affect the instruments' efficiency and environmental effectiveness. If, for example, sources are diffuse and monitoring is costly, charges or permits might be inefficient and technical standards might be preferable. The choice of instruments is, of course, also influenced by the instruments that are already in use.

The shape regulatory changes take can be seen as being dependent on the extent to which different groups are able to articulate their interests. According to

Hahn (1990), one can say that the output of environmental policy emerges from a struggle between key interest groups, or that the political power of the actors affects the results (Downing, 1984). The results of this debate affect the design of environmental policy and the cost efficiency of introducing economic incentives, as well as whether they are introduced at all.

Buchanan and Tullock (1975) suggest that industry might prefer emission standards (in the form of a fixed emission quota for the whole sector) to charges because they result in higher profits. Standards act as a barrier to entry for new firms, thus raising profits for existing ones, whereas charges or emission standards per unit of output do not preclude entry. Moreover, charges or emission standards raise costs because unabated emissions are no longer free. Dewees (1983) argues that shareholders of existing firms should prefer uniform effluent standards (which limit pollution per unit of output) to effluent charges or the sale of tradable effluent rights, because the standards raise prices less. He adds that shareholders of existing firms may prefer tradable permits to charges or standards if these permits are distributed for free. Existing firms prefer stricter standards for new plants than for existing plants. Labor may also prefer tougher standards for new plants to protect existing employment, but because marketable rights are given to firms, they do not have the same appeal for labor as for shareholders of existing firms.

Hahn (1990) identifies two basic interest groups: industry and environmentalists. The former is concerned about the impact of policies on profits and the latter focuses on their impact on the environment. The regulatory agency must balance the interests of both groups in an acceptable policy. A policy can be considered acceptable when it is at least as good as the status quo. Nentjes and Dijkstra (1993) incorporated the above ideas into one political choice model, distinguishing four interest groups: shareholders, labor, environmentalists, and the environmental authorities. Given its objective (profit, employment, etc.), one can derive a preference ordering of instruments for each interest group. Interest groups use connections with political parties to translate their preferences into political choices for instruments. Mainly because of its relative impact on employment, left-wing governments might prefer direct regulation. Under conservative (right-wing) governments "grandfathered" tradable permits might stand a fair chance because of their positive impact (in a closed economy) on the interests of shareholders. Verbruggen (1994, p. 50) argues that the government has more objectives than just environmental policy objectives; among others are its economic policy goal. To meet these often conflicting goals, the government seeks a coalition with producers supporting a regulatory approach.

Hahn argues that because economic incentives promise cost savings, an increase in industry influence will increase the market orientation of environmental policy and tend to decrease the level of environmental quality. This result, however, also depends on the distribution of cost savings within industry. The fact that regulation is more familiar and may offer more ways for existing firms to influence the outcomes in the process of bargaining might make them prefer regulation to economic instruments (De Clerq, 1983). This view is reinforced by the fact that regulation usually shelters existing industry at the expense of newcomers (Verbruggen, 1994, p. 48–49). An increase in industry influence would also tend to result in a reduction in fees. Both industry and environmentalists support earmarking of fees, although for different reasons: industry, because it increases the credit they can claim for reducing pollution and because it might increase their revenues, and environmentalists, because they are in favor of activities that promote environmental quality. The acceptability of tradable permits for the different polluters hinges on conditions of their initial distribution. The auction of pollution rights without a redistribution of the revenues appears to be unacceptable for industry because they would then have to start paying to pollute.

Environmentalists might oppose a free distribution of permits, because it creates windfall profits. They might also oppose using charges as the only instrument because they offer polluters the option of continuing to pollute at the same levels. Furthermore, environmentalists might prefer regulation because it restricts the property rights of the polluter.

Hahn (1990) states that for the regulating authorities the visible impacts on costs and jobs may be important as well. Here, marketable permits and emission charges have the disadvantage that the costs are quite visible, because they rest directly on industry in terms of tax or permits expenditures. The cost savings, however, are more diffuse. The negative impact on jobs might also be more clear with market-oriented instruments. In spite of this, the appearance of economic instruments is more likely if the expected cost savings are relatively high and unemployment levels are low. In favor of charges, however, is the fact that they are a source of government revenue and are more in line with the polluter-pays principle (Bohm and Russell, 1985; Cropper and Oates, 1990). The authorities might prefer regulation to market-oriented instruments because it is more familiar and does not require sophisticated economic knowledge. Charges may be an uncertain source of revenues and are more uncertain in their environmental impacts.

In summary, the role economic incentives can play depends greatly on the extent to which the cost savings they promise exceed the additional costs of changing existing regulation. The form the new policy takes then depends on the distribution of costs and benefits among the different actors, the extent to which they are able to articulate their own interests, and the specific context of the instrument's application. Acceptance of economic instruments is more likely when policy, and especially monitoring and enforcement, is at a higher stage of development. In this respect, economic incentives might face disadvantages, with the exception,

Economic Instruments in Theory 41

perhaps, of "grandfathered" permits. In spite of their cost efficiency advantages, they increase expenditures for industry, especially if charges are not refunded or if permits are not distributed for free. In addition, both the regulatory authorities and industry might prefer regulation, because it is more familiar and the cost configuration is more diffuse.

2.10 Conclusions

This chapter examined the conditions for efficiency and cost efficiency, as well as the administrative practicability, dynamic efficiency, and political acceptability of emission charges, emission trading, and regulation. If the environment is considered a common property resource where free access prevails, private decisions tend to lead to excessive pollution levels. If maximization of welfare is the policy aim, then welfare economics prescribes that negative environmental externalities should be corrected in the form of a Pigovian tax, environmental standards, tradable permits, or Coasian bargaining to obtain a solution that is optimal in terms of maximizing welfare. The welfare impacts of instruments differ and depend on uncertainty about costs and environmental damages.

Implementation of environmental policies that maximize welfare is difficult, however, due to the information requirements on costs and environmental damage and due to market imperfections in other parts of the economy. Furthermore, optimal levels of pollution might not be desirable per se if they have unwanted consequences for the intra- and intergenerational distribution of welfare, or if they imply that the assimilative capacity of the environment will be exceeded and irreversible damage will occur. For these reasons, a more practical object of study is a policy that sets out to reach a set of exogenously fixed environmental standards at minimum costs, accounting for administrative practicability, dynamic impacts, and political acceptability.

Cost efficiency implies the attainment of environmental standards at minimum costs. The conditions for cost efficiency depend on the type of pollutant. With a uniformly dispersed pollutant, the environmental policy can be formulated to meet a total level of emissions in a region. In this case, cost efficiency requires the marginal costs of pollution control to be equalized across all polluting sources. As Chapter 3 shows, this is not true with non-uniformly dispersed pollutants.

Static cost efficiency is not the sole dimension for judging the desirability of a policy instrument. Therefore, *Table 2.1* compares the different instruments along different dimensions based solely on theoretical arguments. One should realize that, to a large degree, the performance of the instrument depends on the circumstances

Table 2.1. Evaluation of instruments.

	Instrument		
Dimension	Emission charges	Tradable permits	Regulation
Cost efficiency	+	+	–
Environmental effectiveness	–	+	+
Administrative practicability	+	+	+
Dynamic efficiency	+	+	0
Political acceptability	0	0/+	+

+ = high; – = low; 0 = neutral.

Table 2.2. Conditions affecting cost efficiency and environmental effectiveness.

	Cost efficiency			Environmental effectiveness		
	Emission charges	Tradable permits	Regulation	Emission charges	Tradable permits	Regulation
Uncertainty about costs	–	0	–	–	0	●
Imperfect markets	–	–	●	–	0	●
Transaction costs	0	–	0	0	0	0
Imperfect enforcement	0	–	●	0/–	–	–
Discontinuous control	0	0	–	–	0	0
Cost-saving technical progress	–	0	●	●	0	0
Economic growth	0	0	0	–	0	–
Inflation	0	0	0	–	0	0

– = negative impact; 0 = no impact; ● = unknown.

under which it is applied, which limits generalizations. Below we discuss each dimension in some detail.

Regarding cost efficiency, emission charges and tradable permits are clearly preferable to regulation. A number of circumstances, however, affect the cost efficiency of the instruments (see *Table 2.2*). Uncertainty about costs (combined with relatively fixed costs), cost-saving technical progress, and collusive behavior might impair the cost efficiency of charges. The cost efficiency of tradable emission permits can be negatively influenced by market power, high transaction costs, or imperfect enforcement. Regulation is generally inefficient, mainly due to a lack of allocative efficiency. Discontinuous control might negatively affect the cost efficiency of standards if control equipment has to operate at less than full capacity.

The environmental effectiveness of regulation and tradable permits is greater than that of emission charges. Uncertainty about costs, collusive behavior, technological indivisibilities, economic growth, inflation, and, in some cases, imperfect enforcement all have a negative impact on the environmental effectiveness of charges. With imperfect enforcement it is also uncertain whether tradable permits meet their environmental objective. Inflation, economic growth, and technical progress do not influence their environmental effectiveness. The environmental effectiveness of regulation is at stake with imperfect enforcement and with economic growth, if the regulation sets emissions per unit of output.

As for administrative practicability, there is little reason to believe that information requirements and the implementation and enforcement costs of charges and tradable permits would be higher than with regulation. An advantage of tradable permits, and to a lesser degree charges, is that no information on costs is required to meet the goal in a cost-efficient way. The ultimate differences in administrative practicability will depend on specific circumstances, such as the availability of monitoring technology, the costs of collecting charge revenues, or the ability to keep track of permits traded.

Concerning dynamic efficiency, both charges and tradable permits have the advantage over regulation in terms of creating higher incentives for innovation and, because they attach a price tag to pollution, of steering the allocation of scarce resources in the correct direction in the long run. Tradable permits do this automatically, even with inflation or economic growth. The innovative push, however, depends on the type of regulation and on which firms implement innovation.

The political acceptability of regulation might be slightly higher than that of charges or tradable permits, because regulation generally is used in existing environmental policy. Whether economic instruments are applied depends on whether the cost savings they promise are high enough to merit a change in existing regulations. This also depends on the specific application context and the stage of development of environmental policy. The type of instrument chosen is affected by the extent to which different actors (industry, labor, environmentalists, government) can articulate their own interests. Industry might prefer regulation, because it increases overall costs less, but might find tradable permits acceptable if they are "grandfathered," and might find charges acceptable if revenues are returned and cost savings are sufficient. Labor might prefer regulation to protect employment. Environmentalists might oppose tradable permits and charges because they create the right to pollute. Government might prefer regulation because it is more familiar, or charges because they promise revenue, albeit uncertain revenue. This multi-actor context and the specific application context will affect the type and the form of the instrument that is ultimately selected.

Chapter 3

Instruments in Theory when Location Matters

3.1 Introduction

When the objective of environmental policy is to attain certain exposure levels (such as ambient concentrations or deposition of pollutants), the control problem becomes more complicated, particularly if more than one receptor is considered (Bohm and Russell, 1985). It is not difficult to see this, because in this case not only the total volume of pollution but also the locations of the polluting sources determine the ultimate environmental impacts. For example, for the acidification of lakes in southern Finland it is clearly more important for emission reductions to take place in Russia, which borders Finland, than in Bulgaria. If total emissions in Bulgaria and Russia were to remain constant but emissions were to increase in those parts of Russia close to Finland, acidification of Finnish lakes would increase.

The purpose of this chapter is to examine if, and under which circumstances, regulation, emission charges, and in particular different emission trading schemes are cost efficient and environmentally effective when location of emission sources matters. In other words, the chapter looks at non-uniformly dispersed pollutants such as SO_2.

In Section 3.2 the conditions for cost efficiency are derived. In Section 3.3 the cost efficiency of charges is assessed, and the cost efficiency of regulation is discussed in Section 3.4. In Section 3.5 the cost-savings potentials of different trading schemes are investigated. Section 3.6 asks whether administrative practicability, dynamic efficiency, and political acceptability differ from the evaluation in Chapter 2 if location is relevant.

3.2 Cost-Efficiency Conditions

When the location of emission sources is of importance, the conditions for an optimum solution are different from those in the simple emission-oriented approach (Tietenberg, 1985). The conceptual framework contains the following additional elements: the level of deposition at receptor j (D_j), the background deposition at receptor j (N_j), the desired level of deposition (the target load) at receptor j (D_j^0), and a (linear) transfer coefficient, which translates emissions of source i to deposition at receptor j (a_{ji}).

The deposition at a specific location is a function of the background deposition plus the sum of the emissions, multiplied by their transfer coefficients:

$$D_j = N_j + \sum_{i=1}^{I} a_j i(\overline{E}_i - R_i) \quad \text{for every } j(j = 1, \ldots, J) \ . \tag{3.1}$$

A cost-efficient solution requires the total costs of emission reductions to be minimized subject to the constraint that the desired deposition levels are met at each receptor point:

$$\min \sum_{i=1}^{I} C_i(R_i) \tag{3.2}$$

subject to

$$N_j + \sum_{i=1}^{I} a_{ji}(\overline{E}_i - R_i) \leq D_j^0 \quad \text{for every } j = 1, \ldots, J \ . \tag{3.3}$$

In addition, $\overline{E}_i \geq R_i \geq 0$. The most relevant of the necessary and sufficient Kuhn–Tucker conditions for a cost minimum are the following (Tietenberg, 1985):

$$R_i \left[\frac{dC_i(R_i)}{dR_i} - \sum_{j=1}^{J} a_{ji}\lambda_j \right] = 0 \quad \text{for every } i \tag{3.4}$$

$$\lambda_j \left[D_j^0 - N_j - \sigma_{i=1}^{I} a_{ji}(\overline{E}_i - R_i) \right] = 0 \quad \text{for every } j \ . \tag{3.5}$$

Equation (3.4) states that for a cost-efficient solution either the emission reduction of the source has to be zero ($R_i = 0$) or the marginal costs of emission reduction for each source have to equal the weighted sum of the shadow prices (λ_j) for each receptor. Weights are the transfer coefficients from source i to each affected receptor j. Equation (3.5) shows that either the required target load (D_j^0) is met exactly or the shadow price (λ_j) is zero. The latter means that the receptor is non-binding. The important conclusion here is that, generally, for a cost-efficient solution the marginal costs per ton emission reduction will be different. Stated another way, the marginal costs per ton deposition must be equalized.

3.3 Ambient or Deposition Charges

When controlling non-uniformly dispersed pollutants, a uniform emission charge generally does not lead to a cost-minimum solution. A uniform emission charge induces polluters to equalize marginal cost per ton emission removed. Equation (3.6) [derived from equation (3.4)] shows the ratio of the marginal costs between two polluters in the optimal allocation (Tietenberg, 1974, 1978):

$$\frac{C'_1}{C'_2} = \frac{\sum_j a_{j1} \lambda_j}{\sum_j a_{j2} \lambda_j} \quad \text{for every } j \;. \tag{3.6}$$

Equation (3.6) shows that, generally, the marginal costs per ton emission must be different. Hence, the optimal emission charge (F_i) for each source must be tailored to the source's location and its impact on the binding receptors, where F_j is the deposition charge (or the shadow price) and a_{ji} denote the transfer coefficients:

$$F_i = a_{1i} F_1 + a_{2i} F_2 + \cdots + a_{ji} F_j \;. \tag{3.7}$$

Without perfect knowledge of the costs and with more than one receptor, attainment of a cost-efficient solution is problematic. If there is only one receptor, the ratio of the optimal emission charges for the two sources depends only on the transfer coefficients for the one receptor. The usual trial-and-error procedure can be applied to achieve the desired ambient concentrations in a cost-efficient way. Equation (3.6) gives the ratio of the cost-minimizing taxes for each source, because with one receptor λ_j would cancel out. Fixing the charge level for one source would then immediately give the charge levels for all other sources. The charge levels could then be set and increased or decreased simultaneously by the same percentage until the desired deposition levels are met.

With more than one receptor, even with complete knowledge of the transfer matrix, it becomes more difficult to determine the optimal charges. Established opinion is that without full knowledge of the cost-minimum solution, the environmental agency cannot know which receptors are binding and what their shadow prices are. Although trial and error might result in a solution that achieves the desired ambient concentrations, the resulting allocation generally will not be cost efficient (Bohm and Russell, 1985). In this case the alleged information advantage of charges would break down because information on costs would be needed to set the cost-minimizing charge levels. However, Ermoliev *et al.* (1995) prove that even with imperfect information on costs, a more complex trial-and-error procedure can be used to determine cost-minimizing charge levels. The procedure consists of two steps. First, emission charge levels are set after conferring with the sources about the marginal cost of control. Second, the charges are implemented. In this stage the agency uses information on the observed difference between actual and desired deposition levels to adjust the deposition charge levels and, using the transfer coefficients [see equation (3.7)], the emission charge levels. If the deposition exceeds the desired level at a receptor, the deposition charge for this receptor is increased stepwise until the desired deposition levels are met at all receptors. If sources minimize costs, they always meet their emission levels at minimum cost, and because the charge levels are based on the difference between observed and actual levels, allocative efficiency is also guaranteed (Ermoliev *et al.*, 1995). In this case the information advantage of charges would be valid again, but it should be realized that the adaptive procedure would take time, and under- or overshooting the targets (and spending more than necessary) might temporarily occur. Adaptation of charges might also create uncertainty and might not lead to the desired results if fixed costs of pollution control are high. Still, these adaptive charge mechanisms would enable industry to meet the deposition goals at lower costs than would a system of standard regulation, without requiring knowledge of costs.

As a kind of fallback from the complexity of ambient charges, uniform charges might be used; depending on the specific regional circumstances, these charges may or may not be more cost efficient than uniform cutbacks (Russell, 1986).

3.4 Regulation and Location

If perfect information on costs were available, the environmental agency could just as easily formulate source-specific regulations to meet a set of deposition targets. This is because the agency would know exactly which emission reductions at which location would be cost-minimal for which source. When information is imperfect (or when cost minimization is not of primary importance to the environmental agency), regulatory instruments, although producing differences in marginal costs among individual polluters, are likely to result in a less than cost-efficient solution. "Regulated" differences in marginal costs do not correspond to optimal differences. The extent to which this occurs depends on regional circumstances and can only be determined empirically. Uniform cutbacks will always be more costly than necessary. However, they might be more or less expensive than uniform emission charges or emission trading in one zone (Spofford and Paulsen, 1990, pp. 2–14).

3.5 Emission Trading

3.5.1 Introduction

This section highlights the use of different tradable permit schemes for non-uniformly dispersed pollutants. After discussing the best alternative (ambient or deposition permits), the section highlights alternatives in the form of trading rules and even simpler schemes such as single-zone and multiple-zone trading and single-receptor deposition trading. Because Chapter 4 (particularly Section 4.4) elaborates on these alternatives, this section discusses them in detail.

3.5.2 Ambient or deposition permits

If the objective of environmental policy is the attainment of air quality goals or deposition targets at certain locations, a system of transferable permits would require the creation of "ambient permits" or deposition permits. Such permits would allow each polluter to deposit a specific amount of a pollutant at certain receptors. The target deposition level would be specified for each receptor. After subtracting background deposition, the remaining deposition at each receptor would then be distributed to each polluter as deposition permits. The only information required would be source-receptor matrices describing the atmospheric dispersion of pollutants. For every receptor a separate market would have to be established. In order to emit one unit, each source would have to keep the appropriate number of deposition permits (according to the source-receptor matrix) for each receptor it affects. If a source wants to increase emissions it must obtain additional deposition permits for each of the receptors its emission reaches.

In a seminal article, Montgomery (1972) provided the proof that under certain conditions the competitive equilibrium for an ambient permit market exists. Moreover, it coincides with the cost-minimum attainment of a set of predefined air quality standards, irrespective of the initial distribution of ambient permits, as long as the license total equals the ambient or deposition standard.

This proof consists of three steps. First, Montgomery characterizes the competitive market equilibrium. L_{ij} represents the quantity of licenses allowing source i to pollute at receptor j. E_i represents the emissions from source i. Each source minimizes costs of controlling emissions (C_i) plus the costs of purchasing licenses, with the condition that no source is allowed to pollute more at any receptor than its "portfolio" of permits allows. Given that L_{ij}^0 is the initial distribution of licenses, the source problem is to minimize

$$C_i(E_i) + \sum_j P_j^*(L_{ij} - L_{ij}^0) \tag{3.8}$$

subject to

$$L_{ij} - a_{ji}E_i \geq 0 \quad \text{for all } j = 1, \ldots, J \tag{3.9}$$

$$L_{ij} \geq 0 \;, \quad E_i \geq 0 \;, \quad P_j \geq 0 \;,$$

with P_j being the price at the jth permit market and P_j^* being the market equilibrium price. A market equilibrium exists when there is a set of prices such that when each firm minimizes costs, the excess demand for licenses is non-positive and excess supply drives the price to zero. In other words, the market clearing conditions hold. Second, Montgomery derives the following conditions for a license-constrained cost minimum, i.e., a vector of emissions minimizing total costs subject to an initial distribution of licenses. If the market equilibrium price (P_j*) equals the shadow price (λ_j) for all i and j, and if the licenses ($L_{ij}*$) held in market equilibrium equal $a_{ji}\, E_i^*$ (E_i^* being the cost-minimizing emission vector), then any market equilibrium is a joint cost minimum. Third, Montgomery proves that the market equilibrium coincides with the cost-minimum solution for attaining a set of air quality standards if the initial license totals equal the desired air quality levels at each point. This result is independent of the initial distribution of the ambient permits.

A simple summary of Montgomery (1972) can be given if one assumes that $L_{ij}^* = a_{ji} \cdot E_i^*$, and that emissions after abatement equal uncontrolled emissions minus emission reduction ($E_i = \overline{E}_i - R_i$). This allows a rewriting of equations (3.8) and (3.9). The source problem now is to minimize costs of controlling emissions and buying permits:

$$C_i(R_i) + \sum_j P_j \left[a_{ji}(\overline{E}_i - R_i) - L_{ij} \right] \tag{3.10}$$

subject to

$$a_{ji} E_i = L_{ij}^* \quad \text{for every } i \ .$$

The first-order condition for a cost minimum is

$$R_i \left[C_i'(R_i) - \sum_j P_j a_{ji} \right] = 0 \ . \tag{3.11}$$

Thus, either the emission reduction of the source is zero or the marginal costs of reducing emissions have to equal the weighted sum of the permit prices of each receptor the source affects (weights are the transfer coefficients). Clearly, if the permit prices for each source at each receptor market (P_j) equal the shadow price (λ_j), the competitive market solution coincides with the cost minimum [see equation (3.4)]. If the license totals (L_{ij}) equal the deposition target (D_j^0) for each receptor, the equivalence of demand and supply ensures that $P_j = \lambda_j$ in each market (Tietenberg, 1985, p. 24).

In summary, the deposition permit system has some desirable properties: it minimizes pollution control costs for any initial distribution of permits, the environment agency does not require knowledge of costs, and the ambient standards are always met.

Several authors, however, have pointed toward some problems with the system's implementation, because transaction costs might be high (Krupnick *et al.*, 1983; McGartland, 1984, pp. 22–25), especially with a large number of receptors. Moreover, thin markets may cause non-price-taking behavior (Hahn, 1986, p. 6). Even under conditions of imperfect competition, however, the free rider problem (McGartland, 1988) is avoided because all property rights are distributed and free access no longer exists. Stated differently, cost efficiency might be impaired, but environmental effectiveness will not be.

Krupnick *et al.* (1983) note that the transaction costs for polluters might be substantial because each cost-minimizing polluter must acquire permits in each of the receptor markets its emissions reach, so that marginal costs of emissions equal the weighted sum of permit prices at each receptor market and the depositions this optimal emission causes must be equal to or smaller than the number of licenses it holds at each receptor. In order to do this, the source would have to collect a portfolio of licenses such that

$$C_i'(R_i) = a_{1i} P_1 + a_{2i} P_2 + \cdots + a_{ji} P_j \tag{3.12}$$

and

$$a_{ji} E_i \leq L_{ij} \quad \text{for every } j \ .$$

If initially $C_i'(R_i)^0 > C_i'(R_i)^*$, the marginal costs of the equilibrium emission reduction, the source will want to increase its emissions and reduce marginal costs. In order to do so it will then have to buy deposition permits for each of the receptors it influences. It must buy in fixed proportions (transfer coefficients) of its desired emission level. This leads McGartland (1984, p. 24) to conclude that as the number of receptors increases, the number of transactions needed per unit of emission reduced increases and obtaining the optimal amount of permits becomes more and more difficult. Transaction costs increase and the incentive for a polluter to reduce emissions and to sell all unnecessary deposition permits is reduced. Consequently, the number of profitable transactions in deposition permits will be limited. The sources might not be able to locate potential buyers of permits quickly, and transaction costs might therefore be high. Depending on the initial distribution of deposition permits, the source might even have to simultaneously buy and sell on different markets (Bohm and Russell, 1985).

A simple example might illustrate the problems. *Figure 3.1* represents a typical airshed with three sources (1, 2, 3) and two receptor areas (A, B). Given the initial distribution, it is assumed to be optimal for source 1 to reduce its emissions and sell its deposition permits for A and B to sources 2 and 3, respectively. In order

Instruments in Theory when Location Matters

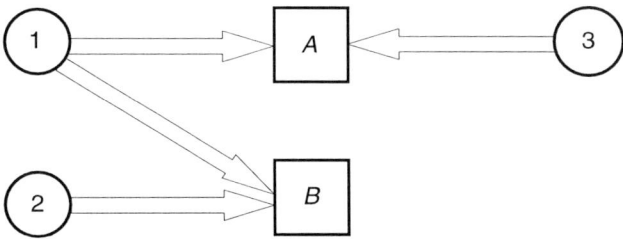

Figure 3.1. Deposition permits.

to arrive at a cost-efficient solution it would also be necessary to sell both types of permits. The cost-minimum condition (3.12) requires that

$$C'_1(R_1) = a_{A1} * P_A + a_{B1} * P_B \ . \tag{3.13}$$

If source 1 is not able to sell its redundant permits for receptor B to source 3 (for example, because of extremely high transaction costs), the bilateral trade with source 2 will result in lower than optimal marginal costs, because $a_{B1} * P_B = 0$. This might not be very problematic, because source 1 can keep its right to deposit at receptor B to sell later.

The point here is that for an optimal trade to occur, source 1 has to trade with both sources 2 and 3. With a perfect competitive market and low transaction costs, there is almost no problem because the deposition permits allow source 1 to keep its permits for deposition at receptor B to sell at a later stage. This is not without problems, however, if source 1 makes the decision without taking the revenues of selling the receptor B permits into account. To arrive at an optimal decision, perfect foresight would be required. Obviously, the fact that all rights to deposit are distributed implies that the common property resource, air, is no longer subject to free access. Because all property rights (the right to cause pollution at certain receptors) are privatized, free riding is no longer possible, and those sources not holding property rights are excluded from depositing waste at the receptors.

In summary, although a system of tradable deposition permits has the distinct advantage of excluding free riding, transaction costs might be high. What is more, the fact that as many markets as receptors will have to be established might lead to thin markets, with just a few buyers and sellers, where price-taking behavior is no longer an appropriate assumption (Hahn, 1986). In this case, strategic behavior and market power may lead to incorrect price setting, as was indicated in the section on emission permits.

3.5.3 Trading subject to rules regarding deposition

In response to the potentially high transaction costs associated with deposition permits, several alternative trading rules were proposed (Krupnick *et al.*, 1983; McGartland and Oates, 1985; Tietenberg, 1985). The common characteristic of all these systems is that emissions (not depositions) are traded under the constraint that deposition targets must be met. In other words, the objective of these rules is to meet a set of deposition targets with a more cost-efficient allocation of emission reductions between the sources. There are three rules:

- Pollution offset (emission trading subject to the condition that the deposition targets are not violated);
- Modified pollution offset (subject to deposition targets and the condition that pre-trade air quality does not deteriorate);
- Non-degradation offset (emission trading subject to non-violation of deposition targets and the rule that total emissions are not allowed to increase).

Figure 3.2 shows how these rules would guide emission trading. There are two receptors, 1 and 2. The lines R_1 and R_2 represent combinations of emissions from both sources for which the deposition targets (concentration levels) for receptors 1 and 2 are met exactly. The starting point (E) is the initial distribution of rights. R_c is the line that represents the pre-trade (current) air quality at receptor 1. The 45-degree line presents combinations of emissions that hold the total level of emissions constant.

Under the pollution offset rule, the trading possibilities are those within the area to the left of line FGB. This is because the ambient standards for receptor 1 and receptor 2 may not be violated. The modified pollution offset rule would restrict trading to the left of line IEB, because neither the pre-trade air quality (R_c) nor the ambient standards for receptors 1 or 2 may be violated. Under the non-degradation offset rule, trading possibilities would be limited to the area to the left of the line $FKEB$, because emissions

- are not allowed to increase (they should remain left of the 45-degree line);
- are not allowed to violate the ambient standards at receptors 1 and 2.

Hence, the pollution offset rule offers the greatest potential for cost savings because it does not make trade contingent on the pre-trade situation of the two sources. Because trading can take place in the whole area below line FGB, the optimum situation can be attained without violating the deposition standards.

The first rule, pollution offset, appears to be the most promising, offering the largest potential for cost savings with the least binding restrictions. The second rule, modified pollution offset, prevents deterioration at receptors where pre-trade air quality is already better than the standards, because neither the deposition

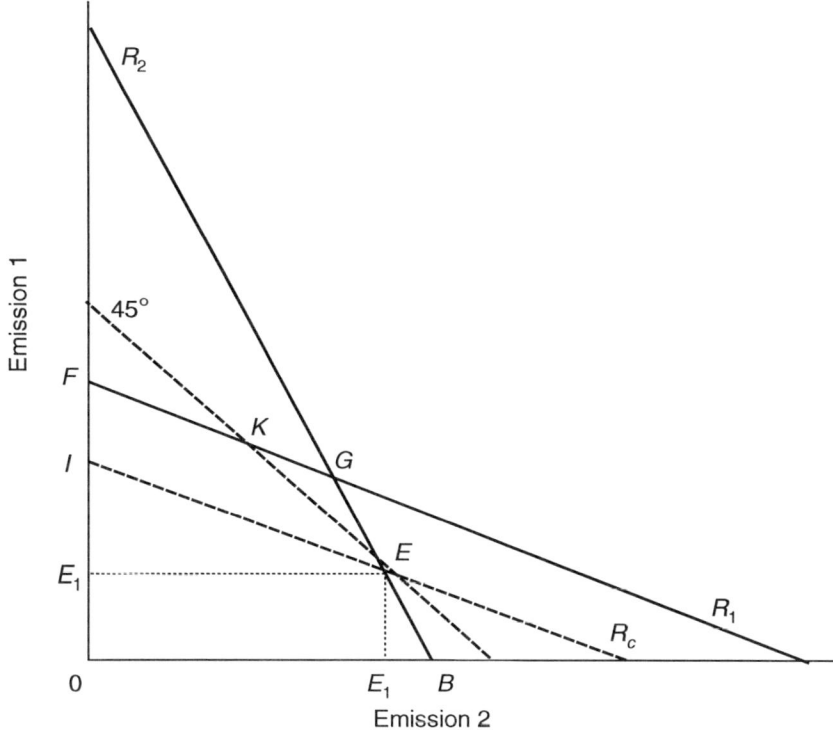

Figure 3.2. Three trading rules.

targets nor the pre-trade air quality are allowed to be violated. Rule three, the non-degradation rule, is similar in practice to the US Environmental Protection Agency (EPA) emission trading program of 1986. With rule three, the ratio at which sources trade (the volume of emissions one source must eliminate to allow another source an increase of one unit) always exceeds one because emissions are never allowed to increase. Even if keeping the ambient standard would allow one source to increase emissions more than another source (the seller) would have to reduce them, this rule would not allow the total emissions to increase. These rules do not force sources to trade according to a fixed rate, such as on a one-to-one basis. In principle, the ratio at which sources exchange their emission permits is not fixed as long as the ambient standards are not violated. This requires the use of a diffusion model that allows verification that air quality standards are not violated before and after each trade.

Because all three rules promise cost savings (perhaps even the attainment of the cost minimum) while maintaining air quality or deposition standards, the next sections will examine them more closely.

3.5.4 Pollution offset trading

Krupnick *et al.* (1983, pp. 240–241) proposed a system of pollution offsets as the most promising approach for achieving regional environmental quality goals for non-uniformly dispersed pollutants – promising in the sense that it minimizes pollution control costs, limits the number of transactions and transaction costs compared with ambient deposition permit trading, and fully guarantees that ambient standards are met. The basic argument for the cost efficiency of the pollution offset system is disarmingly simple. Recall that under this scheme sources are free to trade as long as predetermined air quality standards are not violated at any receptor point. The rule also implies that sources can always obtain additional emission licenses from the environmental agency provided they do not exceed the ambient standards. This effectively implies that interior solutions (such as point I in *Figure 3.2*) that are cost inefficient are ruled out. Furthermore, no point on the frontier (line FGB in *Figure 3.2*) other than the cost minimum (G, for example) can be a trading equilibrium, because from any other point (B, for example) mutually profitable trades exist.

McGartland (1984, pp. 110–116) and McGartland and Oates (1985) proved that under competitive conditions the polluters' problem with pollution offset trading is identical to their problem with the ambient permit system [for which Montgomery (1972) supplied the proof]. This is basically because emission rights can be regarded as implicit rights to deposit waste at certain locations. Furthermore, the regulator's cost-minimization problem is also identical, provided that emission permits are initially distributed so that the air quality constraint is binding at every receptor (McGartland, 1984, p. 116). This implies that if all emission permits are utilized the standards will be exactly met. This could be a difficult task because it is not always possible to distribute emission licenses so that the standards are exactly met everywhere. In general, one can only exactly meet a set of goals (deposition targets in this case) if the number of choice variables (instruments; in this case, emission levels of each source) is at least as high as the number of objectives or constraints (Chiang, 1984, p. 673). The point is, however, that under competitive conditions and in the absence of transaction costs, ambient permit trading and pollution offset trading are both cost efficient.

Unfortunately, pollution offset trading (POS) and ambient permit trading (APS) can only achieve the least-cost outcome after all the gains from trade are exhausted

(McGartland, 1988); that is, when no further cost-saving trades are possible. Two conditions might prevent this: high transaction costs and noncompetitive markets.

Krupnick *et al.* (1983) are of the opinion that transaction costs would be low under POS, because polluters would trade on one market of emission permits instead of operating on a number of deposition permit markets as with APS. McGartland (1988), however, correctly remarks that, under APS, polluters could group deposition permits and sell them as one commodity and concludes that the same number of transactions would be required under both APS and POS to reach the least-cost solution. With perfect information, competitive markets, and vectors of known prices, there would not be any difference between APS and POS. Hahn (1986) believes that with imperfect information, POS might imply high transaction costs because the right to increase emissions is state dependent; that is to say, the deposition at each receptor is a function of the emissions of all other sources. If two sources trade, the deposition pattern is altered and the trading possibilities for other trades are affected as well. Against this, POS might face lower transaction costs than APS, because POS allows polluters to be concerned only with those receptors that threaten to violate the standards since a polluter can increase emissions as long as the ambient standards are met. With APS, a source can only increase emissions if it buys additional deposition permits for all the receptors it affects (McGartland, 1988). However, POS might face higher transaction costs than APS, because a dispersion model must be run before the trade is approved. McGartland (1984, 1988) therefore suggests that POS is only promising if the environmental agency is active and simulates trades beforehand, if dispersion coefficients are made available, and if the number of "binding" receptors is small and stable (the number should not change too much if one trade is implemented). Briefly, under both POS and APS transaction costs might be high and might impair cost efficiency, without, however, affecting environmental effectiveness.

Under imperfect competition, free riding might block the cost efficiency of POS (McGartland, 1988, p. 38). By imperfect competition, McGartland means uncertainty about possible trades and a smaller number of buyers and sellers; in other words, there is no vector of known prices. Recall that under POS, sources can always obtain additional emission rights from the environmental agency if they do not violate ambient standards. This could allow sources to free ride under POS. To give an example using *Figure 3.1*, if source 1, which is considering increasing its emission control, were to sell emission rights to source 3, deposition at receptor *B* would always be reduced to a level below the target (assuming it was binding before trade). This would allow source 2 to increase emissions, at no cost, to the point where the constraint at receptor *B* becomes binding again. Source 2 would not have to buy emission rights from source 1. McGartland (1988) shows that for a cost-minimum solution for meeting the deposition standards, the following conditions for the cost-minimum level of costs of source 1 would be valid:

$$C_1' = \frac{a_{A1}}{a_{A3}} \times C_3' + \frac{a_{B1}}{a_{B2}} \times C_2' , \qquad (3.14)$$

where a_{A1}, a_{A3}, a_{B1}, a_{B2} are transfer coefficients. If source 2 were to act as a free rider, the right-hand side of the equation would not be included in the bilateral trade between sources 1 and 3, and the marginal costs of source 2 would be lower than under the least-cost solution. Under APS this free riding could not occur, because source 1 could keep deposition rights for receptor B and source 2 could only increase emissions if it bought these rights from source 1. This leads McGartland (1988) to conclude that under imperfect competition free riding might prevent realization of the cost-efficient solution.

One might remark that as long as the marginal costs of sources 1 and 2 are not optimal, there is room for more cost-saving trades. In our example, source 1 ends up with marginal costs that are too small. It could increase its emission reductions and enable source 2 to increase emissions further. In this case, however, source 3 could act as a free rider and the game could continue. Briefly, because POS allows free riding, a complicated set of bilateral trades might be needed to arrive at the least-cost solution, if this solution can be reached at all. Still, one should not forget that POS saves costs without conflicting with the environmental objectives.

The information requirements for the environmental agency are more modest than those for the sources and they depend on whether the agency desires to play an active or passive role. The agency would have to maintain a record of the emissions (emission inventory) of each source and would have to inform the sources about the applicable atmospheric transport model. Proposed trades would be simulated beforehand to ensure that the resulting pattern of depositions meets the desired standard at every receptor. Finally, the environmental agency would have to keep an overview of the emissions traded (an emission tracking system) and it would have to define the rules for trading and procedures for acceptance of trades. The agency could also play a more active role, proposing trades and identifying potentially binding receptors [this could actually involve creating the trade ratio function(s) for every set of conceivable trades].

To summarize, POS guarantees that deposition targets will always be met since this is a precondition for every single trade, because in a competitive market POS is equivalent to the deposition permit system and in theory can reach the cost minimum. It functions better with a small, stable number of binding receptors and an active environmental agency that simulates trades beforehand. It implies relatively high transaction costs for the sources compared with ambient permit trading, because the possibilities of trading are contingent on emissions and trades from all sources. The administrative costs, which are basically the information requirements for the environmental agency, are more modest: an emission inventory and a dispersion model are needed to simulate whether a trade will lead to a violation of the deposition standards.

3.5.5 Modified offset trading

In response to some problems with POS, in particular the possibility that deposition will increase at non-binding receptors, McGartland and Oates (1985) proposed a somewhat different scheme, the modified offset system (MOS). The basic characteristic of this system is that it allows sources to trade under two conditions: deposition targets may not be violated and air quality may not deteriorate (so deposition may not become higher than in the pre-trade case).

The modified offset rule restricts trading because neither the pre-trade air quality (R_c) nor the ambient standards may be violated for receptors 1 and 2. Because the standards are redefined, all receptors are initially binding. This is exactly why under competitive conditions the MOS functions as the deposition permit system. The proof is given by McGartland and Oates (1985) and is similar to that for POS in the previous section.

Under imperfect competition, POS and MOS are not equivalent, however. Precisely because the standards are redefined, there is initially no possibility for free riding with MOS. No source can increase emissions without another source having to reduce them. Of course, after some trades have been implemented and before the equilibrium is reached, it may well be that the deposition at some receptors is lower than the redefined standards. Then, of course, just as with POS, free riding may inhibit attaining the cost minimum. There is an additional problem, however. Because initially all receptors are binding, only those sources that affect the same subset of receptors can trade. Clearly, this limits the number of sellers and buyers, reduces competition, and thereby creates conditions for imperfect competition, thus undermining the cost efficiency. The same problem may, however, occur with APS.

3.5.6 Non-degradation offset trading

A third hybrid approach exists: the non-degradation offset or constant emissions offset (COS) approach, whose trading rules are that total emissions are not allowed to increase (constant emission rule) and deposition standards may not be violated. Such a system comes very close to the 1986 emission trading rules of the US EPA (see Borowski and Ellis, 1987; Hahn, 1986; see also Chapter 6). It partly avoids free riding because sources cannot increase emissions without buying "offsets" from other sources due to the constant emissions rule. The attractive property here, as also seen with POS and MOS, is that it guarantees that the deposition standards will be met.

Figure 3.2 can be used to show that whether or not COS can end up at the cost minimum depends on the pre-trade emissions. Let us assume the cost minimum is point E. If point G is the starting point (with total emissions exceeding the cost minimum), the cost minimum E can be reached. Remarkably, total emissions do not remain constant, they decrease. From this point the 45-degree line no longer governs the trade process. In other words, the constant emission rule is not binding. However, if one were to start trading from point B, the cost minimum E would not be attainable. At point B, total emissions are lower than in the cost minimum. The constant emission rule restricts trading to the space on or below the 45-degree line passing through B.

COS is similar to POS because it not only requires emissions to remain at least constant but also prevents violation of the deposition targets. Consequently, the expected transaction costs and information requirements for a polluting source are comparable, if not higher, because sources face one additional constraint. In contrast to POS, COS prevents free riding to a certain degree because no source can increase emissions without buying offset emission reductions from other sources. (Two sources can only free ride together if they make use of the deposition reduction created by a previous trade.) Again, trades are state dependent and contingent on the emissions and previous trades of third parties.

If the environmental agency wants the scheme to be able to attain the cost-minimum solution it must know the optimal emissions. Initial emissions must be equal to or higher than the optimal emissions in order to attain the cost minimum, because COS does not allow emissions to increase. In other respects, the information requirements are similar to those with POS.

3.5.7 Simpler trading schemes

In view of the difficulties encountered in establishing a complete set of properly functioning markets for deposition permits and the difficulties with the three pollution offset schemes, several alternatives have been suggested in the literature to attain deposition targets or ambient standards:

- Trading of emission permits within one zone;
- Trading of emission permits within several zones;
- Single market deposition permit system.

The first system described simply ignores the location of the sources. This "single-zone trading" implies that pollution control costs will be minimized for the initial total level of emissions. However, Bohm and Russell (1985) show that such trading systems might not meet the deposition targets if the total amount of emission permits is so high that deposition targets will be violated at some receptors after trading. One could also determine the total amount of emissions so as to meet the deposition or ambient constraints after trading. In this case, the costs of pollution control would be minimized subject to the condition that the sum of the emissions remains below the total allowable level of emissions. In this case the allowable emission levels would be set so that the ambient standards are not violated after trading and overcontrol of emissions does not take place (Spofford and Paulsen, 1990, pp. 2–10). One way to do this is to set an initial emission level, to determine the cost-minimizing allocation of emission reductions at every source to meet this overall level, and then to examine the impact on air quality. If ambient standards would be violated at this level, the total level of emissions is lowered until, in an iterative procedure (similar to that of setting emission charges), the ambient standards would no longer be violated and emissions cannot be increased without violating the ambient standards.

If the latter procedure is adopted, the costs are always at least as high as those of the least-cost solution for meeting the ambient standards. This is because single-zone trading cannot be cost efficient as it tends to equalize marginal costs, whereas marginal costs generally have to be different if a set of deposition targets is to be met as the environmental objective. The system could be cheaper or more expensive than reductions by a uniform percentage. This depends on the specific regional situation, i.e., the actual transfer coefficients, the cost functions, and the levels of unabated emissions of sources (Russell, 1986). With perfect markets and no transaction costs, single-zone trading will always be more expensive than the ambient permit system. Precisely because it is simple, however, transaction costs can be much lower than in an ambient permit or POS system, as Chapter 6 on practical experience will show. In the case of imperfect information on costs, the regulatory agency does not know with certainty whether ambient standards will be met because it does not know what the ultimate allocation of emissions will be after trade. The only remedy that would remain if ambient standards were violated would be to reduce the total number of permits in the next year until the standards are met. This, however, might create uncertainty on the side of polluters, which would be weighed against the potential cost savings of trading. In summary, single-zone trading does not meet the least-cost solution and can be either more or less cost efficient than regulation. With imperfect information on costs, it might not meet the environmental goals. Its major advantages are its simplicity and its low transaction costs.

Trading emission permits within several zones has a certain surface appeal. It allows the regulatory agency to choose the size and location of zones, taking into account the spatial distribution of sources and possible differences in allowed deposition levels. Therefore, it offers more protection for deposition targets than emission trading in one zone and reduces control costs. However, this is only the case if the environmental agency has complete and correct knowledge of emission control costs. With limited information, the cost will be above the cost minimum because emissions cannot be traded among zones. This is because without knowing the cost-minimum solution, the environmental agency does not know how many permits it should allocate to each zone. In addition, there is no protection against violation of the standards, even in small zones, because it is not exactly known where emissions will take place after trading (Tietenberg, 1985).

No theoretical analysis has been performed on the determination of the optimal number of zones and their size. Chapter 5 will show the results of several empirical studies indicating that cost efficiency depends on regional circumstances and the quality of information on costs (perfect/imperfect).

Single-market deposition permit systems focus on one single "worst case" receptor (a "hot spot"). The permits in this market can be defined only for the most polluted area, the one requiring the greatest reduction in emissions (Atkinson and Tietenberg, 1982). This type of trading might come very close to the cost minimum if there is one dominant receptor. In the absence of multiple binding constraints and with a stable geographical distribution of emitting sources, it also allows a high degree of control of ambient standards. However, if more receptors are binding, trading deposition permits for only one of them is likely to violate the deposition standard for the other binding receptors. If the geographical distribution changes, receptors other than the single "worst case" receptor may become receptors where the deposition targets are violated. Because the focus is on only one receptor, it will inevitably create an incentive for sources to move to regions where the impact on the "hot spot" is restricted. Consequently, new "hot spots" will be created, undermining the selection of the initial "worst case" receptor (Tietenberg, 1985).

3.6 Administrative Practicability, Dynamic Efficiency, and Political Acceptability

Dealing with non-uniformly mixed pollutants requires more information than dealing with uniformly dispersed pollutants (see Spofford and Paulsen, 1990). The information required for the different instruments consists of data on emissions, atmospheric transport, and costs of controlling emissions, and a cost-minimization model. All instruments require an emission inventory and an atmospheric dispersion model in order to be effective. To be environmentally effective, an ambient as well as a uniform charge would also require global cost information to avoid groping in the dark. By contrast, some emission trading schemes (ambient permits and the three trading rules) do not require any information on costs to be cost efficient and environmentally effective. Emission trading in one zone or multiple zones and single-receptor trading need cost information to avoid violation of the environmental targets and to improve their cost efficiency. Uniform regulation does not require any cost knowledge to be environmentally effective. If regulation must be cost efficient, it needs perfect cost information to design nonuniform (source-specific) regulations. In short, several emission trading schemes are as environmentally effective as charges and regulation, but require less information to be more cost efficient. Regarding the administrative costs, there is little reason to believe that, as in the uniform mixing case, there would be significant differences between the instruments.

As with uniformly dispersed pollutants, charges and emission trading can be expected to have a more pronounced impact on innovation than regulation. With nonuniform dispersion, charges might perform better than ambient permits or trading rules; with charges the cost savings of innovation are immediately known, with trading they depend on the permit price and on whether all the fixed deposition permits can be sold.

Moreover, the choice of instruments is likely to influence the location of industry in the long run. Such locational shifts are more likely with instruments that are tailored to the location of the sources: ambient charges, deposition permits, trading rules, multiple-zone trading, single-receptor trading, and source-specific regulation. Such systems tend to create stronger incentives to locate away from areas where permit prices or ambient charges are higher or where regulations are more stringent than uniform or across-the-board regulations (Dales, 1968). This tendency is stronger with ambient charges and ambient permits because new firms would also have to pay for the emissions remaining after meeting the location-specific regulation. From a cost-efficiency point of view, however, this dislocation is perfectly acceptable because it shows that the externalities are location dependent. Politically speaking, location-specific charges, tradable permits, or regulation might lead to protest by the authorities and industry in those regions that are faced with tougher requirements. This might then be an additional stumbling block for accepting location-specific instruments in general and charges and tradable permits in particular.

3.7 Concluding Comments

This chapter has sought to analyze the conditions under which emission trading, charges, and regulation are cost-efficient and environmentally effective instruments for meeting environmental goals when location matters.

Table 3.1. Features of instruments when source location is a factor.

	Cost efficiency	Environmental effectiveness
Charges		
Ambient charges	Yes, but ...	No, unless ...
Uniform charges	No	Yes, if ...
Regulation		
Source-specific	Yes, if ...	Yes ...
Uniform	No	Yes, if ...
Emission Trading		
Ambient permits	Yes, but ...	Yes
Trading rules		
Pollution offset	Yes, but ...	Yes
Modified offset	No, unless ...	Yes
Non-degradation offset	No, unless ...	Yes
Single-zone emission trading	No	No, unless ...
Multiple-zone trading	No	No, unless ...
Single-receptor trading	No	No, unless ...

The main point of the chapter is that with perfect knowledge of costs, ambient charges, source-specific regulation, and ambient (or deposition) permit trading would all be equally cost efficient and environmentally effective. With imperfect information on costs, crucial differences between the instruments occur. In theory, an adaptive procedure is conceivable to set ambient charges so that a set of standards for air quality is met at the lowest costs. In the same way, the environmental agency could issue ambient or deposition permits and allow sources to buy and sell these rights. In the absence of transaction costs and with a perfect market, the ambient standards would be met at minimum costs.

Without knowledge of costs, regulation cannot be designed in a cost-efficient way but can still meet the environmental goals. There is a fundamental difference between ambient charges and ambient permit trading. With ambient charges, the environmental goals are only met when the iterative procedure is successfully completed and the sources react in the expected cost-minimizing way. With ambient permits, the environmental goals are always met, but the least-cost solution might not be attained because transaction costs are potentially high. These costs might be high because polluters would have to trade simultaneously on different markets for ambient concentrations.

For the above, recourse to simpler alternatives is sometimes taken, such as uniform emission charges, emission trading rules, single- or multiple-zone trading,

Instruments in Theory when Location Matters 63

or single-receptor trading. *Table 3.1* summarizes the cost efficiency and environmental effectiveness of the various alternatives.

Of these simple alternatives, uniform emission charges are generally not cost efficient because they ignore the location of the source relative to the receptor. At best they will be environmentally effective after the trial-and-error procedure is successfully completed. Uniform regulation is not cost efficient, but can meet the environmental goals with adequate knowledge of atmospheric dispersion. The three emission trading rules examined all meet the ambient standards because none of them allows emissions to increase when this would violate the standards. Exactly because of this case-by-case approach, transaction costs make these trading rules less cost efficient. Pollution offset trading may not be cost efficient because it allows free riding. The cost efficiency of modified and non-degradation offset trading depends on the starting point for trade. Single-zone trading does not guarantee that environmental goals will be met if costs are uncertain and is not cost efficient but results in lower transaction costs and a thicker market, making market power less of a problem. Multiple-zone trading is not cost efficient, nor is it necessarily environmentally effective. The same holds for single-receptor trading; in a multi-receptor context this might lead to violations of the standards at the other receptors. For all the simpler systems, theory predicts that whether single-zone emission trading, uniform charges, or uniform regulation is the most cost efficient depends on specific regional circumstances such as pollution control costs and atmospheric dispersion patterns, a discussion we will embark on in Chapter 5.

Regarding administrative practicability, dealing with non-uniformly mixed pollutants requires more information than dealing with global pollutants, because atmospheric transport models are needed for an effective policy.

Ambient trading and trading rules have the advantage of not requiring information on costs to be both environmentally effective and cost efficient. Ambient charges, regulation, single- and multiple-zone trading, and single-receptor trading all require some knowledge of costs to be environmentally effective or cost efficient. The administrative practicability is not likely to differ, *a priori*, among the different instruments. The location-specific instruments, especially ambient charges and ambient permits, but also location-specific regulation, trading rules, multiple-zone trading, and single-receptor trading, will have a more significant impact on the location of sources than will uniform instruments (uniform regulation, charges, or single-zone trading). Precisely for this reason they might be politically less attractive. As the next chapter will show, the designing of instruments becomes even more intricate in an international context.

Chapter 4

Economic Instruments in an International Context: Theory

4.1 Introduction

The 1987 Montreal Protocol did not merely restrict the production and consumption of substances that damage the ozone layer, it also established a party's right to transfer its production quota to other parties to improve economic efficiency, an option that has been used extensively by US- and Europe-based chemical companies. This example shows that the international application of economic instruments is no longer a mirage existing only in the minds of economists. In principle, the analysis of the performance of economic instruments in a national context carried out in Chapter 2 is fully transferable to the regulation of transboundary pollution in an international context (Baumol and Oates, 1988). In contrast to the national case, however, no international government exists that can impose emission charges or tradable emission rights, and the cooperation and consent of sovereign nations are required. These countries can only be expected to agree on the implementation of economic incentives if this makes them better off. In the presence of transboundary externalities, taxes or tradable permits can improve the gains from cooperative agreements, although there are also incentives for countries not to cooperate. The use of taxes and emission trading for global pollutants, such as CO_2, is a particularly well-researched issue (e.g., Barrett, 1992; Bohm, 1990, 1994; Hoel, 1992a, 1992b, 1993; Welsch, 1993). Most authors do not make a clear distinction between the application of market instruments as mechanisms for conducting negotiations to arrive at targets and their application as mechanisms to increase the cost efficiency of agreements given a set of targets. Second, with few exceptions (notably Mäler, 1990, 1994; and Nentjes, 1990b, 1994), research has not addressed the theoretical

Economic Instruments in an International Context

niceties of the international application of economic incentives for non-uniformly dispersed or regional pollutants such as SO_2.

This chapter's approach differs from the usual approach because its focus is the application of economic incentives (particularly tradable emission permits, but also emission charges) to realize a set of given targets for non-uniformly dispersed transboundary pollutants. At the starting point, the countries already have arrived at agreements to reduce emissions reciprocally and, therefore, have common targets in terms of emission reductions or reductions in deposition or ambient concentrations (pollution loads). The approach of Chapter 2 is thus continued. The core question posed is under which conditions can emission charges and tradable emission permits be used to improve the cost efficiency of meeting environmental targets in an international context. The perspective adopted in this chapter is mainly theoretical. An evaluation of the practical experience with tradable emission permits (both national and international) is the subject of Chapter 6.

The structure of the chapter is the following. Section 4.2 introduces the conceptual framework and examines whether international accords leave room for improving cost efficiency. Section 4.3 analyzes whether emission taxes could be used as a tool to improve the cost efficiency. Section 4.4 uses the conceptual framework to assess the conditions under which emission trading could be an appropriate instrument for improving the cost efficiency of international agreements.

4.2 Conceptual Framework

4.2.1 Introduction

In this chapter, environmental quality in each country is assumed to be affected by the country's own emissions and by the emissions of other countries. Decisions on environmental policy are made by the national governments. Their perceptions and valuations of the environmental problem are reflected in a generally used, simple separable utility function of the form $U_i = B_i(Z_i) - C_i(R_i)$ (e.g., Mäler, 1990; Hoel, 1991; Nentjes, 1994). B_i represents the benefits from the pollution load reduction (Z_i), and C_i represents the costs of emission reduction (R_i). The concept of environmental benefits does not necessarily imply that benefits can be measured accurately and expressed in monetary terms; however, it is assumed that the political body, in one way or another, weighs the benefits of further reductions in pollution against the additional costs of emission reduction. Environmental benefit and cost functions are twice differentiable with decreasing marginal benefits and increasing marginal costs. Emission reductions and decreases in pollution loads in every country are connected through linear atmospheric transfer relations. In a three-country case, $Z_i = a_{i1}R_1 + a_{i2}R_2 + a_{i3}R_3$ for every $i = 1, 2, 3$, where a_{i1},

a_{i2}, and a_{i3} are linear transmission coefficients. This specification implies that a non-uniformly dispersed (non-global) pollutant, such as acidifying emissions, is being considered. The specification is also valid for global pollutants, because in this case the transfer coefficients are all equal and the benefits are a function of the sum of individual emission reductions. This chapter concentrates on a three-country case because this allows third-party impacts to be taken into account while mathematical notation is kept transparent. The remainder of Section 4.2 will clarify the following concepts:

a. Non-cooperative Nash equilibrium
b. Cooperative Nash equilibrium
c. The full cooperative solution
d. Cost-efficient, Pareto-dominant, joint implementation of emission goals
e. Cost-efficient, Pareto-dominant, joint implementation given certain deposition goals.

Sections 4.3 (on charges) and 4.4 (on emission trading) will then examine how economic instruments can be used to move from (*a*) or (*b*) to (*d*) or (*e*).

4.2.2 Non-cooperative Nash equilibrium

If there is no cooperation and if each government considers its neighbors' levels of emission reductions as given, each country will choose its own best strategy. Each government is assumed to pursue its national interest by maximizing its utility function under the constraint of the relevant transfer equation. This is the non-cooperative Nash equilibrium strategy (Barrett, 1990; McMillan, 1986; Nentjes, 1990a, 1994). For each country the conditions for the non-cooperative Nash equilibrium are derived by maximizing the Lagrange function (here for country 1):

$$H_1 = B_1(Z_1) - C_1(R_1) + \lambda_1 (Z_1 - a_{11} R_1 - a_{12} R_2 - a_{13} R_3) \ . \tag{4.1}$$

Similar functions can be formed for countries 2 and 3. Given that each country pursues its Nash equilibrium, the following first-order conditions are obtained:

$$C_1' = a_{11} B_1' \ , \quad C_2' = a_{22} B_2' \ , \quad C_3' = a_{33} B_3' \ . \tag{4.2}$$

Each country reduces its emissions to the point where marginal costs are equal to the marginal domestic, or national, environmental benefits. This non-cooperative Nash equilibrium is depicted by point N in *Figure 4.1*. The curves $\overline{U}_1$ and $\overline{U}_2$ are iso-utility functions and denote combinations of emission reductions for countries 1 and 2 with similar impacts on welfare. If the countries care about the environment in other countries, part of the deposition in other countries enters the welfare function as well. In this case the Nash equilibrium reductions will be higher than those when this is not a concern (see Nentjes, 1994).

Economic Instruments in an International Context

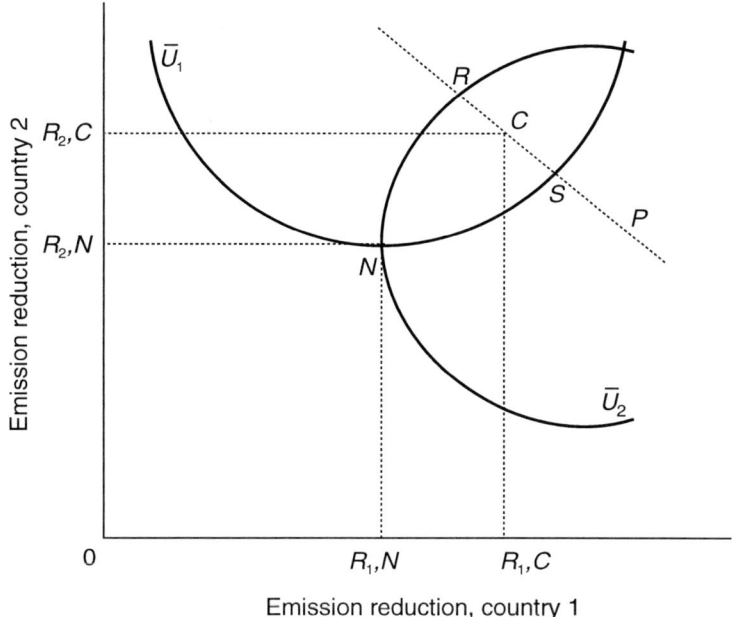

Figure 4.1. Cooperative and non-cooperative Nash equilibriums.

4.2.3 Cooperative Nash equilibrium

Countries can improve welfare by coordinating their pollution control policies. Country 1 can offer to raise its reduction of emissions provided that countries 2 and 3 follow the same line of action (Nentjes, 1994). As a result, country 1 obtains higher reductions in domestic deposition in return for an additional unit of emission reduction than it would have received under a non-cooperative policy regime. The same holds for countries 2 and 3. There is thus room to contract a reciprocal increase in emission reductions from which all countries will benefit.

This is shown in *Figure 4.1*. Any point in the area above $\overline{U}_1$ represents higher welfare for country 1 than any point on the iso-welfare curve and, therefore, is preferable to points on $\overline{U}_1$. Similarly, country 2 prefers points to the right of $\overline{U}_2$ to points on the curve. The contract area, enclosed by curves $\overline{U}_1$ and $\overline{U}_2$, is the set of all possible combinations of R_1, R_2 that are Pareto-superior to N.

In a cooperative Nash equilibrium, countries maximize the net benefits of coordination subject to the condition that welfare levels are at least as high for every country as they are in the non-cooperative Nash equilibrium. In a three-country case, welfare levels in the Nash equilibrium are denoted by 1, 2, and 3. A country then is prepared to increase its emission reductions if the others react and increase their reductions as well. In this case the mutual emission reductions are used as three coordinated instruments. This implies maximizing the following Lagrange function:

$$\begin{aligned} H \;=\;& B_1(Z_1) - C_1(R_1) + \lambda_1 (Z_1 - a_{11}R_1 - a_{12}R_2 - a_{13}R_3) \\ &+ \lambda_2 (Z_2 - a_{21}R_1 - a_{22}R_2 - a_{23}R_3) \\ &+ \lambda_3 (Z_3 - a_{31}R_1 - a_{32}R_2 - a_{33}R_3) \\ &+ \mu_2 \left[-\overline{U}_2 + B_2(Z_2) - C_2(R_2) \right] \\ &+ \mu_3 \left[-\overline{U}_3 + B_3(Z_3) - C_3(R_3) \right] \;. \end{aligned} \quad (4.3)$$

The first-order conditions are

$$\begin{aligned} C_1' &= a_{11}B_1' + a_{21}\mu_2 B_2' + a_{31}\mu_3 B_3' \;, \\ C_2' &= a_{12}/\mu_2 B_1' + a_{22} B_2' + a_{32}\mu_3/\mu_2 B_3' \;, \\ C_3' &= a_{13}/\mu_3 B_1' + a_{23}\mu_2/\mu_3 B_2' + a_{33} B_3' \;. \end{aligned} \quad (4.4)$$

For countries 2 and 3 similar sets of equations hold. Conditions (4.4) imply that in addition to national benefits, country 1 now accounts for part of the transboundary benefits of pollution control when determining its emission level.

4.2.4 Full cooperative solution

In the full cooperative solution each country chooses its pollution level so as to maximize the net benefits of pollution control for all countries, including itself (Barrett, 1990; Mäler, 1990, Nentjes, 1994). The first-order conditions for Pareto efficiency are derived by maximizing the total net benefits of reducing emissions for all countries. This implies maximizing the following Lagrange function:

$$\begin{aligned} H \;=\;& B_1(Z_1) - C_1(R_1) + B_2(Z_2) - C_2(R_2) + B_3(Z_3) - C_3(R_3) \\ &+ \lambda_1 (Z_1 - a_{11}R_1 - a_{12}R_2 - a_{13}R_3) \\ &+ \lambda_2 (Z_2 - a_{21}R_1 - a_{22}R_2 - a_{23}R_3) \\ &+ \lambda_3 (Z_3 - a_{31}R_1 - a_{32}R_2 - a_{33}R_3) \;. \end{aligned} \quad (4.5)$$

First-order conditions are

$$\begin{aligned} C_1' &= a_{11}B_1' + a_{21}B_2' + a_{31}B_3' \;, \\ C_2' &= a_{21}B_1' + a_{22}B_2' + a_{32}B_3' \;, \\ C_3' &= a_{31}B_1' + a_{32}B_2' + a_{33}B_3' \;. \end{aligned} \quad (4.6)$$

Economic Instruments in an International Context

It can be shown that the full cooperative solution can be realized as a Pareto-efficient solution for a case where countries apply money transfers (M) in addition to reciprocal emission reductions. The first-order conditions for a Pareto optimum with side payments are derived by maximizing the following Lagrangian for every country (see Nentjes, 1994):

$$\begin{aligned} H &= B_1(Z_1) - C_1(R_1) + M_1 + \lambda_1(Z_1 - a_{11}R_1 - a_{12}R_2 - a_{13}R_3) \\ &\quad + \lambda_2(Z_2 - a_{21}R_1 - a_{22}R_2 - a_{23}R_3) \\ &\quad + \lambda_3(Z_3 - a_{31}R_1 - a_{32}R_2 - a_{33}R_3) \\ &\quad + \mu_2\left[-\overline{U}_2 + B_2(Z_2) - C_2(R_2)\right] \\ &\quad + \mu_3\left[-\overline{U}_3 + B_3(Z_3) - C_3(R_3)\right] \\ &\quad + \delta(M_1 + M_2 + M_3) \ . \end{aligned} \qquad (4.7)$$

The first-order conditions are

$$\begin{aligned} C_1' &= a_{11}B_1' + a_{21}B_2' + a_{31}B_3' \ , \\ C_2' &= a_{21}B_1' + a_{22}B_2' + a_{32}B_3' \ , \\ C_3' &= a_{31}B_1' + a_{32}B_2' + a_{33}B_3' \ . \end{aligned} \qquad (4.8)$$

A comparison of conditions (4.8) with the conditions for a Pareto-dominant solution [conditions (4.4)] for country 1 shows that the full cooperative solution is a special case of conditions (4.4) where μ_1, μ_2, and μ_3 all equal 1. In other words, a Pareto-efficient solution of emission reduction without money transfers (and only emission reductions as variables) is a range in the Pareto efficiency set with monetary side payments, the special case where side payments are zero. The cooperative Nash conditions (4.4) were derived for a situation where the welfare of countries could only be increased by exchanging reductions of emissions. Pareto-efficient outcomes with larger welfare gains for both countries become feasible by allowing monetary transfers or side payments between parties, because these are used to compensate those countries that end up with additional abatement costs that exceed their additional national environmental benefits. Finding Pareto-efficient solutions with monetary payments can be interpreted as a movement along the R_1, R_2 Pareto efficiency curve, the dotted curve through points R and S in *Figure 4.1*. The welfare of each country now is the sum of the net benefits of reduced pollution plus the net money transfers. Now that international money is involved the Pareto optimum can be a point beyond the disagreement points R and S on the line through R and S in *Figure 4.1* (e.g., point P). This is because a country whose welfare is to be reduced below the welfare of the Nash non-cooperative equilibrium can be compensated with money.

The above analysis makes it clear that even a cooperative Nash equilibrium in terms of reciprocal reductions generally leaves room for additional gains in net benefits for parties by allowing re-contracting using the commitments to reciprocally control emissions as a starting point. In such a second phase, obligations to reduce emissions are exchanged for money.

4.2.5 Cost-efficient joint implementation to achieve an emission goal

Imagine that countries that have already agreed on reciprocal emission reductions and are now considering joint implementation of emission reductions have in mind an objective less ambitious than maximizing net benefits. Instead, the objective could be cost efficiency, with negotiators setting out to reallocate the initial result from reciprocal reduction so that joint costs of pollution control are minimized. The question arises whether an agreement that has emerged from negotiations on reciprocal reductions of emissions leaves room for joint implementation geared to cost efficiency. Two environmental objectives are conceivable: a certain level of emissions and certain levels of deposition. This section examines the merits of the first objective. The next section will analyze joint implementation subject to deposition goals.

Minimization of total costs under the constraint that the total emission reduction equals the reduction initially agreed on $(\overline{R})$ implies minimizing the following Lagrange function:

$$H = C_1(R_1) + C_2(R_2) + C_3(R_3) + \lambda(\overline{R} - R_1 - R_2 - R_3) \ . \tag{4.9}$$

The first-order conditions are

$$C'_1 = C'_2 = C'_3 = \lambda \ . \tag{4.10}$$

A comparison of equation (4.10) and equation (4.6) makes it clear that the cost minimum does not coincide with the full cooperative solution. Note that in the special case of a global pollutant (such as CO_2) this is different; in this case all transfer coefficients are equal and condition (4.6) also implies condition (4.10).

The question arises whether the cost minimum (4.10) is also a Pareto-dominant solution compared with the initial cooperative Nash solution (4.4) and therefore acceptable to all parties. This will be the case if two conditions are fulfilled:

- The total level of emission reduction should be equal to or higher than the sum of individual reductions under the initial agreement $(\overline{R})$.
- The net benefits should be higher than or equal to those of the initial agreement for every country.

Figure 4.2 briefly illustrates this for a two-country case. The initial point C is a cooperative Nash equilibrium of an agreement to reciprocally reduce emissions. This is comparable to point C in *Figure 4.1*. Because C is a point on the contract curve, transactions will only be possible if side payments are used. $\overline{U}_1$ and $\overline{U}_2$ are the welfare levels of the initial agreement. The IC_1 and IC_2 curves are iso-cost curves showing combinations of emission reductions leading to the same overall costs. The 45-degree line depicts combinations of emission reductions that lead to the same total reduction as under the initial agreement. Countries 1 and 2 can jointly implement reductions as long as they end up at a point on or to the right of the 45-degree line. The outcome where aggregate costs are minimized and condition (4.10) is satisfied is point E. In this case total costs would be lower because curve IC_2 is closer to the origin than IC_1. Country 2 would lower its reductions and country 1 would increase them. As a result, welfare in country 2 would increase (point E is inside curve $\overline{U}_2$) for two reasons: it would face lower costs and higher reductions in deposition (point E is to the right of line Z_2, which is the deposition in country 2 due to the initial accord). Welfare in country 1 would be lower, because it would face higher costs and fewer environmental benefits (deposition increases because deposition is below and to the left of line Z_1, which depicts the deposition reduction of the initial accord). In this case, country 2 would have to compensate country 1 to achieve an agreement that not only saves costs and meets the total emission reductions but is also Pareto-dominant. One can imagine solutions where compensation is not possible, or in other words, where the net benefits for country 2 obtained from moving from point C to point E are lower than the net costs for country 1. In such a case cost minimization would not be a Pareto-dominant solution and the solution would not be acceptable for country 1. Complications that occur with more than two parties will be addressed in the sections on taxes and emission permits.

4.2.6 Cost-efficient joint implementation of deposition goals

The environmental objective of an agreement might be to meet certain levels of deposition (deposition standards) or concentrations at a set of receptor points. As in Chapter 3, the cost minimum resulting from the reciprocal reductions agreement is found by minimizing total costs subject to the condition that deposition is lower than or equal to what results from the agreement at every receptor:

$$\begin{aligned} H \ = \ & C_1(R_1) + C_2(R_2) + C_3(R_3) + \lambda_1 \left(\overline{Z}_1 - a_{11}R_1 - a_{12}R_2 - a_{13}R_3 \right) \\ & + \lambda_2 \left(\overline{Z}_2 - a_{21}R_1 - a_{22}R_2 - a_{23}R_3 \right) \\ & + \lambda_3 \left(\overline{Z}_3 - a_{31}R_1 - a_{32}R_2 - a_{33}R_3 \right) \ , \end{aligned} \qquad (4.11)$$

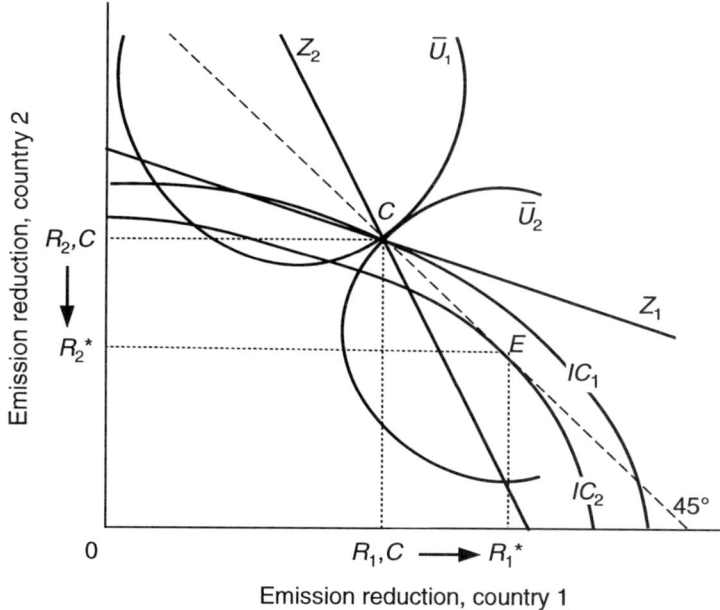

Figure 4.2. Joint implementation of emission reduction.

with $\bar{Z}_1$, $\bar{Z}_2$, and $\bar{Z}_3$ being the agreed on reductions in deposition. The deposition constraints have the form of inequalities. The first-order conditions are

$$\begin{aligned} C'_1 &= a_{11}\lambda_1 + a_{21}\lambda_2 + a_{31}\lambda_3 \;, \\ C'_2 &= a_{12}\lambda_1 + a_{22}\lambda_2 + a_{32}\lambda_3 \;, \\ C'_3 &= a_{13}\lambda_1 + a_{23}\lambda_2 + a_{33}\lambda_3 \;, \end{aligned} \qquad (4.12)$$

where λ_1, λ_2, and λ_3 are the shadow prices of reducing the pollution load in countries 1, 2, and 3, respectively.

In this case we have a system with three equations: the deposition constraints for each country and their relation to the national emission reductions of each country. The system also has three variables in the form of emission reductions. Normally, a system of n equations and exactly the same number of variables yields a determinate solution if one assumes consistency and independence (Chiang, 1984, p. 673). If the initial agreement is a cooperative Nash equilibrium, the deposition constraints are likely to be the depositions that result from the agreement on reciprocal reductions. In the remainder of this chapter we will therefore assume that initially all national deposition constraints are binding.

Economic Instruments in an International Context

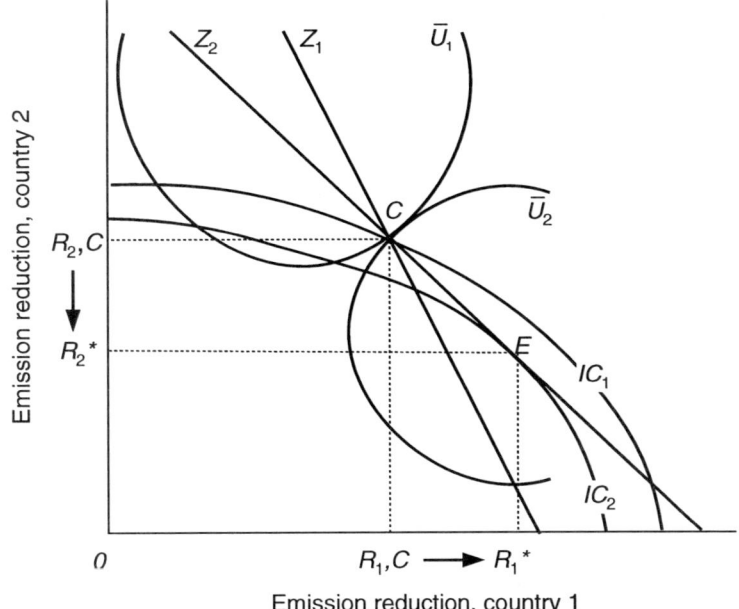

Figure 4.3. Joint implementation of deposition targets.

Figure 4.3 further illustrates the idea of joint implementation subject to deposition standards for a two-country case. Z_1 and Z_2 are the deposition targets of the agreement. Point C is the initial agreement. Initially all deposition constraints are binding. Joint implementation is only allowed as long as reductions result in depositions on or above and to the right of the area enclosed by line Z_2CZ_1. In this situation, country 1 (with relatively low marginal costs) could take over the obligation to reduce emissions for country 2 so as to reduce costs. After joint implementation, these countries could end up at point E. As a result, the total pollution control costs would be lower (a downward shift in the iso-cost curve). Receptor 2 defines the trade ratio $dR_2/dR_1 = a_{21}/a_{22}$. In this example, deposition in country 2 remains constant but costs decrease because emission reductions decline. In country 1, costs are higher but environmental benefits increase because deposition decreases. Welfare in country 1 would be lower (E is below the iso-utility curve $\overline{U}_1$). In order to realize a Pareto-dominant outcome, country 2 would have to pay country 1 to cover part of its incremental abatement costs.

Table 4.1. Overview of equilibrium conditions for marginal costs.

a.	Non-cooperative Nash equilibrium	$C'_1 = a_{11} B'_1$
b.	Cooperative Nash equilibrium	$C'_1 = a_{11} B'_1 + a_{21}\mu_2 B'_2 + a_{31}\mu_3 B'_3$
c.	The full cooperative solution	$C'_1 = a_{11} B'_1 + a_{21} B'_2 + a_{31} B'_3$
d.	Cost-efficient emission goals	$C'_1 = \lambda$
e.	Cost-efficient deposition goals	$C'_1 = a_{11}\lambda_1 + a_{21}\lambda_2 + a_{31}\lambda_3$

To sum up, allowing joint implementation subject to explicit deposition targets will induce a Pareto-dominant improvement of an initial agreement if all national deposition goals are initially binding.

4.2.7 Concluding comments

Table 4.1 summarizes the first-order conditions of the different concepts examined.

An examination of these conditions bearing in mind the discussion in the previous sections allows a concise summary of the main results:

- Neither the non-cooperative (a) nor the cooperative Nash equilibrium of reciprocal reductions of emissions (b) generally meets the conditions for a cost-efficient solution for achieving emission (d) or deposition goals (e). They both leave room for reallocation of emission reduction obligations between countries to reduce costs.
- If countries sign an agreement to meet deposition goals based on the full equilibrium solution, no room for further cost savings is present (c versus e).
- Starting from a cooperative Nash equilibrium and using side payments implies that minimizing costs to achieve the same overall emission reduction level as under the initial agreement is not necessarily a Pareto-optimal improvement for non-uniformly dispersed pollutants, because in some countries benefits may decrease.
- Starting from a cooperative Nash equilibrium and using side payments implies that minimizing costs to meet implicitly agreed on deposition goals implies a Pareto-dominant improvement.
- Neither cost minimizing, Pareto-dominant joint implementation of emission ceilings (d) nor deposition goals (e) in general leads to a full cooperative solution (c). They do, however, enable further improvements in welfare through the reallocation of emissions and the alteration of deposition reductions so that any losses of environmental benefits are compensated by monetary side payments.

The above conditions have been derived for a static case. The fact that international cooperation could be dynamic in the sense that the negotiation "game" would be repeated and negotiators could adopt strategies that depend on past actions of other participants would not, in principle, affect the above conclusions. The difference between non-cooperative and cooperative solutions would become blurred (McMillan, 1986). In a dynamic, or repeated, game the Pareto optimum (full equilibrium) could be sustained by non-cooperative Nash strategies. This is because a dynamic game enables retaliation; countries that deviate from the Pareto optimum can be punished in the next round by other countries that also start to deviate from their Pareto optimum. If the discount rate is sufficiently low the potential gains of choosing the non-cooperative strategy are low and cooperative behavior is enhanced (Barrett, 1990; Mäler, 1993, 1994).

Whereas the previous sections established a conceptual framework, the following sections will show the application of this framework. In particular, the question of the extent to which emission charges and tradable emission permits can improve the cost efficiency of an agreement in a Pareto-dominant way will be addressed.

4.3 Emission Charges

4.3.1 Introduction

The question is whether economic instruments can improve the cost efficiency of international agreements. As in the national context, emission taxes and emission trading promise to cut the costs of achieving environmental objectives. Exactly because of this feature they also increase the chance that more countries will be better off than before. However, they might also raise the compliance costs (pollution control costs in a strict sense plus the net tax or permit payments) for some countries (Barrett, 1992). This section examines the possibilities for an international tax or emission charge to improve on the cost efficiency of a given international agreement.

4.3.2 Cost-efficient and Pareto-dominant taxes

In order to induce countries to participate in a tax agreement, the net benefits of such an agreement should exceed those of the cooperative Nash equilibrium for each country. Otherwise, a country would be better off not signing the tax agreement. Drawing on the work of Hoel (1990, 1992b), in the context of a global pollutant and a uniform tax two types of emission taxes can be distinguished:

- An international tax paid to an international agency plus a reimbursement scheme;

- Harmonized national taxes.

In the case of an international tax there is agreement in the literature that the tax revenues have to be reimbursed for such a tax to be politically acceptable. With a global pollutant and with countries having equal cost and benefit functions, without reimbursement all countries would be worse off with the tax than under the non-cooperative Nash solution (Dasgupta and Heal, 1979, p. 70; Dasgupta, 1982). If countries differ, some countries might face higher pollution control costs with the tax than under the initial agreement despite the fact that overall costs are lower (Barrett, 1992, p. 25). Thus, for a tax to be politically acceptable the revenues will have to be returned to the individual countries.

In the remainder of this discussion we will take a cooperative Nash agreement as an initial solution and examine the extent to which the use of emission charges could improve the cost efficiency of the initial agreement.

If countries participating in international negotiations agree to use taxes as an instrument they first must agree on the appropriate tax levels. Given these tax levels, the rule of the "tax game" implies that every participating country is free to adjust its level of emissions. The adjustment is realized by charging each of the participating countries the agreed on tax rate. As part of the international tax regime, parties must agree on the distribution of the tax revenues among the participating countries. A government that maximizes its net benefits will, of course, take into account the impact of changing its emissions on its costs and on its own domestic environmental benefits. The benefits are determined by the transport equation.

In the case of a global pollutant, the full equilibrium can be realized with a uniform tax rate for all participating countries. The uniform tax is assumed to be charged to citizens of these countries. The tax revenues are distributed so that countries that increase their abatement efforts are sufficiently compensated. In the case of a global pollutant all transfer coefficients are equal. Consequently, the marginal benefit side of the full equilibrium equation (4.6) will be identical for all countries. To realize the full equilibrium, the tax rate (F) has to be set so that

$$C'_i = C'_2 = \ldots = C'_i = F = \sum_i B'_i . \tag{4.13}$$

If the tax (F) is set and charged to firms and consumers in each country, they will adjust so that in each country the marginal costs are equal to the tax rate.

In the case of a non-uniformly dispersed transboundary pollutant, the marginal benefit expression in equation (4.6) differs from country to country. The optimal international tax that brings about the full equilibrium would therefore not be uniform, but country specific. If there are n countries, for country i this full equilibrium tax rate would be

$$F_i = C'_i = a_{1i}B'_1 + a_{2i}B'_2 = + \cdots + a_{ni}B'_n . \tag{4.14}$$

The tax rate is set so that the marginal benefits of all participating countries are internalized. Again, the governments would have to impose this tax rate on firms and households to induce the optimal (full equilibrium) emission reduction. The parameters for redistributing the revenues would have to be fixed so that countries that increase emission reductions are sufficiently compensated for their net additional costs.

If the target of joint implementation is to minimize the total cost of emission reductions, taking the emissions of the initial cooperative Nash equilibrium of reciprocal emission reduction as a starting point then implies that the international tax rate (F_i) is uniform [see equation (4.10)]. For every country $F_i = \lambda$, where λ is the shadow price of the emission constraint. In all countries sources will adjust so that marginal costs equal the tax rate and the shadow price. Again, the tax redistribution parameters have to be set so that due compensation is given and received.

If the goal is to meet the deposition targets of the initial agreement at minimum costs, then equation (4.12) shows that the country-specific tax level would be

$$F_i = a_{1i}\lambda_1 + a_{2i}\lambda_2 + \cdots + a_{ni}\lambda_n \; , \tag{4.15}$$

where λ_i is the shadow price of the ith deposition constraint. Again, sources in each country will adjust so that marginal costs equal the country-specific tax rate. In addition to the agreement on the tax levels for each country, agreement would be necessary on the tax redistribution parameters. Both country-specific tax and reimbursement parameters have to be determined at the outset in such a way that every signatory country is at least as well off with the tax agreement as without it.

The crux of the matter is that setting a tax at a cost-efficient and Pareto-optimal level requires perfect information not only on costs but also on benefits. Recall from Chapter 3 that setting the tax at a cost-efficient level is already difficult when we deal with multiple receptors. Either full knowledge of the costs is required or an iterative procedure is needed, which might be possible in theory but may not be very practical (Ermoliev *et al.*, 1995). Such a procedure takes time to converge, and in the meantime deposition reductions and welfare levels in some countries might sink below their initial levels. In the international context these problems would be compounded, because the environmental agency would also need information on the benefits in order to fix the reimbursement parameters. Not only would perfect information on costs and benefits be required, but these costs and benefits could not change over time, which is difficult to imagine (Mäler, 1993, 43–48). Even if one envisions an iterative procedure where an international agency proposes different tax and reimbursement levels to countries until an agreement has been reached, such an international tax regime faces some deficiencies. First, the collection of revenues by an international agency would shift power away from national governments (Bohm, 1994; Hoel, 1993). Second, if reimbursement is fixed at the outset, the impact of the tax on emissions will be uncertain if costs are uncertain or if costs change over time. As a result, tax revenues are not known beforehand, which may make countries hesitant to sign such an agreement (Hoel, 1993, p. 226). Third, the institutional complexity of the tax is considerable, because it more or less requires all countries that have signed the international agreement to participate from the outset. With a tradable quota system, countries would be able to decide for themselves whether to use this option or not (Hoel, 1993, p. 226). Fourth, in the case of a global pollutant (or minimization of the cost of a given reduction of emissions for a non-global pollutant), the simplicity of a uniform pollution tax that is imposed on sources in all participating countries holds a certain attraction. But even this is lost when the tax rate has to be country specific and, therefore, different between countries.

As an alternative to the international tax, domestic taxes could be harmonized. The main drawback of this approach is that every country has an incentive to act as a free rider and render the harmonized tax as ineffective as possible (e.g., by reducing other taxes) (Hoel, 1993). By making their own domestic tax as ineffective as possible, a country's marginal pollution control costs could be lowered. If the other countries were not to notice this, their emission reductions could remain at the same level. The additional reductions in costs might overcompensate for the loss in welfare due to the reduction in domestic emission reductions. This is certainly so for those small countries where environmental benefits are barely affected by domestic action. In the case of regional pollutants such as sulfur, free riding is less common than in the case of global pollutants, because making the tax ineffective has a price in the form of losses in domestic environmental benefits. The effective domestic tax would be lowered to the point where marginal costs equal domestic marginal benefits.

Although both international and harmonized domestic taxes have disadvantages, they do possess some positive properties as well. The tax has the advantages that the marginal costs of abatement are known and the national emission levels are achieved in a cost-efficient way. The tax might also require a smaller transaction cost than would a system of tradable permits, especially if an emission tax were to take the form of a fossil fuel tax, because such fuels are taxed anyway (Barrett, 1991, p. 86–93). The type of emission tax, however, may depend on the type of pollutant. It may be easier to tax fuels to regulate CO_2 emissions than SO_2 emissions because for SO_2 the sulfur content of the fuel does not determine emissions on its own, the abatement technology adopted does, too.

Setting the appropriate tax might be easy in the case of global pollutants compared with the case of regional pollutants. In the latter case, not only the total level of emissions but also where they are reduced might become uncertain with a tax. Unless the regulator possesses perfect knowledge of costs and benefits in an unchanging world, the distribution of costs and benefits after a tax agreement will be uncertain. A rare example of work on international taxes for regional pollutants can be found in Mäler (1994), who compares the net benefits of a uniform tax on the export of sulfur in Europe with the full cooperative solution. His conclusion is that a uniform export tax of DM 4,000 per ton of sulfur exported generates net benefits close to the optimum solution with similar distributional impacts. He also suggests that the tax revenues are high enough to compensate the losers. Mäler implicitly compares the tax result to the non-cooperative Nash equilibrium, because his benefit functions are calibrated by assuming that the 1985 emission levels constitute a non-cooperative Nash equilibrium. Obviously this export tax is not cost efficient, because in a three-country case the tax level per unit of emission for each country (taking country 1 as example) is

$$F_1 = (a_{21} + a_{31})F_e ,$$

where F_e is the uniform export tax and $(a_{21} + a_{31})$ is the share of the emissions exported.

If the uniform export tax is charged to sources in country 1, they will adjust so that

$$C'_1 = F_1 = (a_{21} + a_{31})F_E , \tag{4.16}$$

and for country 2

$$C'_2 = F_2 = (a_{12} + a_{32})F_E . \tag{4.17}$$

From the above equation it follows that

$$\frac{C'_1}{C'_2} = \frac{a_{21} + a_{31}}{a_{12} + a_{32}} . \tag{4.18}$$

A comparison of this expression with *Table 4.1* shows that it does not correspond to the conditions for the different optima.

It is not obvious that this export tax is Pareto-dominant compared with an initial situation. Halkos (1993), in a repetition of Mäler's exercise with different cost functions, does confirm that this specific uniform tax comes close to the optimum, but he also correctly notes that some countries might prefer a differentiated, country-specific full-equilibrium tax over a uniform export tax. However, none of the authors works out a system of reimbursement parameters to achieve Pareto-dominance over the non-tax case. Mäler (1993, pp. 13–14) is also aware that his benefit functions are educated guesses at best, not in the least because marginal benefits are assumed to be constant and independent of actual deposition levels.

All in all, the sparse results confirm our impression that it might be difficult, if not impossible, to design taxes and reimbursement simultaneously to meet deposition targets not only in a cost-efficient way but also in a Pareto-dominant way.

In conclusion, in the case of an international tax, reimbursement of all revenues to ensure that all countries are better off than under the Nash equilibrium is necessary to obtain a Pareto-dominant improvement. However, full knowledge of cost and benefit functions is required to set both tax and reimbursement parameters so that cost efficiency and Pareto-dominance are achieved. Alternatively, a trial-and-error procedure would be needed, with uncertain tax revenues and environmental benefits. Harmonizing domestic taxes might stimulate free-riding behavior; moreover, tax revenues cannot be used to compensate countries adversely affected. An international tax also requires all countries that signed the agreement to participate, whereas a system of tradable quotas leaves countries the choice of trading or not trading. The tax does, however, have the attractive feature that marginal costs are known beforehand. Where price incentives such as taxes appear to pose problems in an international context, instruments that fix quantities might be a preferable option.

4.4 Tradable Emission Permits

4.4.1 Purpose

This section explores the extent to which different emission permit trading schemes can constitute a cost-saving, Pareto-dominant improvement of international agreements on emission reductions. The following trading schemes are evaluated:

- Emission permit trading in one zone;
- Ambient or deposition permit trading;
- Trading subject to rules regarding deposition.

4.4.2 Emission permit trading

Assume countries have made an initial agreement on national emission reductions. Each country could now be allocated emission permits on the basis of this initial agreement. Permits could be freely bought and sold between all countries under the condition that the sum of the emissions (or, conversely, the sum of the emission reductions) remains the same. Each permit would allow the emission of one unit, irrespective of the location. Hence, we allow straightforward emission trading in one zone. Two possibilities are now conceivable:

- Countries act as trading partners;

Economic Instruments in an International Context

- Countries decentralize decisions and allocate permits to individual firms that trade.

If countries act on the permit market it makes sense to assume that they maximize the net benefits of emission reductions plus the net revenues of permit sales subject to the condition that the emissions remaining after abatement ($\overline{E}_1 - R_1$) do not exceed the number of permits they possess. $\overline{E}_1$ is the level of emissions without any control whatsoever. L_1^0 is the initial allocation of permits. The Lagrangian of this problem is

$$H = B_1(Z_1) - C_1(R_1) - P\left[(\overline{E}_1 - R_1) - L_1^0\right] \\ + \lambda_1(Z_1 - a_{11}R_1 - a_{12}R_2 - a_{13}R_3) \; , \quad (4.19)$$

where P is the permit price. Assuming that countries only care about the impact of their own emission reduction on environmental benefits and ignore the possible repercussions of trade on the emissions of the other countries (analogous to the case of emission charges), we obtain as a first-order condition

$$C_1' = a_{11}B_1' + P \; . \quad (4.20)$$

Equation (4.20) shows that country 1 would reduce its emissions to the point where the net marginal costs ($C_1' - a_{11}B_1'$) equal the permit price. When country 1 considers the reductions of other countries (R_2, R_3) as given, equation (4.20) can be rewritten as the equation $R_1 = R_1(P)$, where emission reductions by country 1 are a function of the permit price. The permit demand function of country 1 will be $(\overline{E}_1 - R_1) - L_1^0$. The demand from country 1 will be positive if $L_1^0 < \overline{E} - R_1$ (that is, if emissions after reduction exceed the number of permits). The demand will be negative (that is, supply will be positive) if $L_1^0 > \overline{E} - R_1$. Similar functions can be derived for countries 2 and 3. Together they form a model of excess demand and supply with R_i and P as unknown variables. To finalize the model the market equilibrium condition must be added, requiring the sum of excess demand and supply to be zero:

$$\left[(\overline{E}_1 - R_1) - L_1\right] + \left[(\overline{E}_2 - R_2) - L_2\right] + \left[(\overline{E}_3 - R_3) - L_3\right] = 0 \; , \quad (4.21)$$

or

$$R_1 + R_2 + R_3 = \overline{E}_1 + \overline{E}_2 + \overline{E}_3 - L_1 - L_2 - L_3 \; . \quad (4.22)$$

From the three excess demand functions and the market equilibrium function the model can be solved. In other words, if individual countries were to regard the market price as given (that is, if they were to act as price takers), then a market equilibrium would exist.

If the market equilibrium exists, equation (4.20) can be written as

$$C'_1 - a_{11}B'_1 = C'_2 - a_{22}B'_2 = C'_3 - a_{33}B'_3 = P \ . \tag{4.23}$$

A comparison with equation (4.10) shows that this is not equivalent to the result of central international planning with the objective of minimizing costs under a constraint on total emissions. However, the result would be equivalent if countries were to neglect the impact of their abatement on national benefits (especially $a_{ii}B'_i = 0$) – in other words, if countries were to individually behave as cost minimizers. The result would also be equivalent if the benefits in any country were dependent only on the total level of emission reduction, because in this case marginal benefits would not change since total emission levels would remain constant. This is the case with CO_2 emissions, for example.

The upshot of all this discussion is that if countries are willing to pay for environmental improvement and if they maximize net benefits, then joint implementation in the form of tradable emission permits (which are traded one to one) will not bring about the joint maximum, nor will it lead to a cost-minimum solution.

The question remaining is whether any trade would occur, and if it were to occur, whether the new equilibrium after trade would be Pareto-dominant compared with the initial cooperative Nash solution. The equations presented here only state marginality conditions and give no definite answer to that question. However, there are arguments that make it plausible that the equilibrium after trade would not necessarily be Pareto-dominant. The most significant one is that the constraints on single-zone emission trading do not exclude negative spillovers to parties not engaged in trading.

Figure 4.4 illustrates what could happen in a three-country case. Two possible trading equilibria are A and B. In case A, country 2 would buy permits from country 1, reduce its emission reduction efforts, and increase its emissions. Although deposition in country 2 would increase (point A is below line Z_2), welfare in country 2 would also increase because the cost savings outweigh the loss of benefits. Country 2 would have to pay for the permits bought from country 1. Country 1 would sell permits, increase its emission reduction efforts, reduce its emissions, and face less deposition. The revenues from the permit sale would have to compensate country 1 for its net welfare loss (A is below curve $\overline{U}_1$). In this particular case, deposition in country 3 (Z_3) would increase because point A is below Z_3, so even without being an actor on the permit market, country 3 would be in a worse situation. Depending on the cost-benefit configuration, permit trading could lead to situation B. In this case, countries 1 and 2 would win (or else they would not make the bargain), and country 3 would face lower deposition, more benefits, and increased welfare than under the initial agreement. So, in spite of the fact that each

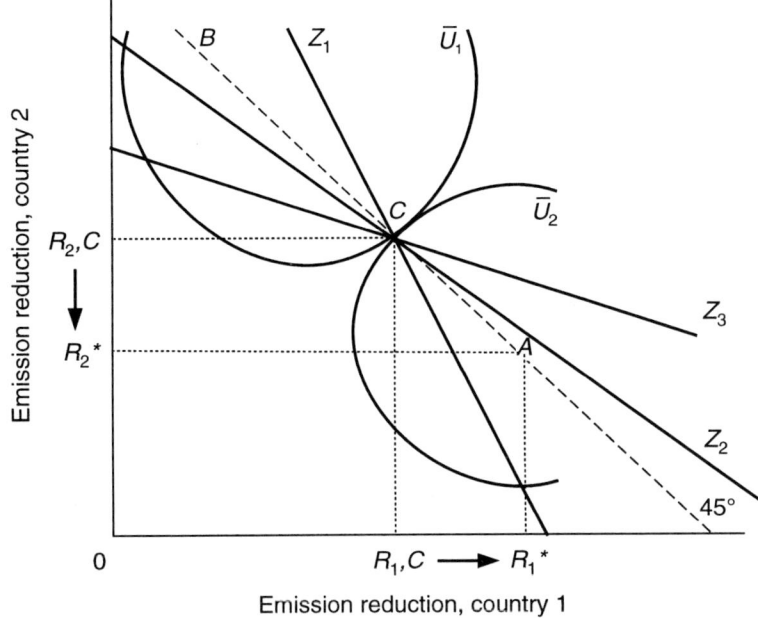

Figure 4.4. Countries trade emission permits.

country acts on the permit market to protect its own welfare, there is no guarantee that a win-win situation will result for all parties.

There are basically two possibilities for Pareto dominance: all countries improve their welfare or some countries improve and all others remain constant. The point is that there is no guarantee that emission trading, putting a constraint on total emissions only, will result in a Pareto-dominant improvement in the case of regional pollutants, because some countries might be confronted with losses in environmental benefits without receiving compensatory reductions in costs. A Pareto improvement only occurs when one of two instances occurs:

- The location of the source is irrelevant (e.g., global pollutants).
- Due to the specific regional cost-benefit configuration, the market equilibrium creates only winners or only winners and non-losers.

The latter situation might occur, but its occurrence can only be ascertained if there is perfect knowledge of both costs and benefits.

It is also possible for individual firms to operate on the permit market. It is important to note that there is a fundamental difference between allowing countries

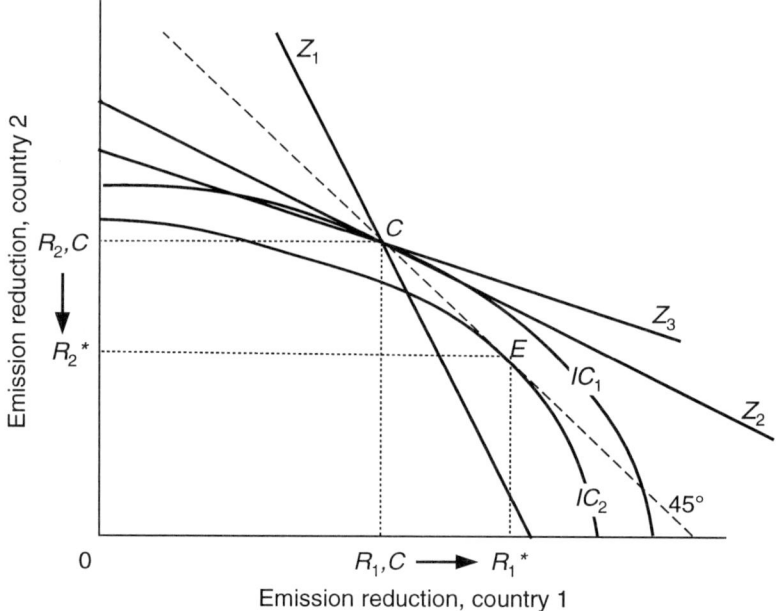

Figure 4.5. Firms trade emission permits.

to trade and allowing individual firms to trade. *Figure 4.5* illustrates this. If firms trade, one can assume that they minimize costs and ignore externalities imposed in the form of higher (or lower) deposition on all countries. In *Figure 4.5*, point E is the market equilibrium. In this particular case, deposition in country 2 has increased and, in contrast to the case where country 2 itself trades, it is now an open question whether the cost savings allowed by trading compensate for the loss of benefits.

A formal analysis supports this. With a perfect market (compare Chapter 2) marginal costs would be equalized among all firms in all countries and the following would hold:

$$C'_i = P .\qquad(4.24)$$

This implies that condition (4.10) is also valid. Stated differently, the costs necessary to keep total emissions below a given ceiling are actually minimized. Again, this in itself does not imply a Pareto-dominant change in the allocation of emissions. Pareto dominance occurs in a number of circumstances. The first of these circumstances is when depositions in each country are independent of which of the countries emit ($a_{11} = a_{12} = a_{13}$, and $a_{21} = a_{22} = a_{23}$, and $a_{31} = a_{32} = a_{33}$). Because the total level of emissions is constant (the license total restricts the total volume of emissions), deposition is also constant. As a result, marginal environmental benefits are always constant for every country. Thus, as long as the total emissions remain constant with emission trading, environmental benefits can be ignored and all countries will be at least as well off with trading as without trading. In this case, the joint maximum would be realized [compare equations (4.24) and (4.10)] if $P = B'_1 + B'_2 + B'_3$; that is, if the license total were set at the optimal level.

Second, Pareto dominance occurs if for every country the expected decrease in pollution control costs plus the net revenues resulting from emission trading compensate for any increases in deposition and associated reductions in environmental benefits. If emission trading is decentralized, knowledge of pollution control costs would be required to foresee what the market equilibrium would be. In the case of market imperfections (high transaction costs or market power), the equilibrium might be different from the perfect market case. In other words, using an emission trading scheme might then be a risky strategy, and those countries that expect little gain from emission trading in terms of cost savings and that have high marginal benefits and relatively low revenues might oppose trading, as is shown in Part II of this study.

That these possibilities are not completely unrealistic has been suggested by Mäler (1993). His simulations suggest that the net benefits of emission trading in one zone (to meet the same level of emissions as the full equilibrium) are twice as high as those of a uniform cutback for all countries together. Unfortunately, he does not analyze the extent to which this is a Pareto-dominant strategy for every country, although he suggests that permits should be auctioned annually and that the proceeds should be used to compensate the losers.

On the basis of the foregoing analysis, it appears that emission permit trading in one zone is not by definition a Pareto-dominant improvement over an initial agreement. However, it can turn out to be so in specific circumstances:

- If we deal with a pollutant where location can be safely ignored, such as with CO_2, and if emissions definitely do not increase;
- If in the permit equilibrium the welfare of all countries improves or is at least maintained, either because the cost savings from trading exceed any losses in environmental benefits, because environmental benefits increase as deposition decreases, or because deposition and benefits remain constant.

Furthermore, if individual firms instead of countries were to trade, there would be less of a guarantee of a Pareto-dominant change, because firms would ignore both domestic as well as transboundary externalities. However, this solution would minimize total control costs.

Although it might thus be possible for emission trading to result in an improvement in welfare for every country, some qualifications are needed. First, in the case of a regional pollutant, countries know that emission permit trading only leads to a win-win situation if they have perfect knowledge of both costs and benefits so that they can forecast the market equilibrium and evaluate its impacts. If firms trade, only cost information would be needed to foresee the equilibrium emissions and depositions. Instead of benefit information, discussion between countries involved could then be used to assess the political acceptability of the results. However, if costs are uncertain or if the market is imperfect, the results may differ from the expectations. As with the tax, this might make countries hesitant to agree to emission trading.

Second, with imperfect enforcement, agents might emit more than they are permitted in order to save costs and permit outlays (Mäler, 1993; Bohm, 1994; Keeler, 1991). If countries trade, this noncompliance is restricted, because if they emit more than permitted they also lose domestic benefits. In addition, the permit market has a built-in stabilizer as too much noncompliance would lower the permit price, rendering noncompliance less attractive. Moreover, whether emission trading would lead to more or less compliance than regulation would depends on the structure of the penalty function (Keeler, 1991; Malik, 1990). A fine at least as high as the expected marginal cost savings of noncompliance could be used to counteract noncompliance, but Barrett (1991, pp. 90–92) questions whether this is realistic in an international context. He suggests other options: if one country fails to meet its obligations, others could fail, as well, or could defect in another context. Another solution could be to agree that cheating countries must reduce emissions below the licenses allocated to them for future years to balance out the shortfall of reductions in the past.

Although tradable emission permits have some problems, they also have advantages. It seems to be easier to allow individual countries to trade than to design complex tax-reimbursement schedules if the initial agreement consists of agreed on national quotas (Hoel, 1993). Moreover, trading is optional and can be introduced gradually. Trading has the advantage that, with perfect monitoring and enforcement, at least national emission levels are known beforehand. Last but not least, emission trading in one zone is simple and can therefore be expected to have relatively low transaction costs, high participation rates, and an absence of market power, especially if individual firms are allowed to trade.

All in all, emission trading might be more acceptable than emission taxes in an international context. However, this section has also shown that whether or not

Economic Instruments in an International Context

emission trading for regional pollutants results in a Pareto-optimal improvement over a given agreement depends on the specific regional context.

4.4.3 Ambient or deposition permits

As an alternative to emission trading, one can think of ambient or deposition permits trading. A permit entitles the owner to one unit of deposition at a certain location. If one source wants to increase emissions, it has to collect a portfolio of deposition permits for each receptor (country) it affects, taking into account the transfer coefficients of the source to the receptor. Again we assume there are two cases:

- Individual countries act on the deposition permit market.
- Individual firms operate on this market.

Individual countries would maximize welfare by striking a balance between reducing emissions and buying and selling deposition permits. The initial allocation of permits for country 1 is represented by L_{11}^0, L_{21}^0, and L_{31}^0, denoting the permits held by country 1 to deposit in receptor country 1, country 2, and country 3, respectively. Each country maximizes welfare subject to the condition that it is not allowed to emit more than the deposition permits it holds. This gives the following Lagrangian:

$$H_1 = B_1(Z_1) - C_1(R_1) - P_1\left[a_{11}(\overline{E}_1 - R_1) - L_{11}^0\right]$$
$$- P_2\left[a_{21}(\overline{E}_1 - R_1) - L_{21}^0\right] + P_3\left[a_{31}(\overline{E}_1 - R_1) - L_{31}^0\right]$$
$$+ \lambda_1(Z_1 - a_{11}R_1 - a_{12}R_2 - a_{13}R_3) \ . \tag{4.25}$$

P_1, P_2, and P_3 are the prices of deposition permits for deposition in countries 1, 2, and 3, respectively. When selling deposition permits, the country is assumed to ignore the impact on the emissions of other countries. This gives the following first-order condition for individually rational behavior on the deposition permit market:

$$C_1' = a_{11}B_1' + a_{11}P_1 + a_{21}P_2 + a_{31}P_3 \ . \tag{4.26}$$

From the perspective of individual rationality, the country would reduce emissions to the level where marginal costs equal the domestic marginal benefits plus the revenues from the sale of deposition permits that becomes possible if emissions are reduced.

In total, three first-order conditions exist, one for each country. From condition (4.26) and the transport equation, emission reductions can be written as a function of the permit prices at the deposition markets:

$$
\begin{aligned}
R_1 &= R_1(P_1, P_2, P_3) \ , \\
R_2 &= R_2(P_1, P_2, P_3) \ , \\
R_3 &= R_3(P_1, P_2, P_3) \ .
\end{aligned}
\tag{4.27}
$$

In addition, the three market clearing conditions must hold, one for each deposition permit market. The condition for permit market 1 reads

$$
a_{11}\left[(\overline{E}_1 - R_1) - L_{11}^0\right] + a_{12}\left[(\overline{E}_2 - R_2) - L_{12}^0\right] \\
+ a_{13}\left[(\overline{E}_3 - R_3) - L_{13}^0\right] = 0 \ ,
$$

stated differently

$$
\begin{aligned}
a_{11}R_1 + a_{12}R_2 + a_{13}R_3 &= a_{11}(\overline{E}_1 - L_{11}^0) + a_{12}(\overline{E}_2 - L_{12}^0) \\
&\quad + a_{13}(\overline{E}_3 - L_{13}^0) \ .
\end{aligned}
\tag{4.28}
$$

If $\overline{E}_1$, $\overline{E}_2$, $\overline{E}_3$ were to be interpreted as the emission levels under the initial cooperative Nash agreement, and if licenses were to be allocated equaling these emission totals, the market clearing conditions could be written as follows:

$$
\begin{aligned}
a_{11}R_1 + a_{12}R_2 + a_{13}R_3 &= 0 \ , \\
a_{21}R_1 + a_{22}R_2 + a_{23}R_3 &= 0 \ , \\
a_{31}R_1 + a_{32}R_2 + a_{33}R_3 &= 0 \ .
\end{aligned}
\tag{4.29}
$$

R_i could then be interpreted as the change in emission reductions from the Nash cooperative solution of reciprocal emission reductions. From the six equations, the six unknown variables R_i and P_i can be solved. Equation (4.26) can be rewritten as

$$
C_1' - a_{11}B_1' = a_{11}P_1 + a_{21}P_2 + a_{31}P_3 \ .
\tag{4.30}
$$

This resembles the condition for cost minimization under a deposition constraint [condition (4.12)]. The difference is the term $a_{11}B_1'$. Therefore, under the assumption that countries account for the impact on domestic benefits, trading of deposition permits does not necessarily minimize costs. In short, with positive marginal benefits, the deposition permit trading will not meet the conditions for cost efficiency. This may not matter too much, however, because total welfare levels might well be higher compared with the situation without trade if countries account for the impact on environmental benefits when trading deposition permits.

Figure 4.6 illustrates the trading of deposition permits. Lines $\overline{Z}_1$, $\overline{Z}_2$, and $\overline{Z}_3$ depict the implicit or explicit deposition goals resulting from the initial agreement C. Point A represents the possible outcome of deposition trading. In this

Economic Instruments in an International Context 89

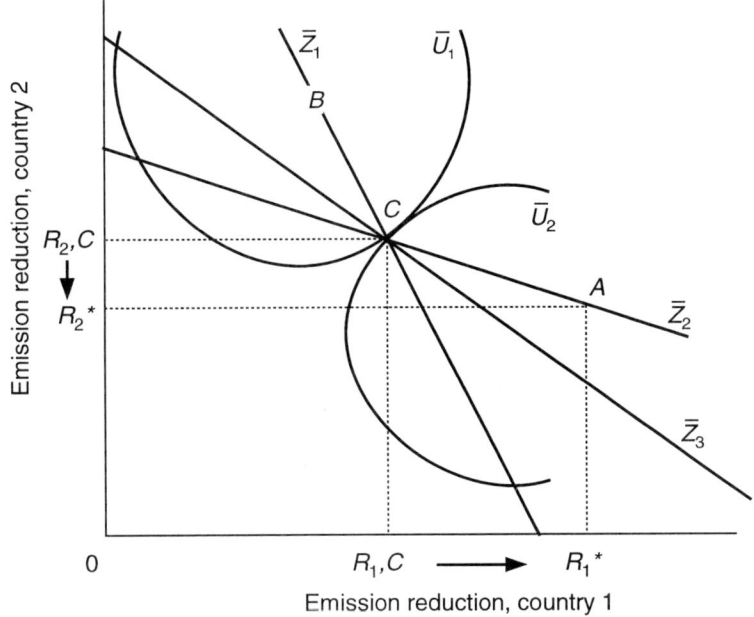

Figure 4.6. Countries trade deposition permits.

case, countries 1 and 2 would strike a deal. Country 2 would decrease emission abatement and reduce costs while deposition (and benefits) would remain constant. Country 2 would have to buy deposition rights from country 1. In country 1, net welfare improvement would consist of an increase in emission reduction efforts and pollution control costs, which would be compensated for by a reduction in deposition at home plus the revenues of selling deposition permits to country 2. Utility in country 3 would increase because deposition in country 3 would decrease without compensatory side payments. Hence, A constitutes a Pareto-dominant improvement. In this particular case there would be an excess supply of deposition permits for countries 1 and 3, driving permit prices to zero.

From a more formal point of view, deposition permit trading is not only individually rational but also Pareto-dominant. The deposition goals for each country can safely be assumed to be those that implicitly or explicitly result from the cooperative Nash equilibrium. More stringent deposition goals would not make sense, because they would not be individually rational. Less strict deposition goals would not make sense either, because the countries' coordination of willingness to pay would result in lower deposition. This necessarily implies that all deposition constraints are exactly met (binding) in every country. If deposition permits are now allocated to each country on the basis of these cooperative Nash deposition levels, deposition permit trading can only result in a Pareto improvement for every country. This is because deposition cannot increase in any country, so benefits cannot decrease either. As a result, net benefits can only increase. Hence, some countries might experience improved welfare because they trade, whereas others might experience constant welfare levels if their deposition levels are not altered and they do not enter trade. The latter, of course, could only occur if it would not be beneficial for these countries to enter trade.

One can conclude that if countries act as welfare maximizers, trading of deposition permits necessarily leads to a Pareto-optimal improvement if the number of deposition permits in each country is based on the actual deposition resulting from the initial cooperative Nash equilibrium agreement.

One could imagine a slight variation of this deposition permit trading if countries are allowed to increase emissions only if they acquire the necessary permits to export deposition. This would then result in the following Lagrangian for country 1 [compare equation (4.25)]:

$$H = B_1(Z_1) - C_1(R_1) - a_{21}P_2\left[(\overline{E}_1 - R_1) - L_{21}^0\right]$$
$$- a_{31}P_3\left[(\overline{E}_1 - R_1) - L_{31}^0\right]$$
$$+ \lambda_1(Z_1 - a_{11}R_1 - a_{12}R_2 - a_{13}R_3) \ . \tag{4.31}$$

Similar functions exist for countries 2 and 3. The first-order conditions for a maximum are

$$C_1' = a_{11}B_1' + a_{21}P_2 + a_{31}P_3 \ ,$$
$$C_2' = a_{12}P_1 + a_{22}B_2' + a_{32}P_3 \ ,$$
$$C_3' = a_{13}P_1 + a_{23}P_2 + a_{33}B_3' \ . \tag{4.32}$$

These functions can be reformulated as demand functions of the deposition permit prices:

$$R_1 = R_1(P_2, P_3) \ ,$$
$$R_2 = R_2(P_1, P_3) \ ,$$
$$R_3 = R_3(P_1, P_2) \ . \tag{4.33}$$

Again, the market clearing conditions (4.28) must hold. This model, consisting of six equations and six variables, can be solved for R_i and P_i.

The first-order conditions (4.32) are a kind of hybrid between a joint welfare maximum (4.6) and cost minimization (4.12). The question arises whether this is

Economic Instruments in an International Context

a Pareto improvement or not. Taking country 2 as an example, the deposition in the form of imported deposition cannot increase because the other countries would need export permits to do so. However, country 1 may itself decide to increase or decrease its domestic deposition, provided that the deposition in countries 2 and 3 does not increase. If countries 1 and 2 were to trade, deposition in country 3 would always have to remain constant. Because countries 1 and 2 will trade only if it increases their individual net benefits, one can expect that trading in export deposition permits will be a Pareto-dominant improvement.

One could also imagine a case where deposition permits are ultimately allocated to individual companies that act as cost minimizers in a perfect, competitive market. In this case, the marginal costs for each firm (in country 1) would become the following (see Chapter 3):

$$C'_1 = a_{11}P_1 + a_{21}P_2 + a_{31}P_3 \ . \tag{4.34}$$

The question is, when is this a Pareto-optimal improvement?

In this case, we have pure cost minimization subject to deposition constraints. With perfect markets for deposition permits and no transaction costs, this type of trading results in the cost-minimum solution [see equation (4.12)]. More specifically, the shadow prices of each deposition constraint (λ_i) equal the deposition permit prices on each market (P_i).

Figure 4.7 illustrates this. IC_1 and IC_2 are iso-cost curves. The starting point for trade is point C. Deposition constraints are $\overline{Z}_1$, $\overline{Z}_2$, and $\overline{Z}_3$. One possible cost-saving trade would be a move toward point B, along line $\overline{Z}_1$. Firms in country 1 would buy deposition permits from firms in country 2. As a result, emission reduction would decrease in country 1 and would increase in country 2. Total abatement costs would be lower, because point B is on an iso-cost curve (IC_2) closer to the origin than curve IC_1 (closer to the origin implies smaller reductions, thus lower costs). As a result, deposition would remain constant in country 1 and would decrease in countries 2 and 3 (point B is above lines $\overline{Z}_2$ and $\overline{Z}_3$ depicting the same deposition reduction as under the initial agreement). This cost-saving trade would therefore be a Pareto-dominant improvement. Again, if the number of deposition permits is set so that the total equals the deposition that results from the initial cooperative Nash agreement, deposition trading between firms can only result in a Pareto-dominant improvement.

In summary, if individual companies act on the deposition permit market, and if these permits are based on deposition levels resulting from the international agreement, there is a guarantee that a Pareto-dominant improvement will occur.

Although deposition permits appear to be a promising route, they have the same deficiencies in an international context as in a national one. Transaction complexity and costs might be high if the number of receptor countries is high,

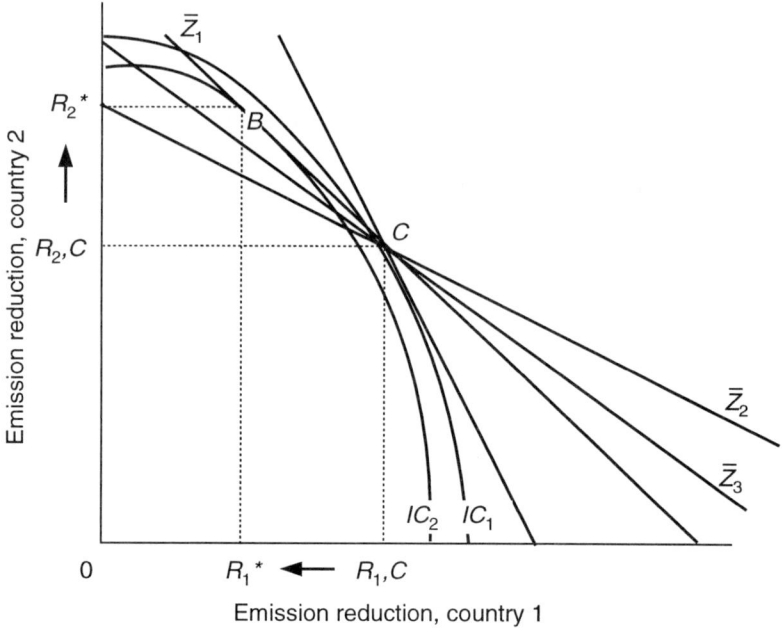

Figure 4.7. Deposition permits with firms trading.

because countries might have to trade with more than one country to obtain the necessary deposition permits to increase emissions. This may tend to limit the number of parties on a particular deposition market, which may imply thin markets where market power might occur. However, this also depends on how well trade is organized. The international agency could, for example, organize regular auctions to establish deposition permit prices for each country and could publish the transfer coefficients. In this way, each firm or country would not only know how many permits it would need to increase its emissions, but would also know whether this would be worthwhile given the prevailing market prices. Achieving the maximum cost savings might still be difficult, but one should remember that a system of deposition permits is Pareto-dominant if it is based on actual deposition. It is also more likely to be accepted than emission permit trading. The Pareto dominance does not depend on uncertain information of costs and benefits. Deposition permits also have the advantage that the environmental effectiveness is great, because no more permits are allocated than are allowed. Ownership and sales of permits have to be registered and monitored; moreover, enforcement mechanisms have to be in

Economic Instruments in an International Context 93

place to prevent countries from emitting more than their collection of deposition permits allows.

4.4.4 Trading rules for case-by-case decisions

As shown in Chapter 3, to avoid the transaction complexity of deposition permits, three alternatives have been proposed in the literature that allow individual trades under certain conditions:

- Pollution offset trading: trading subject to the condition that the deposition targets are not violated in any country;
- Modified pollution offset trading: subject to the condition that deposition targets are met and deposition (if lower than the standards) does not increase in any country;
- Non-degradation offset trading: trading subject to non-violation of deposition targets plus the rule that the total emissions are not allowed to increase.

Clearly, these rules would allow trade only if none of the national deposition goals were to be violated. Trade could be bilateral, sequential, or simultaneous. Throughout this chapter it was assumed that the starting point for trade would be a cooperative Nash equilibrium. Such an agreement implies an implicit or explicit set of national deposition constraints that are all initially binding. Because of this feature, there would be no difference between the pollution offset or the modified pollution offset rule; both rules would allow trade subject to the condition that the Nash deposition targets would not be violated. The non-degradation offset rule requires the same deposition constraint, but would also require that the total amount of emissions did not increase during any trade.

Figure 4.8 illustrates the potential impacts of these trading rules. Pollution offset trading would allow trades as long as the deposition reduction would be above and to the right of line $\overline{Z}_1 C \overline{Z}_2$. This implies that deposition after trade would always be equal to or lower than that under the initial agreement. With the non-degradation rule the same deposition constraint would apply, but on top of this the sum of the emission reductions would have to be at least as high after trade as it was initially. In *Figure 4.8*, the trade has to remain above and to the right of the 45-degree line. In this example the line would not be binding, but it could be binding in other circumstances.

Regarding the modified offset rule, we recall that under competitive circumstances the rule is equivalent to deposition permit trading (see Chapter 3); the same number of transactions would be required to meet the deposition standards at least costs.

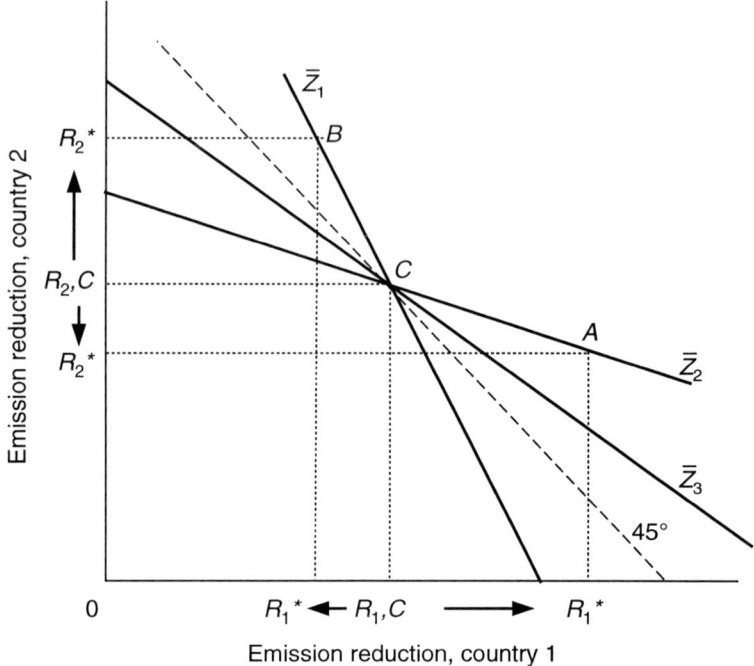

Figure 4.8. Trading rules on deposition.

Stated differently, with a competitive market countries (or firms) could just as easily buy and sell deposition permits for a number of markets as trade emission rights (a bundle of implicit deposition rights). With known prices and full information there would be no difference (see Chapter 3). With noncompetitive markets (no known vector of prices), the system would be fundamentally different. With modified offset trading, positive spillovers and free riding could occur, because some countries might be able to increase emissions without paying. *Figure 4.8* can serve to illustrate this. Assuming that countries 1 and 2 trade and end up at point B inevitably implies that the deposition reduction in country 3 will increase (B is above line $\bar{Z}_3$). If country 3 affects only deposition within its own borders and in country 2, the modified offset rule allows country 3 to increase its emissions to the point where Z_2 or Z_3 becomes binding again without having to buy any emission permits from countries 2 or 1.

In other words, any deposition reductions below the initial agreement are freely accessible common property. Recall that under deposition permits this could not happen because country 3 would not be able to increase its emissions without buying the deposition rights from another country. This has serious implications, especially when countries trade because they account for the impact of trade on domestic deposition and benefits. Country 1 would agree to the trade with country 2 knowing that post-trade deposition would be lower than it was initially. However, country 1 has no property rights to these additional reductions. Under the modified offset rule it cannot prevent country 3 from acting as a free rider and increasing its emissions and deposition. In country 2, deposition might increase again to the original deposition constraint. With ambient permits trading, country 3 cannot increase emissions without buying deposition permits from country 1 or 2. Chapter 3 showed that if firms trade, modified offset trading may no longer be cost efficient when free riding occurs. Analogously, trading between countries might no longer be welfare maximizing with free riding, because part of the spillovers are not internalized. This free riding might not be acceptable in an international context. However, irrespective of whether firms or countries implement pollution offset trading, deposition in each country is always at least as low as under the initial agreement. In other words, modified offset trading and pollution offset trading always constitute Pareto-dominant improvements.

These types of case-by-case rules might be an easy way to improve the flexibility of an agreement, and the requirement that every trade must be accompanied by a deposition impact analysis may not be too complex if the trade is between countries. If individual firms trade, running deposition models to show the impact of trade on deposition might lead to excessive costs. Moreover, as Chapter 6 will show, in practice only a few trades that required atmospheric modeling have been implemented. For our international context, the main advantage of modified offset trading and pollution offset trading is that they guarantee a Pareto-dominant improvement.

Finally, the non-degradation offset rule allows trade if the deposition reductions do not increase anywhere compared with the initial agreement and if the total emissions do not increase. Again, this rule always constitutes a Pareto-dominant improvement. The fact that emissions are not allowed to increase (even if they would not violate the deposition standards) puts an unnecessary constraint on trade at the expense of cost efficiency (see Chapter 3) and welfare maximization. In a noncompetitive context it has the advantage of preventing free riding; no country can increase its emissions without buying emission rights from another country because with every single trade emissions have to remain constant.

In summary, of the three trading rules investigated the pollution offset and the modified offset rules are equivalent. All three rules are Pareto-dominant. Under competitive circumstances the modified offset rule and the pollution offset rule are equivalent to deposition permit trading. With high transaction costs and imperfect information, modified and pollution offset trading might not lead to cost efficiency or a welfare maximum; they enable free riding, which may pose political problems internationally; and they require the use of a deposition model with every single trade, which increases transaction costs.

On the other hand, modified and pollution offset trading are Pareto-dominant and from an environmental point of view can only reduce deposition and improve cost efficiency. Moreover, they increase competition in national permit markets because to a certain degree competition with foreign permit suppliers becomes possible. The number of potential buyers and sellers increases, markets become thicker, and market power becomes less of a problem compared with a situation of pure national permit markets.

One obvious modification of the rule would be to allow two countries to trade as long as the deposition in the territory of third parties is not increased (Nentjes, 1994). Although this appears attractive, under the pollution offset rule any reduction below the standard is a freely accessible, common property resource because every country (or firm) is free to increase deposition up to the standard. Therefore, if two countries were to trade and one country were to also accept the terms of trade on the basis of an expected decrease in domestic deposition (below the standard), a second trade could undo these benefits. That is, of course, unless one were to agree that to avoid this the deposition goals would change after each trade, which does not appear to be a very practical rule.

The case-by-case approach may be too cumbersome if we decentralize trading to individual firms, and simpler solutions such as emission trading in one zone might be requested. This, however, is not a Pareto-dominant improvement under all circumstances, as was shown in this chapter. Another option is to allow trading according to fixed trading ratios in order to reduce transaction complexity. This new option is analyzed in the application part of this study (Chapters 8 to 11).

4.5 Concluding Observations

This chapter sought to identify the conditions under which emission charges and tradable emission permits could be used to improve the cost efficiency of meeting environmental targets for non-uniformly dispersed pollutants in an international context. A number of conclusions can be drawn from the results presented in the previous sections. First of all, agreements taking the form of a cooperative Nash equilibrium of reciprocal reductions of emissions are generally not cost efficient and thus leave room for reallocation of obligations between countries so that total abatement costs are reduced. There is no further room for cost savings only if the countries signed an agreement based on the full-equilibrium solution.

Second, with an international emission tax, reimbursement of all revenues is necessary to obtain a Pareto-dominant improvement. This, however, requires full knowledge of both cost and benefit functions to simultaneously set tax and reimbursement parameters. Without such knowledge tax revenues are uncertain and emission levels, tax rates, or distribution parameters might be set at levels where some countries expect to be worse off; therefore, those countries might opt not to participate.

Third, although emission permit trading in one zone is not by definition a Pareto-dominant improvement over an initial agreement, it can turn out to be so in specific circumstances. This is so if the location of sources does not influence the deposition (as with CO_2) or if the permit equilibrium leads to such a result in the specific regional context. If individual firms instead of countries trade, there is less of a guarantee that a Pareto-dominant change will occur, because firms ignore both domestic and transboundary externalities when trading. However, cost minimization will occur.

Fourth, deposition permit trading is always Pareto-dominant if deposition permits are allocated on the basis of actual deposition levels achieved under the cooperative Nash agreement. This is true irrespective of whether firms or countries trade. However, transaction complexity and costs might be high.

Finally, all three trading rules, the pollution offset rule, the modified offset rule, and the non-degradation offset rule, are always Pareto-dominant because they do not allow increases in deposition anywhere. Such rules are Pareto-dominant and environmentally effective, and they reduce costs. With perfect markets (competitive, with no transaction costs) the pollution offset rule and the modified pollution offset rule are equivalent to deposition permit trading. The non-degradation offset rule would, however, be less cost efficient and would also prevent welfare maximization. With less-than-perfect markets (i.e., no known vector of prices), it might be difficult to find bilateral trades that save costs, increase net benefits, and meet the stringent requirement of non-increasing deposition; free riding is (to a certain degree) possible and deposition modeling of every single trade increases transaction complexity.

As a final point, this analysis has shown that different economic instruments, especially emission trading schemes, not only can improve the cost efficiency of given international agreements but are also politically acceptable because no country would be worse off with trade than without it.

Chapter 5

Empirical Simulation Models and Emission Trading

5.1 Introduction

The overview of economic theory in the previous chapters showed that under certain conditions emission trading can be expected to be Pareto-dominant and to meet given environmental goals at a minimum cost to society. The conditions were found to be dependent on whether the pollutant studied was uniformly mixed or non-uniformly mixed. Stated differently, these conditions depend on whether the environmental objective consists of meeting a total level of emissions or consists of meeting ambient concentration or deposition standards. Evidence from empirical simulation models is useful for examining the extent to which the claims of economic theory hold. Such models simulate emission permit trading in specific regional settings, making use of empirical estimates of emissions, pollution control costs, and atmospheric dispersion of pollutants. These models can provide additional insight into the conditions under which emission trading is cost efficient and meets environmental goals; they can thus help build a bridge between economic theory (Chapters 2 to 4) and practical experience with economic instruments (Chapter 6).

The objective of this chapter is to provide a survey of empirical simulation models of emission trading. The explicit aim is to examine the conditions under which emission trading is cost efficient and meets the environmental objectives (environmental effectiveness). Administrative practicability, innovative impacts, and distributive implications of emission trading are also addressed. The survey is restricted to air pollution studies and to partial equilibrium models. In comparison with previous studies (especially Tietenberg, 1985), this review surveys international applications and includes more recent material, especially related to Europe, and new approaches to modeling emission trading as sequential rather than simultaneous processes.

The chapter has the following structure. Section 5.2 examines the cost efficiency and environmental effectiveness of emission trading when the environmental objective consists of meeting a certain level of emissions. Section 5.3 deals with cost efficiency for non-uniformly dispersed pollutants for those cases where the environmental objective consists of a set of air quality standards. The fourth section summarizes the evidence from simulation models on administrative practicability, innovation, and distributive impacts of emission trading.

5.2 Total Emission Levels as the Objective

5.2.1 The potential cost savings

If a pollutant is uniformly dispersed in the atmosphere, the location of the polluting source is unimportant for its environmental impact; controlling the total amount of emissions is sufficient for protecting the environment. Even if the pollutant is known to be non-uniformly dispersed, as with SO_2, it may make sense in some cases to ignore the location of the source and take the total emission level as the environmental objective. This simplifies policy design and might be quite appropriate under certain regional circumstances, for example, if one deals only with high-stack sources in a small region (see Pototschnig, 1994). This section examines the results of simulation models that analyze whether emission trading can attain a certain level of emissions at lower costs than can be achieved with a command-and-control (CAC) or regulatory approach.

Table 5.1 gives an overview of the results of empirical simulation studies on the potential cost savings of tradable emission permits. The table suggests that emission trading saves costs compared with a CAC allocation of emission reduction and that these cost savings can be significant (up to 85 percent). The international studies that have been performed for CO_2 and SO_2 confirm the potential cost savings. These results are contingent on several assumptions. What these simulation models typically have in common is that they apply algorithms to find the least-cost allocation of emission control measures among the sources in order to meet a given emission ceiling. These algorithms consist of linear, nonlinear, or mixed integer programming models (dynamic or static), sometimes combined with a merit-order dispatch (least-cost production) model for the electricity sector. The results in *Table 5.1* typically rest on the assumption that the emission permit market works perfectly (full competition, cost-minimizing behavior, no transaction costs or uncertainty) and its results match the least-cost solution.

Table 5.1. Potential cost savings of emission trading.

Study	Pollutant	Area	CAC baseline	Cost savings (% of CAC)
National				
Becker *et al.* (1993)	SO_2	Haifa, Israel	Uniform cutback	0–53
Gollop and Roberts (1985)	SO_2	USA	Emission standards	8–75
Hahn and Noll (1982)	SO_2	Los Angeles, USA	Standards	4
ICF (1992)	SO_2	USA	Intrastate trading	45–50
Maloney and Yandle (1984)	HC	USA	Emission standards	7–85
NERI (1992)	NO_x	Ontario, Can.	Emission standards	35–41
Palmer *et al.* (1980)	CFC	USA	Mandatory controls	42
Pototschnig (1994)	SO_2	UK	No interfirm trade	0–32
Toman *et al.* (1994)	SO_2	Poland	Emission standards	11
Welsch (1988)	SO_2	Europe	Emission standards	3–85
Wiersma (1989)	SO_2	Netherlands	Emission standards	5–12
International				
Dean (1993)	CO_2	World	Uniform cutback	6–16
Kverndokk (1992)	CO_2	World	Uniform cutback	20
Rose and Stevens (1993)	CO_2	World	Various	25–50[a]
Klaassen (1995)	SO_2	Europe	Emission standards	4–57

[a]Depends on CAC baseline (Rose and Stevens, 1993, pp. 136, 140).

5.2.2 Why do cost savings differ?

Table 5.1 shows that the potential cost savings of the different studies vary widely and range between 0 and 85 percent. This leads to the question of why such large differences occur. The answer is that cost savings depend on the percentage reduction, the flexibility or stringency of the CAC reference point, and regional characteristics such as the number of sources and control options included. We shall discuss each of these points in turn.

Cost savings tend to be a function of the emission reduction desired. This is clearly shown in *Figure 5.1*, which shows the relative and absolute cost savings of emission trading as functions of the percentage reduction compared with source and plant-specific emission standards, respectively. Because one plant may consist of multiple sources, the plant-specific emission standard (or intraplant bubble) allows plant owners to achieve emission reductions at lower costs. Therefore, absolute and relative cost savings are lower with plant bubbles than with source standards. Clearly, for both CAC reference cases the relative cost savings of emission trading drop considerably when emission reductions increase. This is because the more emissions must be reduced, the fewer the options that are available

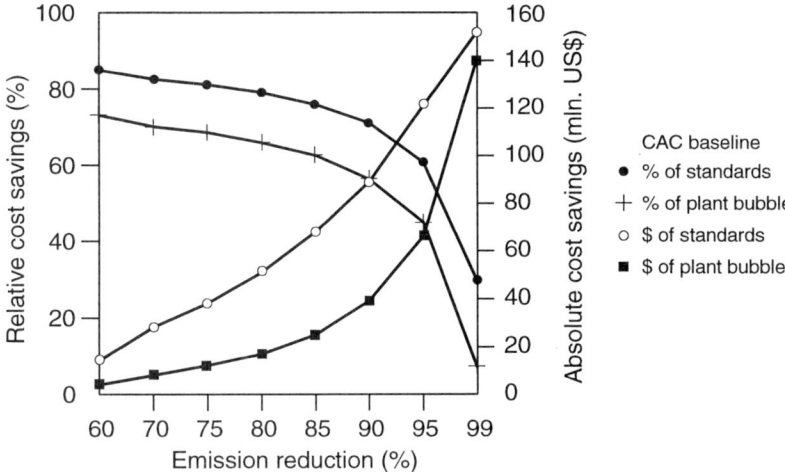

Figure 5.1. Cost savings and percentage emission reduction. Based on Maloney and Yandle (1984).

to meet the desired emission reduction because some technical measures must always be applied to meet the reduction target. This is most obviously the case when the required reduction comes close to the technical maximum.

That the relative cost savings tend to fall when reduction increases is confirmed by Welsch (1989) in a study of the trading of sulfur emission permits among power plants within 12 countries in Europe. In his study, the relation between relative cost savings and the level of emission reduction is not as smooth as in *Figure 5.1*, because specific emission reductions can only be achieved by certain technologies and therefore do not permit cost savings. Welsch (1989) also finds that the cost savings are reduced to 0 to 12 percent if emission reductions go up to 90 percent and correspond to the application of the best available technology. Becker *et al.* (1993) find that for a reduction of 10, 15, or 20 percent their emission permit system saves 53 percent of the costs, compared with a uniform cutback at every source. At a 30 percent cutback, trading saves nothing; at a 37 percent reduction trading saves 31 percent; at a 40 percent cutback trading saves nothing; and at a 45 percent cutback trading saves 21 percent. Although relative and absolute cost savings in this last case tend to be smaller, the relation is not straightforward due to the fact that technological indivisibilities exist in pollution control. Pollution control is modeled as a discrete rather than a continuous process, and installing abatement equipment in response to meeting a fixed emission ceiling may lead to overcontrol. Although emission trading is always at least as inexpensive as the CAC counterpart, for some reductions pollution control costs are the same. In the Becker *et al.* (1993) case, this discontinuous character of pollution control costs is exaggerated, because they deal with only two plants and six sources and continuous options such as low-sulfur fuels are not included.

Whereas the relative cost savings decrease when emission reductions increase, the absolute cost savings may increase or decrease. *Figure 5.1*, for example, shows that the absolute cost savings increase. This stems from the fact that marginal costs increase with increasing emission reduction and total costs of the CAC alternative rise disproportionately with increasing pollution reduction. Similar increases in absolute cost savings are obtained by Rose and Stevens (1993) when the reduction in global carbon emissions moves from 14.9 to 20 percent. However, Becker *et al.* (1993) find that the absolute cost savings tend to decrease with increasing emission reduction. In conclusion, whereas relative cost savings of emission trading tend to decrease with increasing emission reduction, absolute cost savings may increase or decrease. The smoothness of the relation depends on whether the cost functions employed are continuous or discrete.

The cost savings of emission trading also depend on the CAC reference point. *Table 5.1* (column 4) indicates that these reference points may differ. Some studies take a uniform percentage cutback as a reference point (e.g., Becker *et al.*, 1993), others take emission standards required by law (Toman *et al.*, 1994; Gollop and Roberts, 1985; Wiersma, 1989), or plant- or source-specific emission ceilings (Maloney and Yandle, 1984), and still others assume a plant- or region-specific quota and implicitly allow intraplant or intraregion trading in the CAC baseline (ICF, 1992; Pototschnig, 1994). *Figure 5.1* clearly shows that taking plant-specific emission standards (assuming intraplant bubbling) rather than source-specific standards as a baseline implies that the potential cost savings of allowing interfirm trading are reduced. Studies examining the potential cost savings of sulfur trading now taking place in the USA show that allowing intraplant trading would cut costs by 30 to 60 percent compared with the new sulfur regulations. Expanding trading to power plants within a state would allow another reduction in costs by 20 percent, and interstate trading would cut costs by an additional 15 percent (Tietenberg, 1995). Toman *et al.* (1994) show relatively small cost savings for allowing emission trading between Polish power plants, because the emission standards allow for an intraplant bubble.

The international carbon studies show relatively small cost savings; they, however, are based on the assumption that the five to eight (worldwide) regions meet the uniform cutback under the CAC alternative at the least cost within the region. Clearly, cost savings can be expected to be higher if the CAC baseline consists of uniform reductions for each separate country within each region. Hence, for

Empirical Simulation Models and Emission Trading 103

international studies the cost savings depend on the policy assumed for the countries within each region or country. Rose and Stevens (1993) show that the cost savings also depend on the allocation of a non-tradable quota over the regions under the CAC alternative and, as a result, may range between 25 and 50 percent.

To summarize, the potential cost savings of emission trading depend on the CAC reference point and the associated initial distribution of non-tradable emission reductions, as well as the flexibility assumed in the form of intraplant, intrastate, or intraregional cost minimization.

The order of magnitude of the cost savings also depends on regional characteristics such as the type of fuel used and the size of the polluting sources (due to the economies of scale in implementing pollution control equipment). This is clearly shown in the study by Gollop and Roberts (1985) where the range of cost savings, 7 to 75 percent, mainly depends on the differences in control costs in the different regions: the 7 percent is achieved in the Midwest and the 75 percent in the West. The difference is partly caused by differences in the CAC baseline, because Gollop and Roberts assume that intraregional trading meets the same emission ceiling as is expected to be achieved by current, region-specific regulations. The required emission reduction and the intensity with which this reduction is enforced differ between the regions. Clear evidence is also given by Welsch (1989) indicating that similar reductions imply different costs for different countries due to differences in plant sizes and the amounts and types of fuel used. Pototschnig (1994) shows that relative cost savings might be very small if only power plants are allowed to trade, because the differences in marginal control costs would be minor and there would be little to gain from trade. Cost savings increase to around 30 percent in his UK example if sulfur trading is extended to refineries and other industries. That regional characteristics are important was also shown for the Bay of Haifa study where the small number of sources and indivisibilities in pollution control left little flexibility for cost savings (Becker *et al.*, 1993). This contrasts sharply with the study by Maloney and Yandle (1984) in which considerable cost savings occur even with high reductions because the number of individual emission sources (and consequently the differences in marginal costs included is large (548). Therefore, regional characteristics such as the number and diversity of sources and the type of control options (discrete or continuous) affect the potential cost savings of emission trading.

Regarding environmental effectiveness, the models generally assume that enforcement is perfect and that the emission ceilings set as the objective will be met. Some models show that the cost savings obtained under emission trading could also be used to cut emission further than under the CAC approach. NERI (1992, p. S-19) indicates that in Ontario, Canada, NO_x emissions could be cut by an additional 35 percent if the same CAC budget were to be spent under an emission trading program. In an international study, Klaassen (1995) suggests that the same budget expected to be used under a regulatory approach to control SO_2 emissions in Europe could be used in an emission trading program to cut emissions by an additional 40 percent across Europe.

One may ask whether the empirical simulation studies give a realistic picture of the potential cost savings. The answer is that the cost savings are not identical to what would happen in practice (Tietenberg, 1985). First, the models overestimate the cost savings because they typically ignore that an existing CAC system, as well as durable pollution control equipment, is already in place, with sunk costs that limit the flexibility of emission trading. Klaassen (1995), for example, shows that if one combines sulfur emission trading in Europe with existing and recently accepted emission and fuel standards, the cost savings of sulfur emission trading in Europe will drop from 50 to 4 percent. An ICF (1989) study observes that the cost savings of intrastate sulfur trading in the USA would drop if some states were to require the use of domestic fuels and flue gas cleaning instead of allowing imported, cheaper low-sulfur coals. Second, the models tend to underestimate potential cost savings in the long run because they are usually based on accepted technologies, ignoring the positive impact that emission trading has on innovation (see Section 5.4). Moreover, they make use of aggregated, averaged technical and cost data. As a result, only a part of the data available at the firm level is used. This tends to lead to underestimated potential (but undetectable) cost savings. Third, the modelers ignore market imperfections, such as transaction costs, and typically assume frictionless, competitive, regular markets with frequent transactions and stable prices. This neglect of market imperfections tends to lead to overestimated cost savings. One market imperfection that has been given some attention is the issue of market power, to which the next section turns.

5.2.3 Market power and other market imperfections

Theoretical analysis of the influence of market power (exercised on the permit market) on the cost efficiency of tradable discharge permits shows that this influence depends on the following factors: the initial distribution of permits (auction, revenue-neutral or subsidy auction, "grandfathering"); the extent to which the initial distribution of permits, especially to the source exerting market power, deviates from the cost-efficient (competitive equilibrium) distribution; the elasticity of supply or demand of the price-taking sources (the fringe); and the marginal pollution costs of the price-setting firm (Tietenberg, 1985, pp. 125–147; Hahn, 1984). The elasticity of supply and demand of the fringe is a function of the marginal abatement costs and the initial distribution of permits.

A few empirical simulation studies examine the influence of market power. Hahn and Noll (1982) assess the case of a monopsony where the buyer's initial distribution of permits is lower than the number of permits it would hold under a competitive equilibrium for sulfate emissions in Los Angeles. In this situation, the firm with market power would reduce emissions further and would purchase fewer permits than under perfect competition to drive the price below the competitive price. Hahn and Noll find that the buyer could lower the price by 0 to 50 percent, depending on its share of the permit market. The loss of efficiency (the extent to which pollution control costs deviate from the competitive cost minimum) was found to be only 0 to 10 percent and dependent on both the method for estimating pollution control cost functions and on assumptions about ambient air quality standards and pollution control costs.

When the firm acts as a monopoly, its impact again depends on its free initial share of permits. Abatement costs were not seriously affected until the firm received more than 40 percent of the licenses. In that case costs were 10 percent higher than the cost minimum (Hahn, 1984). Hahn suggests that the influence is very sensitive to the level of emission reduction. If the emission reduction is high, the marginal costs of the fringe tend to be high and the demand for permits from the fringe becomes rather elastic. This reduces the market power of the monopoly firm. In the case of hydrocarbon emissions in the USA, Maloney and Yandle (1984) discovered that for both the monopoly and the monopsony the deviation of the permit price from the competitive price increases when the degree of cartelization (the percentage owned by the cartel) increases. With 90 percent of the permits owned by the cartel (monopoly), the permit price is 697 percent higher than under perfect competition and is 17 percent lower in the case of a monopsony. The relative increases in pollution control costs are only 41 percent and 7.7 percent, respectively. Even in the case of a 90 percent monopoly, costs with permit trading were still 66 percent lower (versus 76 percent lower with full competition) than with source-specific standards. In conclusion, the few studies available suggest that although market power might cause significant deviations in the permit price from the competitive price, cost savings relative to CAC may not necessarily be much lower.

Other market imperfections, such as transaction costs, appear not to have received much attention in empirical simulation models. Because transaction costs play a significant role in actual emission trading programs (see Chapter 6), they have also been rediscovered in theoretical analysis (see, for example, Stavins, 1994). The only study that deals with transaction costs in simulation models is by Klaassen *et al.* (1994), who simulated the impact of a fixed level of transaction costs on the cost efficiency of sulfur emission trading in Europe. Their conclusion is that a sufficiently high level of fixed costs reduces the number of cost-saving bilateral trades from 35 to 19 while only insignificantly reducing cost efficiency.

5.3 Air Quality Standards as Objectives

5.3.1 The cost savings of ambient permits

When dealing with non-uniformly dispersed pollutants, both source location and emission level determine the impact on air quality. In order to meet so-called ambient standards, measured as concentrations of specific pollutants at specific locations, both elements must be taken into account. In economic theory, the most appropriate system of tradable permits is a system of ambient permits. An ambient permit allows one source to increase the concentration of emissions at a specific receptor by one unit. In order to emit one unit of emissions, each source has to collect a portfolio of ambient permits for each receptor it affects so that increases in concentrations caused by emitting one unit more are covered by additional ambient permits. Economic theory shows that under certain conditions ambient permits could achieve ambient standards at minimum costs (see Chapter 3).

As is shown in *Table 5.2*, evidence from empirical simulation studies confirms that ambient permit systems might enable sources to meet ambient air quality standards at considerably lower costs than would the CAC type of regulation. Depending on the study, ambient permits would cut costs by 12 to 99 percent. This is somewhat higher than cost savings of emission trading as discussed in Section 5.2. The few international studies confirm the potential cost savings.

These cost savings should come as no surprise, however, because the studies assume that the costs of the ambient permit system are equal to the least-cost solution. They ignore transaction costs and they presuppose a competitive permit market and cost-minimizing behavior. This least-cost solution minimizes costs subject to the condition that air quality standards are met. Before we turn to critical examination of this assumption, attention must be given to what determines the cost savings and why the least-cost solution (and a properly functioning ambient permit system) saves costs.

In addition to the assumption that the ambient permit market is perfect and meets the least-cost solution, the cost savings of the ambient permit system are a function of the ambient standard desired as well as the CAC benchmark.

The relation between cost savings and the ambient standards level is similar to the relation between cost savings and emission levels. More stringent ambient air quality standards tend to reduce the relative potential cost savings, because less flexibility exists to meet the lower standards. The absolute cost savings may well increase, because marginal abatement costs tend to rise with increasing emission

Table 5.2. Cost savings of ambient permits.

Study	Pollutant	Area	CAC baseline	Cost savings (% of CAC)
National (USA)				
Atkinson/Tietenberg (1984)	TSP[a]	St. Louis	State plan (SIP)	72–99
Atkinson/Tietenberg (1987)	SO_2	Cleveland	SIP rollback	25–36
Krupnick (1986)	NO_2	Baltimore	Emission standards	24–96
McGartland (1984)	TSP	Baltimore	CAC requirements	74
Seskin et al. (1983)	NO_2	Chicago	SIP/standards	86–93
Spofford/Paulsen (1990)	SO_2	Delaware	Uniform cutback	16–56
Spofford/Paulsen (1990)	TSP[a]	Delaware	Uniform cutback	67–95
Streets et al. (1984)	SO_2	USA-east	Intrastate trade	77–86
International				
Amann/Klaassen (1995)	NO_x/NH_3	Europe	Current plans	10–44[b]
Klaassen/Amann (1992)	SO_2	Europe	Uniform cutback	51[c]
Kruitwagen (1992)	SO_2	Europe	Uniform standards	43[c]
Welsch (1989)	SO_2	Europe	Uniform standards	12–60[c]

[a]Total suspended particulates.
[b]Cost savings of the least–cost solution for controlling NO_x only, or for simultaneous control of NO_x and NH_3 (ammonia), to meet the same nitrogen deposition targets as would result from current national plans to reduce emissions.
[c]Cost savings of the least–cost solution are given without mention of equivalence to ambient permits.

reductions, so the costs of the CAC benchmark increase as well. *Figure 5.2* shows the relation between ambient standards and cost savings of ambient permits. It shows that the relative cost savings of ambient permits over the CAC baseline are lower when ambient standards become more stringent. The same studies show that although relative cost savings drop, absolute costs may increase or decrease with more stringent ambient standards. These findings appear to be rather robust and independent of the regions and pollutants under study. They are confirmed by Welsch (1989) in one of the few international studies in the literature.

Cost savings again depend on the CAC benchmark. Atkinson and Tietenberg (1984) find that the potential cost savings of an ambient permit system are lowest when the CAC allocation accounts for the impact of the location of the sources on air quality and are highest when the allocation is based on uncontrolled emissions or administrative rules under the state implementation plan. A typical example of the relevance of the CAC benchmark is supplied by Seskin et al. (1983). Costs of the ambient permit systems are 90 percent lower than the CAC benchmark if the CAC strategy is based on technology-based emission standards. The cost savings drop to 86 percent if the CAC alternative consists of a uniform emission limit per source category consistent with air quality objectives. Similar findings are reported

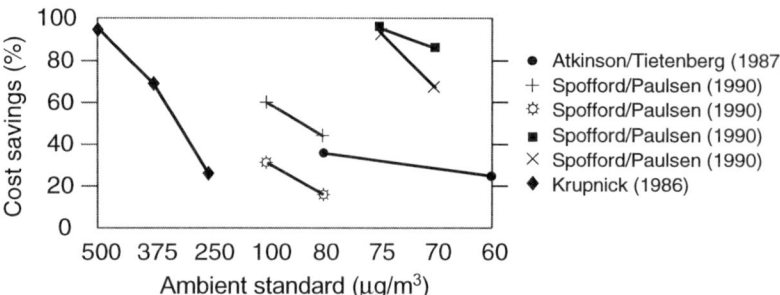

Figure 5.2. Cost savings of ambient permits in relation to ambient standards. Spofford and Paulsen (1990) estimates refer to SO_2, SO_2 only for point sources, particulates, and particulates only for point sources, respectively.

by Krupnick (1986) in his analysis of policies for NO_2 control in Baltimore, MD, USA. For an ambient standard of 250 $\mu g/m^3$, Krupnick's ambient permits are 83 percent cheaper if the CAC regime consists of technology standards [reasonably available control technologies (RACT)]. The cost savings drop to 24 percent if a hybrid policy, consisting of more lenient RACT standards combined with features of incentive-based mechanisms (i.e., emission trading), is used as the reference point. In the international studies, the cost savings depend on the reference point: uniform reductions in each country or uniform emission standards. A clear example is a study by Welsch (1989), which shows that the cost savings of an ambient permit system compared with a uniform cutback in sulfur emissions in 12 countries are not as high as the cost savings of uniform emission standards at every single power plant. This is because the uniform cutback in emissions is assumed to be met at minimum costs within each country. The general conclusion from this is that the potential cost savings of ambient permits depend on the flexibility that the CAC alternative allows and the extent to which the CAC alternative accounts for cost-efficiency options.

There are basically two reasons the ambient permit system might save costs compared with a CAC allocation. First of all, the ambient permit system typically controls fewer emissions than its CAC counterpart (see *Table 5.3*, column 4; see also Tietenberg, 1985). In other words, one reason ambient permits save costs compared with a CAC allocation is that they meet ambient standards with higher levels of emissions than does a CAC allocation and therefore reduce emissions less and save costs. One exception is found in a study by Amann and Klaassen (1995) where total emissions would be lower with ambient permit trading, because under this system expensive measures to reduce a considerable amount of emissions at

Table 5.3. Environmental impacts of ambient discharge permits.

Study	Pollutant	Area	Emissions with ambient permits (CAC = 1)	Air quality (% of CAC)
National (USA)				
Atkinson/Tietenberg (1982)	SO_2	Cleveland	1.86[a]	74[b]
Atkinson/Tietenberg (1984)	TSP	St. Louis	1.2–1.7[a]	NA
Krupnick (1986)	NO_2	Baltimore	1.06–1.48[a]	+/–[c]
McGartland/Oates (1985)	TSP	Baltimore	2.13	Worse
Seskin *et al.* (1983)	NO_2	Chicago	6–7[a]	NA
Spofford/Paulsen (1990)	SO_2	Delaware	1.06–1.9[d]	NA
Spofford/Paulsen (1990)	TSP	Delaware	1.45–1.56[d]	NA
Streets *et al.* (1984)	SO_2	USA	1.27–2.59[e]	Worse[f]
International				
Amann/Klaassen (1995)	NO_x/NH_3	Europe	0.94–9.95	NA
Klaassen/Amann (1992)	SO_2	Europe	2.25	86[g]
Kruitwagen (1992)	SO_2	Europe	2.78	NA

NA = not available.
[a] Emissions removed under CAC divided by emissions removed under ambient deposition permits (Atkinson and Tietenberg, 1987). Otherwise emissions under ambient permits divided by emissions under CAC.
[b] Refers to long-range acid deposition.
[c] Depends on the CAC baseline and ambient standard. Compared with CAC/least-cost baseline, the ambient permit system tends to lead to more receptors with higher average concentrations.
[d] Depends on whether CAC is compared with the least-cost solution or ambient permit trading for point sources only.
[e] Least costs to meet the same deposition of the CAC alternative A: no intrastate trading for two bills (Streets *et al.*, 1984, p. 1189, Table II). Ranges due to different deposition matrices and two different deposition targets. Remaining emissions calculated on the basis of unabated emission of 21.8 million tons.
[f] Deposition increases at all but one (target) receptor area.
[g] Percentage of ecosystems protected.

some locations are replaced by cheaper measures controlling more emissions at more locations. Second, emission reductions are allocated so that the emissions that are controlled have the highest cost efficiency in terms of marginal costs per ton concentration reduced in the region. The ambient permit system takes both location and marginal costs into account when allocating emission reduction responsibilities among the sources. A typical example of what might happen would be the ambient permit system shifting emission reductions from high-stack sources to low-stack sources, which have a more significant impact on local air quality (see Atkinson, 1994). So emission reduction takes place at those sources that have a higher impact on local ambient standards.

In general, ambient permits achieve ambient standards at lower costs by increasing emissions. The next section turns to some problems that may arise as a result of this increase.

5.3.2 Transaction costs and environmental effectiveness of ambient permits

Contrary to what the potential cost savings of ambient permits may lead us to believe, the implementation of these permits can be expected to cause a number of problems including transaction complexity and high transaction costs, increases in emissions, and increases in local concentrations and long-range deposition. Under an ambient permit system, transaction complexity tends to be high because every source has to acquire ambient permits for every receptor it affects. When the number of receptors is large, a complicated set of transactions is required (Tietenberg, 1995). Moreover, the regulatory authority must create markets for every pollutant and receptor (Atkinson, 1994). In their study on particulates and SO_2 in the lower Delaware Valley, Spofford and Paulsen (1990, pp. 4–35) limit the ambient permit system to point sources only, excluding area sources (such as households, for example). In this case, the cost savings are reduced compared with a uniform cutback. Surprisingly, a uniform emission charge applied to all sources appears to be more cost efficient for one of their cases (particulates down to an ambient standard of 70 $\mu g/m^3$). In contrast to what *Table 5.2* suggests, the inclusion of transaction costs may well make ambient permit trading less cost efficient than other strategies, and one might doubt whether a practical ambient permit system really can achieve the least-cost solution. The positive point, however, is that ambient permits always meet the desired ambient standards if enforcement is perfect.

The fact that ambient permits derive part of their cost savings from increasing emissions (see *Table 5.3*) is not without problems and may contradict environmental legislation (see Chapter 6). The obvious solution here is to allow sources to trade as long as they do not violate ambient standards and emissions do not increase. This trading rule is examined in the next section.

A problem related to that of the higher level of emissions under the ambient permit system is that, either inside or outside the region under study, concentrations of pollution might increase compared with the CAC alternative. The few studies that examine the *ex post* levels of air quality (the last column in *Table 5.3*) find that air quality tends to be worse than under the CAC alternative. For example, Oates *et al.* (1988) show that, for the case of Baltimore, the population-weighted average concentrations under the ambient permit system are higher than those under the CAC alternative for every ambient standard and the associated environmental benefits are ignored. Similar observations can be found in Klaassen and Amann

(1992) who found that although the costs of the uniform cutback are higher than the least-cost solution to control sulfur emissions in Europe, the environmental benefits, measured in terms of ecosystems protected from acid deposition, are also higher. Even if emissions are higher under the ambient permit system than under the CAC, local concentrations are not necessarily higher because emissions may be shifted to high-stack sources, which typically have less of an impact on local air quality (Oates and McGartland, 1985). So, although emissions from these sources increase, local air quality may not be greatly affected. Another problem here is that because it implicitly assumes a threshold equal to the ambient standard below which no damage occurs, an ambient permit system may increase concentrations at those receptors where air quality is already better than the ambient standard (McGartland and Oates, 1985). The solution here is another trading rule that states that trading is allowed subject to the conditions that ambient standards are not violated and that concentrations do not increase compared with the pre-trade situation (the modified offset rule, see Section 5.3.3).

The higher emissions under the ambient permit system may also lead to externalities, especially increased long-range deposition of pollutants outside the region for which the costs were minimized (Atkinson, 1994). This is because the emissions tend to be shifted to high-stack sources that have little impact on local air pollution but contribute significantly to deposition at greater distances. What is optimal from a local perspective may lead to a regionally non-optimal solution. Atkinson (1983) solved this problem by adding an additional constraint on the long-range sulfate deposition resulting from the emissions of the region of Cuyahoga County (the Ohio River basin). Adding this constraint reduced the cost savings of the ambient permit system from 33 to 22 percent.

In response to the complexity of the ambient permit system and the associated environmental problems, a number of alternatives have been developed: single ambient permit system, trading rules on deposition and emissions, and emission trading in single or multiple zones. It is to these alternatives that our attention will turn in the next sections.

5.3.3 Single ambient permit trading

One way out of the transaction complexity of ambient permits is to define ambient permits in terms of only one single receptor. Atkinson and Tietenberg (1982) call this the highest ambient permit (HAP) system, because it is based on the receptor that is most polluted (the one requiring the highest reduction in concentration to meet the ambient standard). In their case study for St. Louis, MO, USA, this HAP system is not much costlier than the full ambient permit system. Atkinson (1994) does note, however, that an influx of new sources could create a new "hot spot." Tietenberg (1985) concludes that such a single-permit system guarantees that ambient standards are met at all receptors only if the locations of sources are stable and only in the absence of multiple binding receptors. In the presence of multiple binding receptors, selecting one of them to guide trade implies that ambient standards will be violated after trade. Defining permits in terms of one receptor would also provide long-run incentives to relocate away from that receptor, creating other "hot spots." Atkinson (1994) doubts whether this relocation is a serious problem, because the probability that firms will relocate in response to environmental costs is small, in his opinion. A problem with this approach is that it does not guarantee that the air quality standards will be met at all receptors. An application of the HAP system by Hahn and Noll (1982) shows that the HAP is even cheaper than the least-cost solution. Clearly, this can be so only if the ambient standard is violated at least at one of the receptors for which no ambient permit market is created. Hence, a single ambient permit system might come close to the cost minimum, but does not guarantee that environmental objectives will be met at every location.

5.3.4 Emission trading rules

In response to potential losses in environmental quality due to ambient permit trading, several alternative trading rules have been proposed in the literature and examined in simulation studies. These trading rules treat emission trading on a case-by-case basis, allowing trades to continue if certain conditions are met:

- The pollution offset rule (POS) allows emission trading as long as ambient standards are not violated.
- The modified offset rule (MOS) permits emission trading as long as ambient standards are not violated and, at those places where air quality is better than the standards, air quality does not get worse.
- The non-degradation offset rule allows emission trading as long as ambient standards are not violated and emissions do not increase.

Typically, trading under these rules requires a calculation of the impacts on air quality of each and every trade and requires case-by-case approval (compare Section 6.2).

The evidence for the potential cost savings of these rules is summarized in *Figure 5.3*. The studies depicted can be divided into the following two groups:

- Those that simulate trading as a simultaneous process and assume that it will meet the least-cost solution (POS, MOS, and non-degradation);

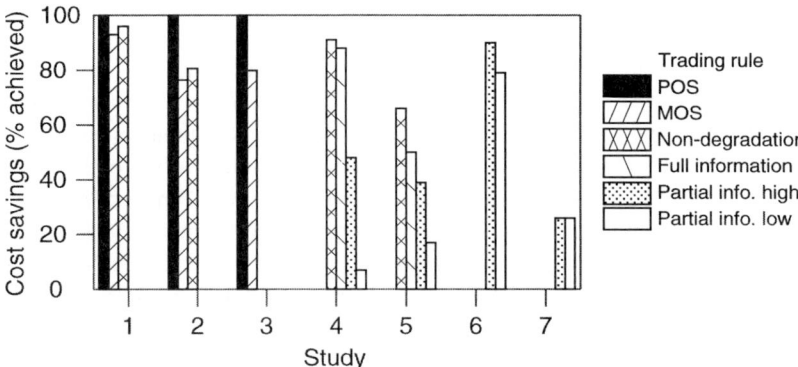

1,2 = Atkinson and Tietenberg (1987), SO$_2$ and particulates, respectively.
3 = McGartland and Oates (1985).
4,5 = Atkinson and Tietenberg (1991), primary and secondary ambient standard, respectively.
6 = Burtraw et al. (1994).
7 = Kruitwagen (1992).

Figure 5.3. Cost savings of trading rules.

- Those that simulate trading as a bilateral, sequential process with traders having full or partial information on the potential cost savings and feasibility of each trade (full information, partial information high, and partial information low).

Clearly, all studies show that the modified offset system and the non-degradation offset system save fewer costs than the ambient permit system or the pollution offset system relative to CAC. This is no surprise because both systems build in an additional constraint in the optimization, and as long as this constraint is binding it attracts a cost penalty.

The studies that simulate trading as a simultaneous process are the first three studies (sets 1, 2, 3) from the left-hand side on the x-axis of *Figure 5.3*. The Atkinson and Tietenberg (1987) case studies (sets 1 and 2 in *Figure 5.3*) show that the non-degradation trading rule reduces the potential cost savings further than the modified offset rule. McGartland and Oates (1985) (set 3) confirm that the additional constraint under the modified pollution offset system has a price in terms of reduced cost efficiency. In spite of these additional limitations, both the modified pollution offset and the non-degradation offset are expected to save costs compared with the CAC allocation.

The nature of the actual trading process may differ considerably from the implicitly assumed simultaneous trading simulated in mathematical optimization (Atkinson and Tietenberg, 1991). An alternative is to model trade as a bilateral sequential procedure where two sources trade, and in the next step another two sources trade, and so forth. In the first study to model trading as a bilateral sequential process, Atkinson and Tietenberg (sets 4 and 5 in *Figure 5.3*) find that this type of modified offset trading may achieve only 7 to 88 percent of the potential cost savings of the least-cost solution. For two levels of ambient standards they simulated the non-degradation rule as a simultaneous process (the non-degradation bar in sets 4 and 5 of *Figure 5.3*) and as a bilateral sequential process (full information and partial information bars in sets 4 and 5 of *Figure 5.3*). If trading is simultaneous, the non-degradation rule will achieve 65 to 90 percent of the potential cost savings of the least-cost solution over the CAC alternative. The cost savings are lower simply because the non-degradation rule does not allow emissions to increase (even if an increase would not violate ambient standards). The results of bilateral sequential trading depend on the information traders have on the potential cost savings of each trade, on how they select trading partners, and on whether they know beforehand whether or not trades will violate standards. If the traders have full information on costs (that is, if they know which trade will generate which level of absolute cost savings) and if the order of trading is such that trades with the highest cost savings are consummated first, the cost savings of the bilateral trading sequence are only slightly less than under the non-degradation rule (full information bar in *Figure 5.3*). If traders have only limited information on costs and select their partners either at random or in such a way that the first seller is the one with the lowest (initial) marginal cost, the potential cost savings are lower and may vary from 40 to 50 percent (partial information high) to 10 to 20 percent (partial information low). These findings lead Atkinson (1994) to conclude that treating the cost minimum found through mathematical programming as if it were the emission trading equilibrium may be a serious misrepresentation of the emission trading process.

These results are supported by Burtraw *et al.* (1994) (set 6 in *Figure 5.3*), who analyzed the potential cost savings from implementing non-degradation offset trading between six countries in Europe as a bilateral sequential process. Their main result is that cost savings of bilateral sequential trading may vary between a high of 94 percent and a low of 79 percent of the potential cost savings achievable under a joint cost-minimization problem. The cost savings are higher if trading partners base the choice of their trading partner on the initial difference in marginal pollution control costs rather than select partners at random. They also observe that cost savings are higher if trade is decentralized and if instead of one country, two or three firms within a country are allowed to trade. They offer no explanation for this simulation result. The reason is probably the following. Burtraw *et al.* (1994) simply divide the cost function of each country into two or three parts. They call these parts firms, but they could also be regions within the country. Because of this, the number of trading partners is simply doubled or tripled (and the number of trades is increased by a factor of four to nine). This increases the chances that a partner can be found for a cost-saving trade such that the deposition constraints are met. In itself, the procedure has nothing to do with allocating permits to firms or regions. The only conclusion is that further decentralization within a country can improve cost efficiency.

Kruitwagen (1992) (set 7 in *Figure 5.3*) performed a similar simulation for the pollution offset system in the context of a set of sulfur deposition standards in Europe. Her results suggest that if trading is treated as a bilateral sequential process the cost savings of the pollution offset may not even come close to attaining the theoretically conceivable cost minimum.

In summary, trading rules restricting deposition or emission increases limit the cost savings of emission trading. The cost savings are lower if trading is treated as a bilateral sequential process rather than a simultaneous process, because trade might then be path-dependent and one trade might preclude the consummation of another trade with higher cost savings. The cost savings depend on the quality of cost information available to the traders and whether countries or firms are assumed to trade. We will also analyze this in Chapter 10 for the case of sulfur trading in Europe.

5.3.5 Is single-zone emission trading a solution?

To avoid the potentially high transaction costs of case-by-case trading rules, one can also ignore the spatial distribution of the sources and use emission trading (on a one-to-one basis) in the whole region as an instrument (single-zone emission trading). In this case the total level of emissions is determined in an iterative procedure so that when the competitive trade equilibrium is achieved the ambient standards are met. This is formally equivalent to the problem in Chapter 2 [equations (2.7) and (2.8)]; the only difference is that the maximum allowable level of emissions is set so that ambient standards are met after the completion of the emission trading process.

The evidence on the cost efficiency of using emission permit trading to achieve ambient air quality goals is collected in *Table 5.4*. The table shows that no unique statement is possible on whether or not emission trading is more cost efficient than its CAC counterpart. Several studies show that emission trading is cheaper (e.g., Streets *et al.*, 1984), but others indicate that emission trading might be considerably more expensive than its CAC counterpart. All studies show that emission permit trading is more expensive than the least-cost solution. But there clearly is no guarantee that emission permit trading is more cost efficient than CAC for meeting a set of ambient standards.

Table 5.4. Costs of single-zone emission trading to meet ambient standards.

Study	Pollutant	Area	CAC baseline	Cost ratio (CAC = 1)
Atkinson (1983)	SO_2	Cleveland	Uniform cutback	1.10–2.80
Atkinson/Tietenberg (1982)	TSP	St. Louis	SIP	0.15–0.85
Atkinson/Tietenberg (1984)	TSP	St. Louis	Various rules[a]	0.42–0.45
Krupnick (1986)	NO_2	Baltimore	Emission standards	0.15–6.50[b]
Oates/McGartland (1985)	TSP	Baltimore	SIP[c]	0.41
Roach et al. (1981)	SO_2	Four Corners	Uniform cutback	0.41
Seskin et al. (1983)	NO_2	Chicago	SIP/standards	2.34
Spofford/Paulsen (1990)	SO_2	Delaware	Uniform cutback	0.95–1.04[d]
Spofford/Paulsen (1990)	TSP	Delaware	Uniform cutback	0.10–0.78
Streets et al. (1984)	SO_2	USA	Intrastate trading	0.36–0.43[e]

[a] SIP or uniform cutback compared with past levels or CAC allocation, taking into account impacts on air quality.
[b] Results of uniform charge are taken as being representative for emission trading.
[c] State implementation plan (SIP).
[d] Only point sources trade, area source controlled on an ad hoc basis.
[e] Based on Streets et al. (1984, p. 1191, Figure 2) for reduction in deposition in the Adirondack Mountains between 10 and 40 percent. CAC allocation based on proposed acid rain bills.

Tietenberg (1985, 1995) thoroughly analyzes the reasons for this, perhaps surprising, result. He finds that the cost efficiency of CAC and emission permit trading, both being inadequate instruments for meeting ambient standards, is related to two components:

- The equal marginal cost component;
- The degree-of-required-control component.

Regarding the first component, the fact that emission trading equalizes marginal abatement costs implies that for the same level of emission reduction emission trading is always cheaper than its CAC alternative. The second component, however, tells us that the degree of emission reduction required to meet the ambient standard is not always the same for each of the alternatives. This is because both instruments are based on different spatial allocations of control responsibility over the various sources in the region. Under emission permit trading, the spatial allocation results from the equalization of marginal costs. Under the CAC regime, the spatial allocation depends on the specific type of regulation adopted. Under the emission trading scheme, the total allowable amount of emissions (the number of permits) would be lowered (and permit price, marginal costs, and emission reduction would be increased) so that, after trade, ambient air quality would be met everywhere. Under the CAC, policy standards would be tightened and emission reductions would be increased so that ambient standards would be met. The second component (the degree of emission reduction) can favor either the CAC allocation or the emission permit trading allocation. If the CAC alternative requires more control than single-zone emission trading, the CAC alternative is always more expensive. If the CAC alternative requires less control, the outcome is uncertain and depends on the spatial distribution of the sources. To give an example, if a region consists of a few low-stack sources and a few high-stack point sources, emission trading would require both types of sources to have the same marginal control costs even if the high-stack sources have less impact on local air quality. A carefully designed CAC policy might control the low-stack sources to a higher degree than the high-stack sources and end up with lower costs. If the atmospheric impact of the sources on the receptors is similar, or if emission trading would reduce emissions exactly where needed to protect sensitive areas (see Streets *et al.*, 1984), then an emission trading system could be expected to reach air quality standards at lower costs. Whether or not single-zone emission trading is more cost efficient than a CAC alternative thus turns out to depend on regional characteristics such as the atmospheric dispersion pattern and pollution control costs. It also depends on whether or not the CAC alternative takes the location into account and designs location-specific regulations.

A drawback to using single-zone emission permit trading is that there is no guarantee that the environmental quality standards will be met unless the environmental agency has full knowledge of the costs and unless the permit market works perfectly. If pollution control costs are not known with certainty, or if imperfections prevail on the permit market, the emission trading equilibrium may not coincide with what is expected. In this case, the ambient standards might be violated after emission trading. This is because the location of the actual emission reductions differs from the modeled locations. In this case, the environmental agency would have to lower the number of permits until the ambient standards are met.

The final drawback of single-zone emission trading stems from its inability to steer the location of new sources because there is only a single permit price. In the long run the single zone gives too little protection because the source location can be crucial to meeting the ambient standards (Tietenberg, 1985, 1995).

Major advantages of single-zone emission trading are that the design is simple, and precisely because there is only a single permit market the number of market participants is potentially large, trading is relatively straightforward, transaction costs are low, and the conditions for a perfect market appear to be more likely than with ambient permit trading or trading rules.

5.3.6 Multiple-zone emission trading as an alternative

A final option is multiple-zone emission trading. In this case, the region is divided into a number of zones and emission trading is allowed only within each of the zones. Superficially, this zoning has attractive features because compared with single-zone emission trading it appears to reduce the risk that air quality standards will be violated. This is because permits can no longer move from one corner of the region to another; they can only be traded within each of the zones. However, it also appears that multiple-zone emission trading attracts a cost penalty because trading between the zones is not allowed. The extent to which this alleged cost penalty and favorable impact on environmental quality occur depends foremost on how emission ceilings are initially allocated over the zones. Strictly speaking, there are two possibilities:

- Full information case: emissions are allocated over the zones so that the total regional pollution control costs are minimized in order to meet the ambient standards given the knowledge that marginal costs within each zone are equalized.
- Incomplete information case: emissions are allocated without taking costs into account, for example, using uniform cutbacks, similar emission standards, or more stringent cutbacks in the most polluted area such that the ambient standards are met initially, before trading takes place.

In the full information case the first step is to determine the zones. The choices of the number and size of the trading zones are based on political boundaries such as states or counties (Spofford and Paulsen, 1990, Chapter 5, p. 1) or on geographical location (McGartland, 1984, p. 57). Zones may also be chosen on the basis of ambient receptors. In this case, sources may be located in the receptor-based zone they affect most (McGartland, 1984, p. 58). Given the choices of the number and size of the zones, an optimization model can be used to determine the cost-minimizing volume of emissions in each zone necessary to meet the ambient standards. This cost-minimization model reads (McGartland, 1984, p. 58)

minimize
$$\sum_i C_i(R_i) \tag{5.1}$$

subject to
$$(\overline{E}_i - R_i) \leq D_j^0 \qquad \text{for all } j \ ,$$

and
$$C_i'(R_i) = C_k'(R_k) \qquad \text{for all sources } i, k \text{ placed in the same zone.}$$

With imperfect cost information, emissions are initially allocated to the zones on the basis of a CAC type of regulation. After the trading in each zone, emissions are reallocated and the deposition can be calculated. If trade violates the deposition standards, emissions in each zone are reduced (for example, by a uniform percentage) until trading takes place again and deposition is compared with deposition standards. This procedure is repeated until deposition standards are met everywhere and no excess reduction takes place. In reality, imperfect cost information might imply that deposition after zonal trading could be higher or lower than the standards. In this case, the regulator would have to adjust the total number of permits until the standards are met everywhere. Such a procedure would be similar to the adaptive procedure for setting ambient charges (see Ermoliev *et al.*, 1995).

Figure 5.4 summarizes the evidence from several empirical studies based on full information. The figure depicts the relative costs of zonal emission trading as a function of the number of zones. Clearly, if the number of zones approaches the number of sources, each source is treated as a separate zone and with full information can be allocated the least-cost number of permits. Without exception, the studies show that increasing the number of zones reduces the cost penalty compared with the least-cost solution, because the different impacts of the sources allocated in each zone on air quality at each receptor are accounted for and the initial CAC allocation is based on cost minimization. Similar results were obtained by Roach *et al.* (1981). By increasing the number of zones from 1 region to 4 states and then to 35 airsheds, their cost penalty was cut from 4.4 to 2.5 and then to 1.2. With full information and a sufficiently high number of zones, these zonal policies might come close to the least-cost solution. Spofford and Paulsen (1990) show that their 11-zone system is more cost efficient than are uniform emission reductions. With full information, however, the environmental agency might just as well set the optimal emission reduction for every single source; then there would not be any need for emission trading. With full information of costs and a sufficiently high number of zones, emission trading in zones (or, for that matter, uniform charges per zone) might come close to the least-cost solution. In this case, however, designing a least-cost CAC solution would also be possible.

In the incomplete information case, the agency may determine the allocation of emission reductions in the different zones, not by using cost information, but by using political rules of thumb, existing practice, uniform emission standards, or the relative contribution of the zones to air quality problems. In this case, increasing the number of zones may lead to increased, rather than reduced, costs (see *Figure 5.5*). The fact that costs are minimized within each zone for the allocated emission level may or may not compensate for the loss of cost efficiency caused by the fact that the wrong levels of emissions are allocated to the zones. Furthermore, depending on the region and the pollutant, the costs of zonal emission trading may be higher

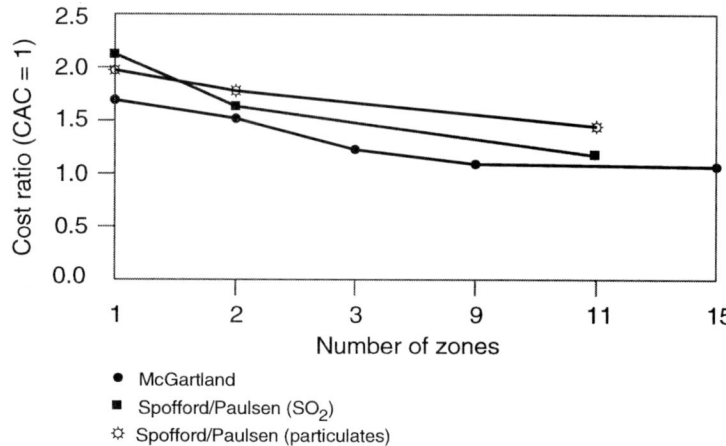

Figure 5.4. Multiple-zone trading with full information. Sources: McGartland (1984); Spofford and Paulsen (1990).

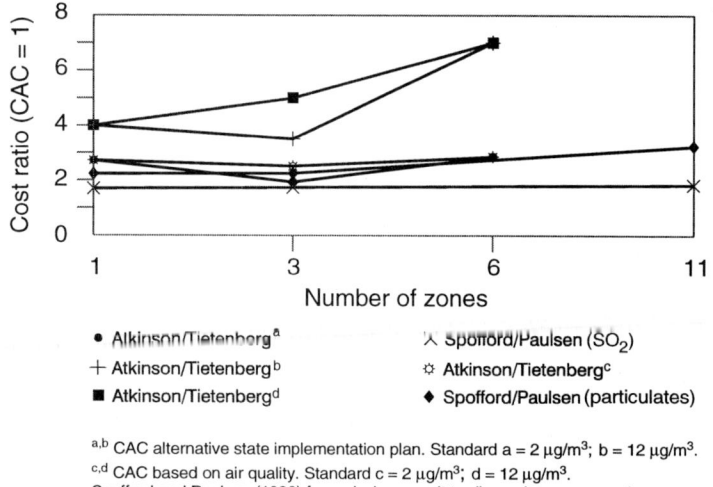

[a,b] CAC alternative state implementation plan. Standard a = 2 μg/m³; b = 12 μg/m³.
[c,d] CAC based on air quality. Standard c = 2 μg/m³; d = 12 μg/m³.
Spofford and Paulsen (1990) for emission permit trading point sources only.

Figure 5.5. Multiple-zone trading with imperfect information. Sources: Atkinson and Tietenberg (1982); Spofford and Paulsen (1990).

or lower than CAC types of allocations (Spofford and Paulsen, 1990; Atkinson and Tietenberg, 1982). Finally, there is no guarantee with these zonal emission trading schemes that the air quality standards are met *ex post*, in spite of the fact that they were met *ex ante*, before trading started. Atkinson and Tietenberg (1982) find that emission trading in three as well as in six zones leads to *ex post* ambient concentrations exceeding the ambient standards for nearly every standard and initial allocation examined. Spofford and Paulsen (1990) conclude that multiple zones cannot be relied on to control the spatial redistribution of emissions during trading. In the case of SO_2, increasing the number of zones did not prevent violation of the ambient standard; emission ceilings would have to be reduced more to achieve this (at extra costs). In the case of particulates, the additional zones, and the additional cost, were not necessary to meet the ambient standards and one zone would have sufficed. In summary, in the case of incomplete information on costs, zonal emission trading is not cost efficient and in effect might be more expensive than its CAC counterpart, nor does it guarantee that ambient standards will be met.

5.4 Administrative Practicability, Innovation, and Distributional Issues

5.4.1 Administrative practicability

The administrative practicability of policy instruments consists of two elements, the information intensity and the ease of monitoring and enforcement. Theoretical examination (Chapter 2) suggests that emission trading requires less information to be cost efficient and environmentally effective than regulation or charges do, although administrative costs could be higher because an emission tracking system is required. One of the few studies on this issue estimated the administrative costs of implementing sulfur emission trading among utilities under the US Clean Air Act Amendments (CAAA) of 1990 (ICF, 1992). The administrative or implementation costs consist of the costs of transactions, the costs of the emission tracking system (registering who holds permits and comparing permits with actual emissions), the costs of organizing auctions and sales, and the costs of continuous emission monitoring and issuing permits. The ICF study does not discuss the administrative costs of enforcement. *Table 5.5* shows that the total administrative costs of emission trading are expected to be slightly higher than those of the regulatory alternative. The administrative costs for the regulatory authorities consist of the costs of the allowance tracking and transfer system, roughly half of the costs of auctions and direct sales, an insignificant part of the monitoring costs, and the administrative burden of implementing permits (roughly 70 percent of permit costs). The regulated

Table 5.5. Administrative costs of sulfur trading in USA (1993–2010) (mln. US$, 1990 value).

	Regulatory approach	Emission trading
Tracking system and transfers	–	2–6
Transaction costs	–	200–400
Auctions and sales	–	2–7
Monitoring	2,512	2,395
Permits	–	68
Total	2,512	2,665–2,876

sources have to pay the transaction costs of buying and selling permits, part of the costs of the auctions and direct sales, the monitoring costs, and around 30 percent of the permit implementation costs (ICF, 1992, Chapter 4, pp. 8–37).

A major component of the costs is the expected transaction costs of US$200 to 400 million for the utilities. Transaction costs are based on a traded volume of 1 to 2 million tons/year and a transaction cost of 1.5 percent per trade (ICF, 1992, pp. 4–12). They could well be higher because in practice brokers' commissions are around 5 percent (see Section 6.5) rather than 1.5 percent; although in that case transaction volumes would be lower. A major component of the implementation costs, however, is the cost of monitoring. Monitoring costs are expected to be lower under the trading program than under a traditional program if one assumes that the EPA does not require continuous monitoring for sources with very low emissions (ICF, 1992, pp. 4–26). This seems rather artificial because the same rational approach could probably be adopted under a traditional CAC approach, as well. In that case, no differences in monitoring costs would occur. Whatever assumption is made about the monitoring costs, the total additional administrative costs are only a small fraction (2 to 4 percent) of the potential cost savings. Differences in enforcement costs, if any, are not included here. In summary, the administrative costs of emission trading could be slightly higher than those of a regulatory alternative, mainly because of the associated transaction costs. These additional costs are only a minor part of the expected cost savings.

5.4.2 Innovation

In theory, regulation, emission charges, and tradable permits do not all have the same impact on innovation (see Chapter 2). Emission charges and tradable permits tend to promote innovation more than regulation does, because the expected cost savings are higher whether the regulator responds to innovation or not.

Surprisingly, empirical simulation models have not addressed this issue. An interesting exception is work by Wiersma (1991), which examined the impact of learning effects in the case of sulfur abatement in the Dutch electricity sector. A learning curve for flue gas desulfurization was assumed, thus implying that the fixed costs of flue gas purification are reduced when the installed capacity increases. The model study compares the abatement costs and emission reduction of four instruments: existing emission and fuel standards, a uniform emission charge, an annual emission bubble (intrafirm emission trading), and a cumulative bubble for five years. The regulator responds to cost-reducing innovation by reducing the emission charge from 1,500 to 1,300 Dutch guilders per ton of SO_2 emitted to keep the emissions at the desired level. Wiersma finds that innovation reduces the abatement costs for all policies, but especially increases the cost savings (compared with the standards) of the pollution charge (24 percent instead of 12 percent lower costs) and the cumulative bubble (23 percent versus 12 percent). To a lesser degree, innovation increases the cost savings of the annual bubble (5 percent instead of 4 percent). The increased cost savings are due to the fact that both the charge and the cumulative bubble lead to the accelerated application of flue gas purification at the expense of using low-sulfur fuels. This is because flue gas purification becomes cheaper over time as a function of installed flue gas desulfurization capacity due to the learning-by-doing effect. It is remarkable that the additional cost savings of economic incentives are not due to the higher incentive to innovate as such, as one might expect on theoretical grounds. The savings are caused by the fact that as cost-reducing innovation occurs, economic instruments offer polluters the flexibility to allocate pollution control resources to those control options that become relatively cheaper over time.

5.4.3 Political acceptability and distributional impacts

Policy changes in the direction of an increased use of economic instruments such as emission trading might occur if the benefits of making these changes exceed the costs and if the distribution of costs and benefits leads the different actors to opt for economic incentives as a supplement for, or complement to, regulatory approaches. An important item in this respect is the initial distribution of emission permits, because costs of compliance for the polluters not only consist of the resource costs of controlling pollution but also of the transfer payments associated with the initial distribution of permits. If polluters are required to pay for the initial permits their financial burden might be higher with emission trading than with traditional CAC approaches in spite of the lower pollution control costs. The following are the two basic approaches to allocating emission permits among polluters (Tietenberg, 1985, p. 97):

- Those involving financial transfers to or from the government: auctions or subsidies (where polluters buy permits from governments to reduce pollution);
- Those involving financial transfers among polluters only: "grandfathering" (free allocation) or zero-revenue auctions (where auction proceeds are returned to the polluters).

Of course, mixed strategies are also possible, such as "grandfathering" combined with an auction of part of the permits.

The remainder of this section discusses the financial burden of tradable permits and its distribution for the case of point sources in a national context. The section then examines the financial burden and its distributional impacts in an international context.

The empirical simulation models for point sources within one nation yield some insight into the relevance of permit expenditures compared with pollution control costs. The total financial burden is defined here as the sum of pollution control costs plus the expenditure on permits. It is assumed that polluters have to buy all permits at an auction and that auction proceeds are not returned. *Table 5.6* indicates that for those studies that looked at emission trading for uniformly mixed pollutants (that is, trading in one zone), in nearly all cases the financial burden of emission trading will be higher than the costs of the CAC alternative if the polluters must buy the permits at the equilibrium price (column 5). In some cases (Palmer *et al.*, 1980) the burden might be more than eight times higher. In other cases the burden is not significantly higher (ICF, 1992; Toman *et al.*, 1994; Wiersma, 1991) and the expenditures on permits are low compared with the pollution control or resource costs (column 6). If the emission reductions required are high, permit expenses become less relevant compared with pollution control costs. If reductions are low, the uncontrolled emissions are high and the potential permit outlays might be substantial, as may be the case for CO_2 emissions. Koutstaal (1993, p. 132), for example, finds that for a 10 percent reduction in CO_2 emissions in the Netherlands pollution control costs would be less than 5 percent of the permit expenditures if permits were to be auctioned.

In a review of studies dealing with emission trading within one nation, Tietenberg (1985, pp. 104–105) finds that for non-uniformly dispersed pollutants the financial burden of ambient permits is lower than the abatement costs of the CAC alternative because the estimated cost savings exceed the expected permit expenditures. In most studies this aspect has not been examined, however. If single-zone emission permit trading is used to meet ambient air quality goals, the financial burden may or may not be higher than the resource costs of the CAC alternative. This also depends on whether the resource costs of emission trading were lower than the CAC alternative to begin with. The empirical evidence indicates

Empirical Simulation Models and Emission Trading 125

Table 5.6. Financial burden and permit expenditures when permits are sold.

Study	Pollutant	Area	CAC baseline	Financial burden (CAC = 1)	Permit expenditures/ control costs
National					
ICF (1992)	SO_2	USA	Intrastate trading	0.9–1.2[a]	0.78–1.0[a]
NERI (1992)	NO_x	Ontario, Canada	Emission standards	1.49[b]	1.53
Palmer *et al.* (1980)	CFC	USA	Mandatory controls	8.4	15.4
Pototschnig (1994)	SO_2	UK	No interfirm trading	2.25[c]	1.66
Toman *et al.* (1994)	SO_2	Poland	Emission standards	1.04	0.16
Wiersma (1991)	SO_2	Holland	Emission standards	1.19[d]	0.26
International					
Klaassen (1995)	SO_2	Europe	Emission standards	1.66–1.78[e]	0.86–0.99
Rose and Stevens (1993)	CO_2	World	Intraregion trading	10.5–15.9[f]	18.9

[a]Results depend on the year (1995/2000) and high or low scenario. Permit prices based on ICF (1991).
[b]For auctioning under the lowest-cost case.
[c]For the year 2000 for England plus Wales bubble.
[d]Based on a permit price of 1,599 guilders/ton (charge level).
[e]Based on equilibrium permit price of DM 700/ton SO_2 (Klaassen, 1995). The alternative consists of emission and fuel standards.
[f]Depends on initial distribution of permits. Based on total unabated CO_2 emissions of 9.1 million tons (1990 level minus 20 percent) and a permit price of US$39/ton (Rose and Stevens, 1993, p. 21).

that if ambient permits are used, a free distribution of deposition permits might not be necessary, because the total financial burden of ambient permits could be lower than the cost of the CAC alternative. With single-zone emission trading, "grandfathering" or zero-revenue auctions are more likely because the pollution control cost savings of single-zone trading may not be high enough to bring the total financial burden below that of the CAC alternative.

Not only is the total financial burden for all polluters relevant, the distribution of compliance costs among the polluters is as well. In an overview of national studies, Tietenberg (1985, p. 108) states that the few studies dealing with the auction of emission permits suggest that auctions might increase the differences in financial burden among polluters. When looking at permit markets for non-uniformly dispersed pollutants, Tietenberg (1985) observed that if permits are auctioned off an additional source of cost variability is introduced, because existing sources located in areas requiring more emission control face higher emission control costs and higher permit expenditures due to their location. "Grandfathering" permits on the basis of the levels of pollution in the past might create a secondary permit market where a few large sources would sell permits if the allocation of permits were to ignore economies of scale. Whether or not market power becomes an issue also depends on the differentiation of standards and marginal costs in the pre-trade allocation. Market power depends not only on the initial distribution of permits but also on the equilibrium distribution, because it is excess demand and price elasticities that determine the potential market power (Hahn, 1984). "Grandfathering" permits on the basis of the CAC allocation would ensure that no source would be worse off than under the CAC approach (see also NERI, 1992, p. 123). If the CAC alternative is not sufficiently stringent to meet the air quality goals, "grandfathering" might not make every source better off, but it might come very close (Atkinson, 1994). Auctioning permits, even if they are least-cost ambient permits, negatively affects far more firms and discourages the innovation that improves the situation for every source. In summary, the auctioning of permits runs the risk that some sources will be much worse off than under the CAC alternative. With grandfathering or with the zero-revenue auction of permits, the initial permit distribution can be used to improve the situation compared with the CAC situation for existing polluters.

The data from international studies (Klaassen, 1995; Rose and Stevens, 1993) confirm that the auctioning of permits (and, similarly, charges whose revenues are not returned) can cause opposition, because the financial burden might be considerably higher than under the CAC alternative. This seems problematic in view of the fact that in an international context countries cannot be forced to accept such an initial allocation.

The distribution of the financial burden might be even more relevant in an international context than in a national context, because countries can only be expected to sign an agreement if it is beneficial to them. The international studies usually assume that emission permits or carbon quotas are allocated initially for free. The studies on tradable carbon permits generally show that irrespective of the initial allocation all countries benefit from trade, because the costs for those that engage in trade are lowered whereas the environmental benefits are not affected. This is because the benefits depend on the global carbon emissions, which are not altered. Rose and Stevens (1993, pp. 136, 140) show that regardless of the initial distribution criterion all countries benefit from emission trading if the worldwide goal is a 20 percent cutback, although some countries benefit more than others. Some regions (USA and Western Europe), however, would benefit with a smaller reduction (14.9 percent). Kverndokk (1992, pp. 106–107) compares the cost savings of tradable carbon permits with a uniform 20 percent cutback. He concludes that "grandfathering" permits on the basis of the 1990 level of emissions seems to be the most politically acceptable distribution, because the other rules examined (GDP, population) imply that some regions will end up with higher costs than under a uniform percentage cutback. The problem here is that there is little reason to take the uniform cutback as reference point, because it would imply a limit on economic development in developing countries (Edmonds *et al.*, 1993, p. 299). This makes Edmonds *et al.* conclude that developing countries might be better off if the initial allocation of rights takes place so that all their expected emissions are covered than if the rights are "grandfathered" on the basis of past emissions. In case of SO_2 emission trading, the problem is that the net benefits of emission trading do not depend only on the costs, because trading shifts the spatial pattern of emissions and influences the distribution of benefits. As a result, sulfur deposition may increase in some countries and benefits may be reduced (Klaassen, 1995). In this case, the "grandfathering" of permits on the basis of a CAC agreement may imply that after trading some countries will be faced with less environmental protection and similar or lower pollution control costs. To induce these potential losers to participate, they could be allocated more permits initially so that net revenues from permit sales increase. These additional permits could then be taken away from countries with large expected gains.

In conclusion, as with national trading schemes, the total financial burden of international tradable permits might be higher than with the CAC alternative. International studies usually assume a free distribution of rights and disagree on whether "grandfathering" rights on the basis of past emissions is the most acceptable political solution. In the case of international SO_2 trading, even "grandfathering" on the basis of an initial agreement without trading might not be without distributional problems, because after trading some countries might have lower environmental benefits and these welfare losses may exceed the cost savings from trade.

5.5 Conclusions

This chapter surveyed the cost efficiency and environmental impacts of emission trading in empirical simulation models with either emission levels or ambient standards for the air quality as objectives. Moreover, administrative practicability, innovation, and distributional implication were examined.

The national and international studies collected suggest that if the environmental objective consists of meeting a certain emission level, the following can be concluded:

- The potential cost savings of emission trading compared with those of CAC types of instruments are larger, *cet. par.*,
    - The lower the percentage reduction;
    - The less the CAC alternative accounts for cost differences;

    - The larger the number of control options;
    - The larger the number of sources.
- The relative cost savings tend to decrease when emission reduction increases, but the absolute cost-saving levels might increase or decrease.
- The cost savings can also be used to reduce emissions further than would have been possible with the same budget under a CAC alternative.
- The actual cost savings might be lower than the modeled cost savings because the models
    - Ignore that existing regulations and durable pollution control equipment are in place;
    - Assume frictionless, regular markets with frequent and stable prices so that the result equals the least-cost solution.
- The actual cost savings could be higher than the models predict for the following reasons:
    - Models underestimate the potential cost savings in the long run because they are usually based on accepted technologies and do not account for the fact that trading stimulates cost-reducing innovation;
    - Models are based on less complete information on control options and costs than individual trading parties possess.
- Market power can have a substantial impact on permit price if the market share of the price-setting firm is high without necessarily significantly affecting the cost efficiency of permit trading.

Furthermore, evidence from simulation models indicates that if (ambient) concentration or deposition standards are the objective, the following can be concluded:

- Ambient permit systems can meet ambient air quality standards at considerably lower costs than CAC regulation if such ambient permit systems equal the least-cost solution. The cost savings of the ambient permit system, however, are typically achieved at the expense of increased emissions and increased local concentrations or long-range deposition that do not violate the ambient standards.
- Trading rules restricting deposition and emission increases in order to safeguard environmental quality reduce the potential cost savings of emission trading, especially if such rules are modeled as bilateral sequential processes with partial information; however, they still promise cost savings compared with any initial allocation of permits without decreasing environmental quality.
- Single ambient permit trading might come close to the cost-efficient solution but is not able to prevent the violation of ambient standards.

Empirical Simulation Models and Emission Trading 129

- If the environmental agency has perfect information on costs, single-zone emission permit trading may or may not be more cost efficient than CAC for meeting a set of ambient standards. This depends on regional characteristics. In the case of imperfect information on pollution control costs or unexpected market imperfections on the permit market, attainment of the environmental goals is not guaranteed.
- Given full information on costs, with multiple-zone emission trading zones can be delineated in such a way that the market equilibrium might come close to the least-cost solution. In the case of incomplete information on costs, there is no guarantee that zonal emission trading is cost efficient. It might be more expensive than its CAC counterpart and might violate ambient standards.

Finally, on administrative practicability, innovation, and distributional issues the empirical studies suggest that the additional administrative costs of emission trading mainly consist of transaction costs, which are borne by the polluters. The administrative costs are likely to be small compared with the potential cost savings and are not likely to be significantly higher than the administrative costs of CAC. Innovation further increases the potential cost savings of trading because it tends to induce learning by doing.

The auctioning of emission permits is likely to imply that the financial burden (pollution control costs plus permit expenditures) of emission trading exceeds that of the CAC alternative. The "grandfathering" of permits or zero-revenue auctions pose less opposition and might be more profitable to polluters than would CAC. The "grandfathering" of permits on the basis of the CAC alternative creates opportunities for Pareto-dominant improvements in a national context. The next chapter will show that "grandfathering" is the preferred allocation in the actual implementation of emission trading.

Chapter 6

Emission Trading for Air Pollution in Practice

6.1 Introduction

The previous chapters have sought to examine the properties of economic instruments in economic theory and in empirical simulation models. This chapter turns to the practical implementation of emission trading. Overviews of the use of economic incentives, and marketable permits in particular, have been provided by Anderson *et al.* (1990), Carlin (1992), Elman *et al.* (1992), Hahn (1989), Hahn and Hester (1989a, 1989b), Opschoor and Vos (1989), Opschoor (1994), and Tietenberg (1990), among others. These overviews are biased because they mainly rely on experience in the USA and they ignore recent steps in Europe and international experience. Moreover, they do not deal with the most recent and radical attempt to implement marketable permits: sulfur emission trading in the USA after the Clean Air Act Amendments (CAAA) of 1990. This overview attempts to bridge this gap.

The objectives of this overview are twofold:

- To analyze how a selected number of emission trading schemes for air pollution control are designed and how they operate in practice;
- To evaluate their cost efficiency, their environmental effectiveness (the extent to which environmental goals are met), their administrative practicability (information requirements and costs), and their dynamic efficiency (impact on innovation), and especially the circumstances affecting their cost efficiency.

The chapter is limited to applications for air pollution. The description of the tradable permit schemes focuses on background and objectives; pollutants covered; definition of a permit, sources involved, and the time dimension; the initial

Emission Trading for Air Pollution in Practice 131

allocation of permits; rules governing permit trading and the ways to obtain permits; the tasks of the environmental authorities; and monitoring and enforcement.

The overview in this chapter contains a wide variety of emission trading schemes. This reflects the fact that in practice the choice is not strictly between regulation or tradable permits because mixtures of instruments are implemented. Some of the examples in this chapter (for example, sulfur allowance trading in the USA) are more pure marketable permit systems and include a clear allocation of property rights to individual polluters as well as transferability of these rights. In other cases (for example, bubbles in Denmark and the Netherlands), property rights are allocated to a whole sector rather than to individual firms, and thus the situation leans toward flexibility in existing regulation and intrafirm "trading" rather than to the creation of markets for tradable permits.

The structure of the chapter is as follows. Sections 6.2 to 6.5 describe the experience in the USA. Section 6.2 discusses the EPA's emission trading policy; Section 6.3 evaluates lead trading; Section 6.4 analyzes the transfer of ozone-depleting substances; and Section 6.5 takes a close look at the sulfur allowance trading program. Sections 6.6 to 6.8 shed light on European initiatives, evaluating Danish (Section 6.6), Dutch (Section 6.7), and German (Section 6.8) experiences. Section 6.9 looks at international transfers of ozone-depleting chemicals.

6.2 The EPA's Emission Trading Policy

6.2.1 Design

The basic statute governing air quality in the USA is the Clean Air Act (CAA). The CAA is directed at implementation of emission control strategies to meet the national ambient air quality standards (NAAQS). These standards specify maximum allowable concentrations for specific pollutants in the air (Liroff, 1986; USEPA, 1990). Standards exist for CO, NO_x, lead, particulates, ozone, and SO_2, and are set by the EPA. For non-attainment areas, where one or more of the standards are violated, individual states must develop state implementation plans (SIPs) showing the measures that will be taken to meet the standards. For attainment areas, where air quality is better than the standard, prevention of significant deterioration (PSD) standards specify the allowable increments in concentrations. Moreover, technology-based emission standards [new source performance standards (NSPS)] were introduced in 1971 for both new and modified sources. For non-attainment areas, new and modified sources have to comply with the most stringent standards, the lowest achievable emission reduction (LAER), and existing sources have to apply reasonable available control technology (RACT) accounting for technological and economic feasibility (see *Table 6.1*). In attainment areas, new and modified

Table 6.1. Trading options and emission limits.

Type of source	Type of area	Trading option	Optional or obligatory	Emission limit	Can trade avoid emission limit?
New	Attainment	Offsets	Optional	BACT	No
	Non-attainment	Offsets	Obligatory	LAER	No
Modified	Attainment	Netting	Optional	BACT	Yes
	Non-attainment	Netting	Optional	LAER	Yes
		Offsets	Obligatory	LAER	No
Existing	Attainment	Bubbles	Optional	State limits	Yes
	Non-attainment	Bubbles	Optional	RACT	Yes

Source: Hahn and Hester, 1989b, p. 370.

sources have to apply best available control technology (BACT), prescribing the best technique while accounting for economic considerations (Liroff, 1986). Existing sources in attainment areas have to meet state limits, which typically are no more stringent than RACT (Hahn and Hester, 1989b, p. 380).

The development of emission trading began in 1972. In response to the formulation of NSPS, smelter operators proposed to avoid these standards if the additional emissions could be netted out by other, cheaper measures at the same plant. This was officially allowed in 1975, but a court decision in 1978 struck down this rule. Meanwhile, it became clear that states could not meet their SIP deadlines and the offset policy was born. This policy allowed new and modified sources to enter non-attainment areas as long as they applied LAER technologies and any additional emissions were offset by other sources in the same area. In 1979, banking of emission reductions was allowed and the bubble concept was introduced. The bubble allowed emission trading between existing sources in attainment areas (Liroff, 1986). In 1982, the bubble concept was revised, and interim general trading rules were formulated and the bubble was extended to non-attainment areas (USEPA, 1982). In 1986, the EPA published its final emission trading policy statement, setting out general principles for emission trading (USEPA, 1986; Borowski and Ellis, 1987).

The objectives of the emission trading policy are to stimulate more economically efficient means of control and to promote flexibility in order to meet air quality standards more quickly (USEPA, 1979, 1982, 1986). The pollutants covered are all the air pollutants for which national ambient air quality standards exist as well as hazardous pollutants.

A tradable permit, or emission reduction credit (ERC), is defined as an emission reduction that is surplus, enforceable, quantifiable, and permanent. Surplus means the reduction is not currently required by law. This might pertain to an emission reduction below

- The baseline emissions required for attaining and maintaining the NAAQS;
- The applicable emission standards for new and modified sources.

A number of emission reductions are not eligible for use in emission trading, such as reductions resulting from plant shutdowns. All existing sources and major new, stationary sources in both attainment and non-attainment areas can trade (Borowski and Ellis, 1987).

The initial distribution of permits is based on "grandfathering" and depends on the definition of a source's baseline emissions. These baseline emissions depend on three factors: the emission rate, capacity utilization, and operating hours. They are typically determined on a case-by-case basis. For attainment areas, the baseline emissions generally are the lower of the actual or allowable emissions. For non-attainment areas with approved SIP plans, baseline emissions are the emissions used in the SIP. For similar areas without approved SIP plans, baseline emissions are whichever is the lowest of actual, SIP-allowable, and RACT-allowable emissions, for each of the three baseline factors for each source involved in the trade.

There are four ways sources can trade:

- Netting enables a modified source to avoid the most strict emission standard, which normally would be applied for modified (or new) sources, by reducing emissions from another source at the same plant provided that it does not increase total emissions (netting is, by definition, intrafirm).
- Bubbles allow existing sources to apply different levels of emission reductions as long as the total emissions level of the sources does not increase. This can be within one firm or between different firms.
- Offsets can be used by new sources or in the case of major modifications in existing sources in non-attainment areas. These sources can obtain offsets from existing sources to show progress in meeting the NAAQS. In attainment areas, these sources can use offsets to demonstrate protection of air quality standards or visibility.
- Banking is the use of created ERCs for future use (either in the firm or for later sale).

Table 6.1 shows that these trading options are not always mandatory and cannot always be used to avoid compliance with emission standards (Hahn and Hester, 1989b, p. 370).

The final general rules governing trading as formulated in 1986 are manifold:

- Trades must be for the same pollutant.

- For particulates, SO_2, CO, and lead, trades must satisfy ambient tests. For volatile organic compounds (VOC) and NO_x no such test is required. An ambient test is a demonstration to show that the trade has no significant negative impact on air quality. Trading is not allowed if it violates the NAAQS or PSD increments in attainment areas and it cannot create new violations or delay meeting the NAAQS in non-attainment areas. Generally, this requires air quality modeling unless the emission increases are below certain minimum levels or unless sources are located within 250 meters and certain conditions are met.
- Interstate trading is allowed as long as the requirements of the more stringent state are met.
- Emission credits from existing sources cannot be used to avoid NSPS for new sources, BACT in PSD areas, or LAER in non-attainment areas.
- Bubbles are allowed in non-attainment areas as long as the baseline emissions of the sources reflect the lowest of actual, SIP-allowable, or RACT-allowable emissions, as long as there is a net air quality benefit (an at least 20 percent cutback in emission below baseline levels), and as long as ambient tests are satisfied.

Finally, states may adopt alternative, general trading rules that assure attainment and maintenance of the NAAQS. Such state rules are not permitted to increase emissions above the baseline. States have interpreted these rules to guarantee the NAAQS in the following three ways: requesting offset rates exceeding one to ensure that emissions are reduced, limiting trading to relatively small zones to avoid "hot spots," or requiring dispersion modeling (Hahn, 1986).

6.2.2 Cost efficiency

There is no dispute that the EPA's emission trading policy resulted in considerable cost savings, although it is also evident that the number of transactions that took place was lower than expected. *Table 6.2* gives an overview of the traded volumes and the estimated cost savings of the four ways to trade (Hahn and Hester, 1989a; Dudek and Palmisano, 1988). The table shows that the total expected cost savings range between US$1 and US$13 billion. This is maximally 4 percent of total air pollution control expenditures over a similar period of time (Anderson *et al.*, 1990). Annual cost savings are somewhere between US$100 and US$1,400 million.

Netting can be regarded as the most successful part of the EPA's emission trading policy. The large range in estimated trades reflects the lack of data. Cost savings of netting consist of savings in pollution control costs and savings in costs of permitting procedures. The average control cost savings are estimated to be

Emission Trading for Air Pollution in Practice

Table 6.2. Trades and cost savings under the EPA's emission trading policy.

Type	Volume (no. of trades)	Cost savings (mln. US$)	Period	Annual cost savings (mln. US$)
Netting	5,000–12,000	525–12,300	1974–1984	53–1,230
Offsets	2,000–2,500	20–25	1977–1986	2.5–6
Internal	1,800–2,250	–	–	–
External	200–250	–	–	–
Bubbles	132	435–570	1979–1986	60–80
EPA	42	300	–	–
States	90	135–270	–	–
Banking	<120	Small	1979–1986	–
Total	–	980–13,135	–	116–1,435

Sources: Hahn and Hester (1989a) and Dudek and Palmisano (1988).

US$0.1 to US$1 million per case. Firms also save permit costs, because netting avoids the use of permit procedures for major sources. These permit cost savings range between US$5,000 and US$25,000 per case (Hahn and Hester, 1989a). Of the 2,000 to 2,500 offsets that took place, 90 percent were intrafirm. Dudek and Palmisano (1988) estimated cost savings of US$10,000 per offset. By mid-1986, around 130 bubbles had been approved and around 100 were pending. Only two of these were interfirm. The estimated cost savings of the 130 bubbles range between US$435 and US$570 million. The lower value uses lower average cost savings for state-approved trades (less bureaucracy), and the higher value uses EPA estimates of an average cost savings of US$3 million per bubble (Hahn and Hester, 1989a). The cost savings of banking are not available but are believed to be small in view of the small number of trades. Remarkably, most of the cost savings have been realized by intrafirm or internal trades. This appears to be due to the fact that transaction costs are higher for trades between different firms and the fact that there is a greater chance that created rights will be confiscated (Hahn and Hester, 1989b, p. 380).

Various explanations have been offered for the paucity of the EPA's emission trading. The following circumstances have been indicated in these explanations:

- Regulatory restrictions that limit the set of feasible trades;
- High transaction costs which restrict trades that in principle are allowed;
- Uncertainty about the value of property rights.

Regulatory restrictions on trades existed in the form of emission standards. The demand for permits was limited, because trading was grafted onto an existing regulatory framework (Boland, 1986). New sources still had to meet tight emission standards, which restricted their demand. Offsets were only mandatory for new sources and sources with major modifications in non-attainment areas (see *Table 6.1*). Most sources could avoid this obligatory use of offsets by applying BACT. The remaining emissions would then be too low for all but very large sources to classify as having major modifications (Palmisano, 1993). Neither new nor modified sources could use external trades to avoid emission standards, although modified sources could use internal trading to circumvent these limits. This made netting more attractive than external offsets. Demand for trading from existing sources was also limited because durable pollution control equipment had already been installed (Hahn and Hester, 1989a, Tietenberg, 1989). The message is that combining trade with emission standards not only restricted trading but also created a bias in favor of internal trading because modified sources could use netting to avoid emission limits.

The second explanation is that transaction costs may have been high. Transaction costs consist of the following two elements:

- Finding a trading partner and negotiating, coming to an agreement, and implementing the agreement;
- Obtaining approval from the authorities.

Searching for sellers appears to have been a formidable task because market information was scarce, typical permit prices and clear price signals were absent, and the outcome of the search was unpredictable (Vivian and Hall, 1981; Hahn and Hester, 1989a; Foster and Hahn, 1994). Brokerage fees, which depend on the value of the trade and the complexity of the transaction, varied from 4 to 30 percent (Dwyer, 1992; Foster and Hahn, 1994). Other elements were the cost and the length of the approval procedure. The administrative fees and the preparation of supporting material might typically cost $25,000, with $10,000 as lower bound (Foster and Hahn, 1994), although netting reduced these costs by avoiding major source review procedures. Federal approval is much more costly and lengthy than state approval for emission trading (Hahn and Hester, 1989a). It is important to note that air quality modeling is expensive and tends to raise new questions; moreover, only a few trades requiring such models have been implemented (Tietenberg, 1989).

Another element of transaction costs existed in the form of offset rates exceeding one. These offset rates specify the number of emission credits the buyer must acquire to increase its emission by one unit. With the 1986 policy change, a 20 percent additional cutback was required (an offset rate of 1.2), which further reduced the already low interest in bubbles (Elman, 1993). Similarly, in several states offset rates exceeding one for distant sources restricted demand (Dwyer, 1992). The use of tougher definitions of baseline (pre-trade) emissions belongs in the same category. With the 1986 policy the lowest of actual, SIP-allowed, or

Emission Trading for Air Pollution in Practice 137

RACT-allowed emissions must be used as the baseline for trading (Elman, 1993). If actual emissions exceed this pre-trade level, the implicit offset rate exceeds one.

Finally, uncertainty about the value of the property rights has also limited emission trading. Uncertainty about whether sellers would achieve their reductions, about the buyers' baselines, and about whether the trade would be accepted has limited the interest in trading (Hahn and Hester, 1989a). Moreover, trading policy has been uncertain and has changed frequently, thus inhibiting trading (Boland, 1986). There has also been too much debate and controversy, making states afraid to touch the issue (Elman, 1993). The conflicting interests of environmentalists and business have led to the creation of policies with no explicit definition of the nature of the property rights (Hahn, 1989). Furthermore, for strategic reasons firms would rather hoard permits for their own future use than sell them (Vivian and Hall, 1981). Responsible managers are believed to avoid risk rather than act as textbook profit maximizers (Palmisano, 1993). Finally, obtaining approval for a trade is not only costly, its outcome is uncertain, which is perhaps more important. Data for California suggest that, on average, out of 100 trades proposed, 50 fall through during brokering. Of the remaining proposals authorities accept 10 immediately and reject 20, leaving 20 to be revised before being accepted (Foster and Hahn, 1994).

In summary, there are three major reasons for the poor performance of the EPA's emission trading policy: regulatory restrictions in the form of obligatory emission standards; high transaction costs arising from difficulties in finding partners and obtaining approval or from offset rates exceeding one; and uncertainty about the value of the property rights created in the form of emission reduction credits.

6.2.3 Environmental effectiveness, administrative practicability, and innovation

Although the EPA's emission trading program was primarily designed to promote cost efficiency, it is clear that it achieved cost savings while generally having a neutral impact on both the level of emissions and on air quality. In some cases bubbles led to an acceleration of compliance and faster achievement of emission reductions. In other cases partial deterioration of air quality may not have been prevented. Dudek and Palmisano (1988) maintain that emission trading has not been hostile to air quality objectives because institutional safeguards exist for maintaining environmental protection. Although offsets by definition should reduce emissions, Hahn and Hester (1989a) remark that they may not protect air quality in the way intended because of inadequate emission inventories and because allowable emissions may be higher than actual emissions. The ultimate impact of trading on environmental quality was insignificant because the number of offsets was small, regulators carefully analyzed trades, and trading rules contained provisions to prevent emission increases. Bubbles and netting had little negative impact on emissions and air quality due to the review process and the participation of environmental and public interest groups that promoted the disapproval of bubbles with negative environmental impacts, and due to the fact that netting is intrafirm and at the same location (Hahn and Hester, 1989a; Liroff, 1986). Banking has had a neutral to positive impact because banked permits can only be used temporarily, if at all (Rehbinder and Sprenger, 1985). In conclusion, emission trading generally has had a neutral impact on emissions and air quality, although in some specific cases positive or negative impacts may have occurred.

The administrative burden of the emission trading policy is considerable. The long preparation time for trading policies, the frequent policy amendments, and the lengthy approval process for individual cases (4 to 29 months), including the creation of an emission reduction credit, imply not only a substantial workload but also high administrative costs (Opschoor and Vos, 1989). In the case of netting, however, savings in administrative costs may have occurred, because netting allowed sources to avoid a major source review process (Hahn and Hester, 1989a).

The emission trading program is believed to have encouraged technological progress in pollution control to a limited degree. The availability of relatively cheap emission control opportunities combined with the lagged industrial response to the emission trading program in general reduced incentives for broad-scale invention and innovation (Dudek and Palmisano, 1988). In some cases, bubbles allowed industries to benefit from innovative solutions (Opschoor and Vos, 1989). One example of such an innovation is the substitution of water-based solvents for solvents containing VOCs (Tietenberg, 1991).

6.3 Lead Trading

6.3.1 The institutions

The lead trading program was instituted as part of the regulations to reduce the lead content in gasoline. This lead phasedown program started in 1973. Until 1983, individual refineries had to meet standards expressed as the average lead content measured in grams of lead per gallon of gasoline produced in each calendar quarter. The reason for the averaging was that continuous monitoring of the output of every single refinery, as well as meeting the standard for every gallon, was impractical (Nussbaum, 1992). In 1982, the EPA imposed new, lower limits on the lead content. Interfirm lead trading was now introduced, partly in response to the fear that smaller producers would have difficulties meeting the lower standard of

Emission Trading for Air Pollution in Practice

1.1 g/gallon. In 1985, standards were decreased again (to 0.5 g/gallon). The EPA specified that lead trading would be stopped and the standard would be lowered to 0.1 g/gallon, but later in 1985, the EPA allowed the banking of lead rights.

The rights to add specified quantities of lead to gasoline could be traded among refineries and importers. The initial distribution of rights was determined by the amount of leaded gasoline a refinery produced and the standard per gallon. If one refinery added, on average, less lead per gallon in a calendar quarter it could sell the difference between the actual and the allowed quantities. Prior to the introduction of banking, these rights expired if they were not used in the same quarter. From the beginning of 1985, a company's rights could be banked through 1987 for sale or for the company's use in the future (Nussbaum, 1992). The rules governing trades were simple (Hahn and Hester, 1989b; Peeters, 1992):

- Every calendar quarter, lead rights had to be in excess of, or equal to, actual lead use.
- Gasoline production, lead use, and any trades had to be reported to the EPA for every quarter.
- Trading could not be used to avoid the more stringent California standards.
- The lifetime of the right was one calendar quarter; after the introduction of banking, rights could be banked only until 1987.
- Small refineries, subject to less stringent standards, could not sell rights to large refineries.

Although reporting of data was needed every quarter, the EPA did not have to respond to every trade or approve it beforehand.

6.3.2 Cost efficiency of lead trading

Lead trading successfully saved costs. The market in lead rights was very active and trading increased during the lifetime of the program. Whereas in 1983 less than 1 percent of the rights was traded, the beginning of 1986 showed a sharp increase in trading activity. The number of lead rights traded as a percentage of all lead used increased from 20 percent in the first quarter of 1986 to 60 percent in the second quarter of 1987 (Hahn and Hester, 1989b). The price of the rights rose from three-quarters of a cent to slightly over four cents per gram of lead after banking was allowed (Nussbaum, 1992). Although cost savings were not evaluated *ex post*, the EPA estimated that the emission trading provisions adopted in 1982 saved US$65 million (Nussbaum, 1992) and that the three years of banking cut costs by US$226 million (USEPA, 1985a). Because more banking took place than expected, estimated cost savings of close to US$300 million appear reasonable. This roughly corresponds to cost savings of 15 to 20 percent compared with the

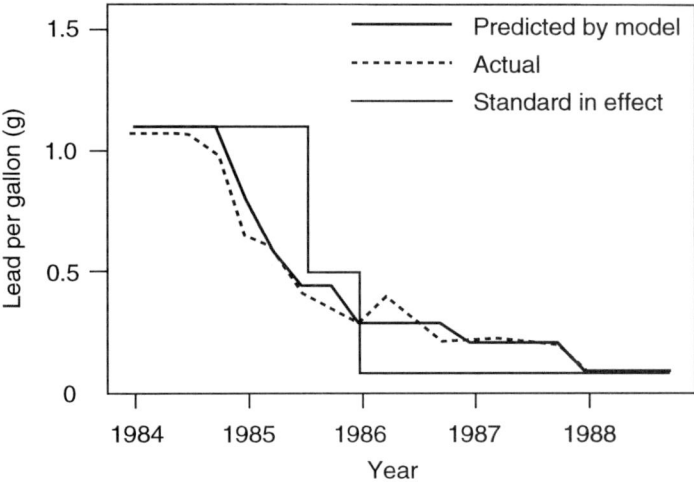

Figure 6.1. Lead banking and lead standards (Nussbaum, 1992).

costs to refineries of meeting lead regulation without banking (USEPA, 1985b; Portney, 1990, p. 64). A major reason for banking was that refineries could use this option to delay compliance with the stringent lead standards for 1987 (Hahn and Hester, 1989b), as is shown in *Figure 6.1*.

Significant factors explaining the success were the minimal trading restrictions and the lack of burdensome administrative requirements, as well as the existence of well-established markets in refinery feedstocks and products including gasoline additives (Hahn and Hester, 1989a). The latter implies that refineries were already used to carrying out transactions with each other, so transaction costs in terms of searching for a partner could be much lower than under the EPA's emission trading program. Hahn (1989) adds two additional features: first, monitoring the amount of lead could easily be done within the existing regulatory apparatus; second, agreement existed on the environmental goal (the phasing out of lead) of the program.

6.3.3 Environmental impacts and administrative costs

Lead trading, and especially banking, has led to a sharper, more rapid decrease in the average lead content of leaded gasoline than would have been the case without banking (see *Figure 6.1*), although the total use of lead in the period from 1985 to 1987 was the same (Nussbaum, 1992). Despite the fact that the lead standards were met, the actual use of leaded gasoline was not reduced as much as expected because the price of leaded gasoline was lower than the price of unleaded gasoline (Boland, 1986). Trading did not result in the emergence of "hot spots" (high concentrations at certain locations), although lead rights were transferred from modern refineries on the West Coast to older refineries on the East Coast. This migration of lead rights would not have had a significant impact (a 5 percent increase above the standard) in the East even if all West Coast rights had been shipped to the East Coast. The gasoline distribution system, mainly the vast pipeline, tends to mix gasoline from different producers in the marketplace, which mitigates environmental concerns about the occurrence of "hot spots" (Nussbaum, 1992).

The administrative practicability of lead trading was less than expected. Monitoring and compliance proved more labor intensive due to the fact that the number of small gasoline blenders, who mixed leaded and unleaded gasoline, swelled. Many of them did not know how to fill out the forms and, frequently, credits that could not be claimed were quickly sold. Furthermore, the computer system had to be adapted to track the lead rights of the small blenders. This led to an increase in personnel and computer requirements (Nussbaum, 1992). In addition, enforcement proved to be less straightforward than without trading. The number of possible violations increased because the value of lead created an incentive to cheat. Violations included trading improperly generated rights, banking rights that were needed for current compliance, and selling the same rights twice (Nussbaum, 1992). The EPA had not established a requirement to verify lead rights data reported by the participants. They had a backlog in reviewing these reports, which delayed enforcement actions, and concern was expressed that this could lead to the use of lead rights in amounts exceeding allowed levels (GAO, 1986). Tracking the possible violations proved to be even more labor intensive. Not intervening in sales or certifying rights beforehand, however, also reduced administrative costs, especially for the industry (Nussbaum, 1992). In short, administrative information requirements and costs seem to have been higher with trading than without.

6.4 Chlorofluorocarbons and Halons

6.4.1 Designing the market

The USA has implemented a market in tradable permits to meet the requirements of the Montreal Protocol of 1987. This protocol aims at protecting the stratospheric ozone layer and limits the production and consumption of ozone-depleting substances such as chlorofluorocarbons (CFCs) and Halons. In the original Protocol, both production and consumption of CFCs and Halons had to be frozen in 1992 at their 1986 levels and had to be reduced by 50 percent in 1998 (USEPA, 1992a).

Subsequent amendments sharpened these reductions. To meet the requirements, the USA implemented marketable permits because they promote efficiency and offer more certainty of meeting production and consumption quotas than a fee (Peeters, 1992).

The tradable permit system covers both CFCs and Halons. The EPA created production and consumption allowances for each group. Allowances are defined as granted privileges and are specified in kilograms for one year (USEPA, 1992a). Banking of the allowance is not allowed. Both production and consumption allowances were allocated to producers and importers on the basis of their 1986 production and import levels. They were not allocated to consumers, because the large number of consumers (5,000–10,000) would have increased the administrative costs (Hahn and McGartland, 1989). Production allowances were allocated to seven producers. Seventeen producers and importers received consumption rights (Peeters, 1992, p. 269). The main reasons for "grandfathering" instead of auctioning the rights were political. Producers feared that auctions would be complex and would create uncertainty about permit price and the possibility of obtaining permits; they may also have been against redistributing revenues to the government (Hahn and McGartland, 1989).

The rules governing trades reflect the relation between consumption and production allowances . In the Protocol, consumption equals production plus imports minus exports. To produce CFCs and Halons, producers need both consumption and production allowances . This is related to the fact that the Montreal Protocol sets separate targets for consumption and production. Because consumers are not allocated consumption rights, consumption in the USA can only be controlled by allocating consumption rights to producers (for that part of production being consumed inside the USA) and to importers. Imports require only consumption allowances . Exporters can obtain additional consumption allowances for exports. Producers can obtain additional production allowances for exports to developing countries or for parties to the Protocol or upon the transfer of production rights from another Protocol party (USEPA, 1992a). An example might clarify this. If a US producer has 20 production rights and 15 consumption rights, he is only allowed to produce 15 kilograms. If, however, he wants to export 5 kilograms to another Protocol party, he is allowed to produce these 5 kilograms. The exported 5 kilograms have to be added to the actual level of consumption of the other party and should lead to less domestic production by the importing party.

The EPA must be notified of the intention to transfer allowances within the USA. The notification has to include information on the parties involved, the type of right, the type of pollutant, the control period, and the remaining allowances after the transfer. The EPA only checks if the seller has sufficient allowances to justify the transfer and responds within three working days. If production allowances are bought from other countries that are parties to the Protocol, the EPA can grant these additional allowances to the buying party in the USA. In this case, the embassy of that country has to declare that the country has reduced its production rights by the amount transferred (Peeters, 1992).

To monitor compliance, producers and importers have to report their production and import levels to the EPA every quarter. The EPA checks whether the companies have sufficient allowances to carry out transfers, responds to requests for additional allowances, resolves policy issues, carries out inspections, reviews quarterly reports, and guides and carries out compliance and enforcement activities (USEPA, 1992b).

In 1990, the Clear Air Act was amended to further restrict ozone-depleting substances (USEPA, 1992b). This also led to a change in the trading rules. Transfer of allowances is permitted provided the remaining allowances held by the seller are reduced by the amount transferred plus 1 percent. This was introduced to ensure that any trade will result in greater total reductions in production than would occur in the absence of trade. The rules regulating the transfer of allowable production levels to other Protocol parties are more strict. Transfers of production allowances to another party are allowed if the USA revises its aggregate national production limit so that the maximum allowable production level equals the lowest of the following levels:

- The maximum production level permitted for the USA in the Montreal Protocol minus the amount transferred outside the USA;
- The maximum production level in the applicable US law minus the amount transferred outside the USA;
- The average of the actual national production level in the USA for the three years prior to transfer minus the amount transferred abroad.

In effect, this implies that the EPA will subtract from the trading company's balance the greater of two quantities: the amount transferred, or US allowable production minus US average production over past three years plus the amount transferred. The latter requirement actually implies that if the actual production is below the allowed level (the production rights), the unused rights are not transferred to other parties, thus increasing production there (thereby, keeping the total production constant). Instead, these rights are confiscated to ensure that production is reduced more quickly to attain more rapid protection of the ozone layer. This actually implies an effective offset rate exceeding one for production transfers abroad. In reality, this creates a bias in favor of domestic rather than international trade. Regarding the transfers to other parties, the EPA will review the transfer request and approve it if it is consistent with the Montreal Protocol and with US domestic policy. Again, the USA may also acquire production allowances from another party if the other party has revised its domestic production limit (USEPA, 1993).

The allowances represent a significant economic value. To redistribute the associated rents, a tax was installed. In 1991 the tax level was US$1.37 per kilogram; it increased to US$2.65 in 1994. It is supposed to rise by US$0.45 every year (Vos *et al.*, 1992, p. 52).

6.4.2 Cost efficiency and environmental effectiveness of CFC trading

The extent to which the market has been efficient is difficult to judge because such information is considered confidential. From July 1989 until October 1990 fewer than 15 trades took place, corresponding to 0.5 percent of the consumption allowances (Peeters, 1992). Through mid-1991, 34 market participants had consummated 80 trades (Stavins and Hahn, 1993). As of September 1993, the traded amount was roughly 10 percent of the allowances . Permit demand increased somewhat, because in 1994 production caps had to be met (Voight and Lee, 1993). No estimates are available of the cost savings of these trades. Earlier model studies suggest that the costs of a tradable quota policy could be more than 40 percent lower than those for mandatory controls (Palmer *et al.*, 1980, p. vi).

Stavins and Hahn (1993, p.12) confirm that it is not easy to evaluate the efficiency of the market because studies on the trading pattern and costs saved are lacking. Acceleration of the phaseout of the ozone-depleting chemicals and the introduction of the tax complicate such an evaluation. They report that in 1992 firms were producing below their allowable levels. Because banking was not allowed, there was an excess supply of permits. They suggest that this might have been because the tax pressed actual levels below the allowable amount and hence was the binding instrument. However, excess supply might also have been due to the fact that firms had to respond to the accelerated phaseout of CFCs and Halons.

One might question the extent to which CFC trading has been cost efficient. Relatively low transaction costs suggest a somewhat cost-efficient outcome (Stavins and Hahn, 1993). Production is highly concentrated; only two companies produce 75 percent of the CFCs in the USA (Hahn and McGartland, 1989). It is not known to what extent this has caused the market permit price to deviate from a competitive market price, leading to a less than cost-efficient outcome. This might happen if firms with market power act as permit sellers and depress their supply in order to increase the permit price and increase their own revenues. Or it might occur if they act as buyers and suppress demand to depress the permit price and reduce their permit outlays (Hahn, 1984). Clearly, US regulation that subtracts the difference between actual and allowable production from the international transferred quota has negatively influenced international trades and cost efficiency, because this creates a bias in favor of domestic production. Marginal costs of domestic production and production abroad are no longer equalized, because transferring production abroad is subject to an additional penalty. This impact will diminish when allowable production is reduced and comes closer to actual production (Voight, 1993).

The environmental effects of the trading program seem not to have been problematic. Actual production levels are lower than allowable levels. There have been 10 to 11 violations due to ignorance of the rules. The noncompliance fines are so high [US$25,000/kg according to Peeters (1992)], however, that they are working as a deterrent (Voight and Lee, 1993). No data are available on the administrative practicability and the impacts on innovation.

6.5 Sulfur Trading under the 1990 Clean Air Act Amendments

6.5.1 Harnessing market forces to curb acidifying emissions

In November 1990 the Clean Air Act Amendments (CAAA) became law (US Congress, 1990; USEPA, 1990). Title IV of these amendments contains provisions to control the acid deposition caused by sulfur and nitrogen oxide emissions. The objectives of the acid rain program are (USEPA, 1992c)

- To reduce the adverse impacts of acid deposition through a 10 million ton reduction in the annual SO_2 emissions and an NO_x reduction of around 2 million tons from the 1980 levels;
- To achieve these reductions at the lowest costs by employing traditional methods and an emission allowance trading system.

For SO_2 the acid rain program introduces an emission trading scheme. The objective of this emission or allowance trading is to allow utilities to choose the most cost-efficient way to reduce emissions. Each allowance permits a unit to emit one ton of SO_2 during or after a specific year (USEPA, 1992d). An allowance is defined as a limited authorization to emit SO_2. It does not constitute a property right and does not limit the authority of the USA to terminate or limit such authorization (US Congress, 1990, pp. 2591–2592).

Total SO_2 emissions were around 23 million tons in 1985. In this same year, electric utilities emitted 16 million tons. The acid rain program establishes a national cap on the SO_2 emissions of all utilities of 8.95 million tons/year. This is a reduction of around 10 million tons, comparable to the 1980 level of the electric utility emissions. This cap is to be implemented in two phases. Phase I starts in

1995 and requires the 110 highest emitting electric utility plants in 21 eastern and midwestern states to meet an interim ceiling of 5.7 million tons. Phase II begins in 2000 and will include not only these 110 units but also smaller, cleaner plants throughout the USA. With phase II, a national cap of 8.95 million tons of SO_2 is placed on the number of allowances.

The initial allocation of allowances is such that allowances are allocated for each year beginning in 1995. The baseline for allocating allowances is the average fossil fuel consumption from 1985 to 1987 and an emission rate [in phase I, 2.5 pounds of SO_2 per million Btu (British thermal units) of fuel input]. Additional allowances are given to various units, including units in Illinois, Indiana, and Ohio.

The rules governing permit trading are straightforward. At the end of each year, each unit must hold allowances at least equal to its annual emissions. Furthermore, regardless of the number of allowances a unit holds, it is never entitled to exceed the ambient air quality standards for public health. The unit is free to buy the allowances, but it might not be able to use them to increase emissions if this threatens to violate ambient standards.

There are different ways to obtain permits, apart from the initial allocation. First, allowances may be sold, bought, or banked by any person – not only utility representatives, but also individual corporations, brokers, municipalities, environmental groups, and private citizens. Second, allowances can be obtained from three EPA reserves (USEPA, 1992d, 1992e)

- For installing technologies that remove at least 90 percent of the emissions (phase I);
- For implementing energy conservation or renewable energy;
- From a set of allowances kept for auctions and direct sales.

Newcomers (those who begin operating in 1996 or later) will not be allocated any allowances. They will have to buy on the secondary market or from EPA auctions and direct sales.

The annual auctions are being organized to help create a price signal and to create an additional avenue for utilities to purchase allowances. The main object of the direct sales is to ensure that new independent power producers always have a way to buy permits so that existing utilities cannot withhold allowances to reduce entry of these producers and thus reduce competition. The allowances that are sold at auctions and by direct sales come from a special reserve of 2.8 percent of the annual allowances. The reserve was created by taking away 2.8 percent of the annual allowances from every plant that obtained these allowances with the free initial distribution. Auctions and direct sales both started in 1993 and will be held every year.

The auctions consist of two markets:

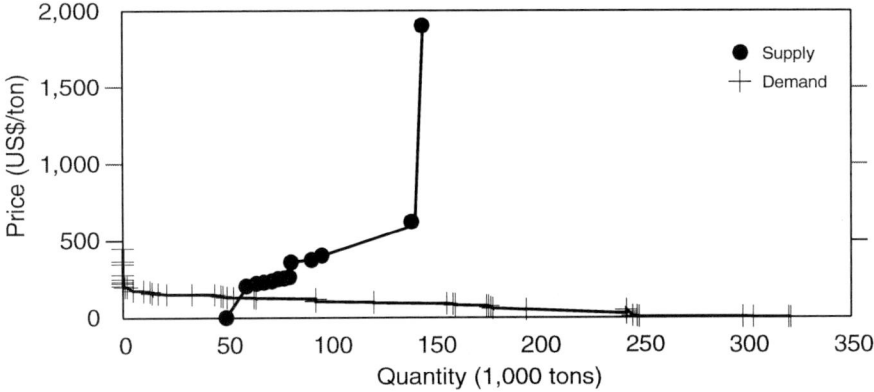

Figure 6.2. The spot market for SO_2 allowances in 1993.

- A spot market were allowances that can be used in the same year (or banked for later use) can be bought;
- An advance (forward) market for the sale of allowances that can only be used for compliance no sooner than seven years in the future.

In addition to the allowances offered from EPA reserves, private allowance holders can offer their allowances for sale at the EPA auction. The Chicago Board of Trade administers the auctions and sales.

The auction proceeds in the following manner. Bidders submit sealed offers for the number and type of allowances they want to buy, listing the maximum price they are prepared to pay. Allowances from EPA reserves are offered at a supply price of zero. Private suppliers have to indicate their minimum supply price as well as the type and quantities they want to supply. Ranking supply and demand in ascending order, starting with the lowest supply and demand price, results in demand and supply curves. *Figure 6.2* shows this for the 1993 spot auction. The figure shows that supply starts at a level of 50,000 tons for a zero price. This is the EPA supply. The remaining part of the supply curve shows private suppliers.

Normally, allowances would be sold at the market clearing or equilibrium price (around US$130 in *Figure 6.2*). However, at the EPA auction the lowest supply prices are always coupled with the highest demand prices. The bidders do not pay the market clearing price but the actual price they are prepared to pay. EPA allowances are always sold first, because they have the lowest supply (asking) price. After EPA allowances have been sold, allowances of private holders are sold, again at the price the bidders are prepared to pay. This continues until all allowance supplies have been sold or the minimum supply price for the allowances exceeds the maximum demand price, or until all bids are used up. The reason for selling allowances at the demand (bid) price and not at the market clearing price is political. It is related to the fact that the auction revenues (and any unsold allowances) are returned on a *pro rata* basis to those units from whom the EPA withheld allowances to create the special reserve. The coupling was expected to ensure that midwestern utilities (expected to act as sellers) would get most of the revenues (McLean, 1993).

At direct sales, allowances are sold for US$1,500 (indexed to inflation) on a first-come-first-served base. Independent power producers, however, have priority in purchasing allowances for a fixed price of US$1,500 (indexed to inflation) in these sales (USEPA, 1992e). If allowances are not sold, they are offered in the next year's auction. Proceeds from the sales are again returned to those units that contributed to the special reserve (USEPA, 1992c).

Monitoring and enforcement imply that the EPA records all permit transfers and ensures that a unit does not emit more than the number of allowances it holds. For this purpose, the EPA must maintain an allowance tracking system. Allowance transfers require the submission of transfer forms to the EPA, signed by the representatives of both parties (USEPA, 1992d). Compliance is determined at the end of the year. Utilities are granted a 30-day grace period at the beginning of the next calendar year during which allowances can be purchased to cover emissions. Excess allowances may be banked or sold. Each unit must continuously measure and record emissions. This implies that most plants are equipped with continuous monitoring equipment to guarantee the confidence in allowance transfers. If a unit emits more than allowed, a penalty of US$2,000 must be paid per excess ton of emissions. Moreover, the excess emissions must be offset in the following year (USEPA, 1992c).

6.5.2 Cost efficiency, or how well the market operates

Obviously, it is too early for an *ex post* analysis of this sulfur trading, because phase I only started in 1995. It is possible, however, to indicate the expected cost savings, to compare expected market activity and prices with reality, and to analyze the circumstances that have influenced the market so far. *Table 6.3* shows that the sulfur trading program has the potential to cut costs by US$9.4 to US$13.4 billion over the lifetime of the program (ICF, 1992). This is a reduction in costs of 40 to 45 percent. Costs consist of direct pollution control cost measures and the administrative implementation costs. The pollution control cost estimates are based on a linear programming model and assume a perfect permit market. Implementation costs consist of allowance system, monitoring, and permit costs.

Table 6.3. Expected cost savings of sulfur trading (1995–2010) (mln. US$, 1990 value).

Type of costs	Traditional control	National trading program	Cost savings
Pollution control	19,100 to 30,900	9,500 to 17,100	9,600 to 13,800
Implementation			
Allowance system	0	207 to 416	−207 to −416
Monitoring	2,512	2,395	117
Permits	0	68	−68
Total costs	21,612 to 33,412	12,170 to 19,979	9,442 to 13,433

Sources: ICF (1991, 1992).

Major components of the allowance system costs are the expected transaction costs of US$200 to US$400 million for the utilities. Minor cost components are the allowance tracking system, the auctions, direct sales, and the conservation and renewable energy reserve. Transaction costs are based on a traded volume of 1 to 2 billion tons/year and a transaction cost of 1.5 percent (ICF, 1992). Major elements of the implementation costs are the costs of monitoring. Monitoring costs are expected to be lower under the trading program than under a traditional program because the EPA does not require continuous monitoring for sources with very low emissions (ICF, 1992, pp. 4–26). This seems rather artificial considering the same rational approach could have been adopted under a traditional program. In short, the sulfur trading program is expected to lead to considerable savings in overall costs.

Achievement of the potential cost savings depends on how well the market operates in practice. *Figure 6.2* shows the supply and demand observed at the EPA's spot auction in March 1993 for permits to be used in phase I. *Table 6.4* gives the estimates of prices and volumes based on model calculations (column 2), an overview of actual prices and volumes on the secondary markets (column 3), and the official data on the EPA allowance auctions held in March 1993, March 1994, and March 1995 (columns 4 to 6). The table suggests the following:

- Prices and trade volumes of the auctions (spot and forward) are far below prices and trade volumes on the secondary market.
- Actual prices on the spot market tend to equal actual prices on the forward market.
- Actual trade volume (auction plus secondary market) is higher than the volume estimated with the model.

Table 6.4. Market prices and traded volumes.

	Model	Secondary market[a]	Auction 1993	1994	1995[b]
Prices [US$ (1993 value)][c]					
Phase I (1995–2000)	100–130[d]	180–270	157	159	132
Phase II (2000–2010)	261–504[e]	160–250	136	149[f]	128[f]
Volume (1,000 ton/yr)					
Phase I (1995–2000)	310–370[d]	>420	50[g]	50[g]	51
Phase II (2000–2010)	681–1075	–	100[g]	126[h]	126

[a] Price data: Wall Street Journal, 24 August 1993, Compliance Strategies Review (1993), Lipskey (1993), and Rico (1995). Price data are not available for all trades. Volumes on secondary market based on Rico (1995; Table 4). This refers to the period from May 1992 to October 1994 and only includes trades exceeding 10,000 tons. Total volume traded is 2.3 million tons, of which 0.2 million tons pertain to reselling through brokers. The remaining 2.1 million tons are traded between utilities. Most trades pertain to a five-year contract (Weissman, 1993). This gives 2.1/5 = 0.42 million tons per year. This is a minimum estimate because smaller trades are not included and more phase I trades are possible as phase I lasts until 1999. Of the 2.1 million tons, 0.420 million tons are intrastate; therefore, the minimum annual interstate trading is 0.340 million tons.
[b] USEPA (1995).
[c] Prices in 1993 US$; auction prices in current prices.
[d] ICF (1991) [ICF (1994) estimates no trading in phase I, only banking].
[e] US$261 in the year 2000 increasing to US$504 in 2010 (ICF, 1994) [ICF (1991) predicted US$200 to US$420].
[f] Weighted average prices paid at the six- and seven-year advance auctions.
[g] All offered by the EPA (USEPA, 1993). Prices are the average prices paid.
[h] Of which 100 were offered by the EPA and 25 were left over from the 1993 direct sales (USEPA, 1994).

A first question is why prices and quantities on auctions are so much below the level of the secondary market. A number of explanations are given for this phenomenon.

First, some authors expect that the pay-what-you-bid format of the auction is an incentive for buyers to state a lower demand price than their true willingness to pay, which would depress the market clearing price and the volume traded (Rico, 1995; Cason, 1993). However, one should bear in mind that the outcome depends on the bid strategy of the buyers at the margin. To avoid the risk of not getting the allowances, they must state a price that is not below the expected market clearing price. Therefore, it is by no means certain that the pay-what-you-bid scheme tends to depress the auction price.

Two other arguments to explain a too-low auction price and trade volume seem more relevant. Some brokers believe that the EPA's auction is inappropriate for the type of market, because the most significant demand for permits occurs when a utility has to decide whether or not to install a flue gas desulfurization unit. This is more easily handled in a few large trades ahead of the compliance dates (Weissman, 1993). In addition, the timing of the first auction (March 1993) was not optimal, because most companies had just finished their phase I planning and were just starting to plan phase II. In short, prices and sales at the March 1993 auction were depressed due to the timing of the auction and the structure of the market. The downward bias in market prices might limit the ability of the auction to set cost-efficient price signals. It should be noted, however, that at the second auction (March 1994) the market clearing auction prices (on the spot and forward markets) were higher than the March 1993 auction prices and approached the price in the secondary market. Given the opportunities for profitable arbitrage between primary and secondary markets in 1993 and 1994, we could expect the 1995 auction price to approach the lower prices on the secondary market. The 1995 auction results show that the auction prices have dropped compared with 1994 results. Prices on the secondary market have gradually come down, as well (Hahn and May, 1994, p. 33). Data for February 1995 show that the secondary market price for a phase I allowance is around US$137 (Lobsenz, 1995). Primary and secondary market prices thus have gradually converged.

Table 6.4 suggests that actual prices on the spot and forward markets tend to be equal, with forward prices somewhat below the level of spot prices. This may look surprising at first sight given the necessity of higher emission reduction in phase II combined with higher electricity demand and higher fuel input, which induce higher marginal control costs; but actually the tendency toward equalization of prices shows that the market works as it should. What happens is that firms bank part of their allowances and professional investors (ICF, 1992, 1994) possibly buy allowances in the 1995 spot market and store them for use or for selling in 2000 or later, when emission targets will be more stringent and marginal abatement costs and spot prices of allowances will be higher. An indication of the activity of professional investors is the bid at the 1994 spot auction of the Allowance Holding Corporation, which acquired 26 percent of the 50,000 allowances sold (USEPA, 1994). The possibility of intertemporal trade causes future scarcity to already be signaled in today's price and reduces present use of the good. Consequently, emissions in the period from 1995 to 2000 will be reduced below the 1995 target level, whereas from 2000 on emissions will be somewhat above the target level. The official prognoses have been late in recognizing this impact of the market on the development of emissions over time. It was only in 1994 that *Figure 6.3* was published (ICF, 1994). *Figure 6.3* clearly shows the difference between legally required and expected emission reductions and the importance of banking for accomplishing early reductions.

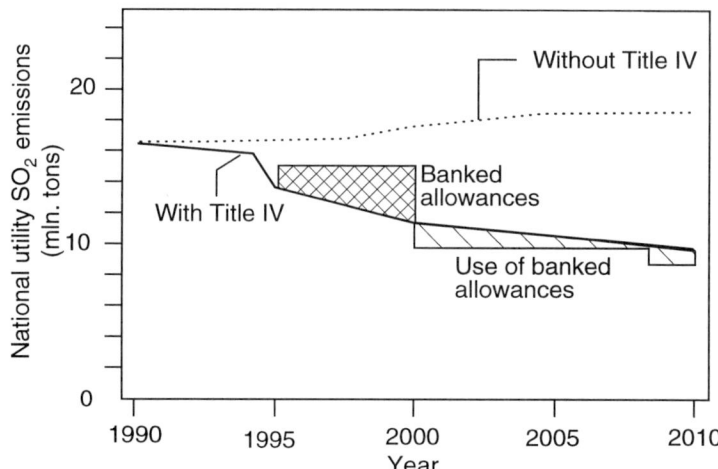

Figure 6.3. Sulfur emissions and banking.

In a market with perfect foresight and certainty, the forward and spot price will coincide. In reality, the forward price is lower than the spot price (at auction and in the secondary market). The explanation for this phenomenon is that in a world with imperfect foresight, buying phase I allowances in the spot market gives the added option of using them in the period from 1995 to 2000 in addition to the possibility of storing them for use in phase II (Hahn and May, 1994). The price margin is the premium paid for the extra liquidity.

The model predicts increasing real spot prices for allowances – close to a doubling of the price in the decade between 2000 and 2010. This would boil down to a price increase of 7 percent per year on average (but about 9 percent in the period between 2000 and 2005 and about 6 percent from 2005 to 2010). Extrapolating back in time to 1994 gives us a spot price of US$180 (at 7 percent discounting) or US$140 (at 10 percent discounting), which is in the range of actual spot prices in 1994. The model, however, overestimates future prices, because the linkage that banking provides between forward and spot prices has been ignored (Hahn and May, 1994).

A third striking feature in *Table 6.4* is that actual interstate trade on the primary and secondary markets has been above the model estimate up to now. In the model the volume of interstate trade is between 800 and 1,100 kton per year in the period between 2000 and 2010. In reality, interstate trade over the period from 1992 until May 1994 was about 1,500 kton per year. This might be interpreted as an indication that the allowance market works for reallocating emissions between producers. This is contrary to the rather pessimistic expectations of a number of observers who have stressed the factors that reduce demand and supply in a way that is detrimental to the environmental effectiveness and the cost efficiency of the program. These factors include reduction in capacity use of boilers in phase I, uncertainty about property rights, and the regulated character of the electricity industry. These points will be subsequently discussed.

One factor that might have depressed permit demand, at least for phase I permits, is the existence of a loophole in the legislation. Many utilities have plants that fall under the phase I cap, but they also have other plants that fall only under the phase II cap, not under the phase I cap. This gives utilities an incentive to run these phase II boilers at higher generation rates, reducing the electricity output and emissions produced by phase I boilers and reducing the demand for allowances. The EPA is currently implementing regulation to prevent such shifts (Rico, 1995).

A minor element that might have led to a thin market is the uncertainty on the status of the property rights. The CAA makes it clear that an allowance is not an impregnable property right and hence can be limited, revoked, or otherwise modified in the future without compensation. Changes in federal and air pollution regulations could also occur. The EPA, for example, is considering the introduction of a short-term standard for SO_2 to avoid health problems. This could lead to limits on trades. Furthermore, a major topic of discussion is whether emission standards have to be applied to new or modified sources. Some state that the application of BACT or NSPS in PSD areas should be a case-by-case decision of the individual state. Others are of the opinion that compliance with standards cannot be avoided (Endres and Schwarze, 1993).

The negative public attitude in some cases and a lawsuit by New York State in its effort to prevent utilities from selling permits may also have depressed market activity (Weissman, 1993). The EPA, however, believes that the New York State lawsuit did not have much effect on trading, although some utilities in New York are awaiting the outcome nervously (McLean, 1993).

Another factor influencing proper functioning of the permit market is the tendency in some states – Bohi (1993) counted six – to promote options other than buying allowances, mainly by granting favorable cost recovery of the capital cost of flue gas desulfurization equipment. This in itself could encourage overinvestment in pollution control and selling or banking of allowances. A potential market failure of a more general nature is the regulated character of the electric utility industry. Most US electric utilities have monopolies provided by state or local authorities (ICF, 1992). Utility rates and revenues are subject to state or federal control and regulations. Rates for privately owned utilities are set to allow them to recover all "prudently incurred" costs and make limited profits (ICF, 1992). This implies that the utilities' rates of return (profit divided by the asset or capital base) have to be below a certain allowed rate of return. If the rate of return allowed exceeds the cost of capital, this will lead the profit-maximizing firm to substitute capital for other factor inputs and to operate at an output level where costs are not minimized (Averch and Johnson, 1962). In empirical studies evidence has been found that overinvestment actually may occur. This type of behavior has consequences for the utilities' reactions on a permit market. What is decisive is how the initially "grandfathered" allowances are valued and how revenue from allowances that have been sold and expenditures on permits that are bought are treated. Public utility commissions (PUCs) in most states use the "original cost method" (Bohi *et al.*, 1990; Endres and Schwarze, 1993). Strict application of this rule implies that allowances that have been grandfathered have a book value of zero, whereas abatement capital will be valued according to investment expenditure. Consequently, abatement capital will enter the rate base and initially "grandfathered" allowances will not. As far as revenue is concerned, the "Uniform System of Accounts" of the Federal Energy Regulatory Commission (FERC) requires net revenue from selling capital assets to be transferred to electricity consumers (the rate payers). Tschirhart (1984) has shown that this set of rules leads to a bias in favor of capital-intensive pollution control equipment, such as flue gas desulfurization equipment, compared with emission permits. (By the same token, utilities have an incentive to overinvest in abatement capital relative to reducing emissions by substituting low-sulfur coal for higher-sulfur coal.) The results of both rules are that regulated utilities overinvest in flue gas desulfurization equipment, that prices of emission permits are depressed, and that unregulated private electricity producers underinvest in abatement capital. This distortion in the allocation of abatement capital reduces the cost efficiency of the trading scheme. To avoid any bias in favor of flue gas desulfurization equipment, the initial allocation of emission permits and the costs of purchased allowances would have to be required to show up in the rate base and would have to be treated in the same way as investments in flue gas desulfurization equipment (Bohi *et al.*, 1990).

Although there is the critical possibility of overinvestment, it is by no means clear how important the Averch–Johnson effect is in general, or even more, whether and to what extent it has induced overinvestment in flue gas desulfurization equipment in regulated utilities and has depressed prices and trade volume on the permit market. In a summary of regulatory policy and utility compliance plans for 11 states (85 percent of phase I permits), Bohi (1993) observes that two-thirds of the utilities switch to low-sulfur fuels, one-sixth install flue gas desulfurization equipment, and only one-third hold allowances. Such figures in themselves do not, however, say anything about the allocative efficiency of the decisions. For the moment, the most important issue is the uncertainty about what the calculation rules are going to be. Many utilities have shown concern about buying and selling permits without knowing how their costs and revenues will be treated by PUCs (Rico, 1995). This inhibits trades because the PUCs wait for the utilities to trade and the utilities wait for the PUCs to formulate the rules for treating costs (Weissman, 1993). Although the majority of PUCs were not contemplating limiting the activities of utilities in an allowance trading market, most of the commissions had not yet decided how to treat costs and revenues of trades for rate-making purposes (Solomon and Rose, 1992). Uncertainty, in particular about the treatment of permits, might well induce utilities to install pollution control capacity and to hoard allowances for later use rather than selling them. This tendency is strengthened by the utilities' reputed conservatism and risk aversion (Hausker, 1992). As of March 1995, most PUCs have decided what they are going to do but are providing few incentives to engage in trading. Gains of trades are hardly shared with shareholders and are typically passed on to ratepayers in the current period. The costs of acquiring allowances are not placed in the rate base but are treated like fuel costs. This is not to say that uncertainty about the allowance market is not important. It appears to be less important, however, than the uncertainty related to the regulatory reform in the form of emerging wholesale and retail competition. This uncertainty makes utilities rather hesitant to undertake new capital investments (Burtraw, 1995). This seems to fit with what happens in practice where much fuel switching (to low-sulfur coal and to gas) is taking place and little new pollution control equipment is being built.

Transaction costs in terms of finding a trading partner and gaining approval seem not to be a problem. Although brokerage fees are higher than expected [around 5 percent according to the Wall Street Journal (24 August 1993)], they are much lower than under the EPA's emission trading policy, where ranges were between 5 and 30 percent (Foster and Hahn, 1994; Dwyer, 1992). Administrative approval of trades also does not seem to be a distorting factor.

In summary, there are indications that the market works well. Future tighter emission targets from 2000 onward are signaled by a rising price on the spot market for emissions and induce extra reductions of emissions until 2000. Some major remaining problems seem to be uncertainty about the treatment of permit costs and permit capital for regulated utilities and the possibility that the rules, insofar as they are clear, might induce overinvestment in flue gas desulfurization equipment in regulated utilities and underinvestment by independent electricity producers.

6.5.3 Environmental impacts, administrative practicability, and innovation

The question is to what extent the allowance trading program meets its objective of reducing the adverse impacts of acid deposition. This depends on both the level of the emissions reduced and their locations, because some areas in the USA are more sensitive to acid deposition than others. Regarding the total emission cap for phase I, there exists a loophole problem. Utilities can shift electricity production to plants that fall under phase II only and thus unexpectedly increase the emission of phase II units (USEPA, 1992c). As a result, SO_2 emissions could be 1 million tons higher than the goals of the CAA. To avoid this, the EPA now requires phase II boilers to report their generations and emissions during phase I and requires an offset of emission reduction to these shifts (Rico, 1995).

The risk of exceeding ambient air quality standards is nil. First, at present, even before the reductions planned under the 1990 CAAA, these standards are only exceeded in 4 out of 3,000 regions (Leaf, 1993). Second, the new CAA cuts emissions even further. Finally, although allowances can be freely bought or sold, their use can be prevented if ambient air quality standards are exceeded (USEPA, 1992d).

Whether the allowance trading results in the expected improvement in acid deposition in the sensitive areas is another question. First of all, the inclusion of the emission trading programs ensured that instead of a 7 or 8 million ton cutback a 10 million ton cutback in emissions could be realized (Goffman, 1993). This large-scale reduction, and the fact that sources must still do what states require to meet ambient standards, allowed the assumption that regardless of where the acid deposition occurs a large improvement in sulfate deposition will occur (Kete, 1992). Furthermore, studies suggest that given the distribution of marginal costs over the sources, trading will reduce emissions and deposition where they are needed to protect sensitive areas (NAPAP, 1991; Goffman, 1993). The latter, however, is questionable. Bohi (1993) suggests that none of the states plans interstate trading matching the output of the optimization model due to the constraints on trading. The three states that have interstate trading plans will take positions (buy/sell) opposite to those expected. A comparison of the interstate trading observed in Rico (1995) and the simulation by the US National Acid Precipitation Assessment Program (NAPAP, 1991) sheds some light. NAPAP (1991) suggests that nine states would show noticeable changes in emissions due to interstate trading: Indiana, Pennsylvania, Illinois, Kentucky, and Ohio would sell; Texas, Michigan, Florida, and Massachusetts would buy. Rico (1995), however, shows that of the five states expected to sell, three were net sellers in practice (Pennsylvania, Kentucky, and Ohio), and two (Indiana, Illinois) bought permits. Of the four NAPAP buyers, three did not trade at all (Michigan, Florida, Massachusetts) and one (Texas) was a seller. Consequently, reductions in deposition might not necessarily occur where they are most beneficial for the environment.

The trading program, however, has also enhanced the environmental effectiveness in a way that was not foreseen by the authorities. The banking of allowances results in an extra reduction in the period between 1995 and 2000 in exchange for higher emissions from 2000 on (see *Figure 6.3*). This can be valued positively from an environmental point of view, because it slows down the rate of acidification in sensitive areas.

Administrative practicability depends on information requirements and costs. *Table 6.3* shows that implementation costs, i.e., monitoring costs, are expected to be considerable. These costs would also have to be borne under a regulatory regime. Major additional administrative costs are the costs of the emission allowance tracking system (US$200 to US$400 million) and the permit costs. Implementation costs might be underestimated in view of the additional measures taken to close the loophole as this leads to additional, cumbersome paperwork associated with previous trading programs (Rico, 1995).

There is some evidence that the sulfur trading program stimulates innovation. The average removal efficiency of flue gas desulfurization equipment has increased and the design has become simpler because back-up installations, used when the desulfurization unit does not operate, are now no longer needed as allowances can be bought if emissions are higher due to maintenance problems (McLean, 1993; Burtraw, 1993). Moreover, increased cross-competition between low-sulfur coal and flue gas desulfurization equipment has decreased the costs (Burtraw, 1993). This is not only the result of decreased desulfurization equipment costs; it is also due to the increased availability of other options such as dispatching electricity, more conservation and efficiency management, and the supply of low-sulfur coals (Weissman, 1993).

6.6 Power Plants Quota in Denmark

6.6.1 The quota arrangements

By 1984, Denmark had accepted legislation that imposed a national bubble of 125 kton SO_2 on the emissions from its power plants to be attained in 1995. In 1987, a ceiling for the year 2005 was set at 85 kton (a 60 percent cutback from the 1980 level). For NO_x, a cap for the year 2005 was fixed at 60 kton. On the basis of ongoing international negotiations the original ceilings have been lowered. As of June 1993, the SO_2 quota for 1995 was 116 kton and for 2000 it was 73 kton. The NO_x quota is 85 kton for 1995 and 61 kton for the year 2000 (Elkraft, 1993). [The regulation guiding this is no. 885 of 18 December 1991 from the Environment Ministry (Miljoministeriets bekendtgorelse).] The objective of these quotas is to provide more flexibility to the power plants to meet emission reductions (Sørensen, 1993).

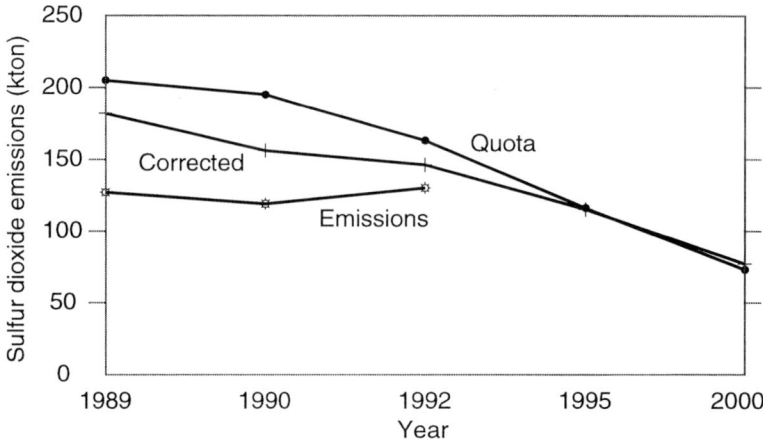

Figure 6.4. Sulfur dioxide quota for power plants in Denmark.

The Ministry of Environment fixes the quotas each year for the coming eight years after a review of plans submitted by the power companies. Two power plant consortia, Elsam and Elkraft, are involved. Quotas for the first four years are binding and quotas for the remaining four years are provisional. The quotas pertain to all plants with a capacity of 25 MW(E) or greater. Although quotas are fixed on an annual basis, the annual ceilings may be exceeded by a maximum of 10 percent as long as the cumulative emission ceilings for the four-year periods (1993–1996 and 1997–2000) are not exceeded (Elkraft, 1993). The allocation of the annual national ceilings between both power plant companies is left to the two companies. They can rearrange emission reductions (Sørensen, 1993). This could be seen as a first step toward an emission trading program: the setting of a quantitative goal for the emissions. The second step, allocating quotas to individual firms, is actually not made. The rules governing these quota arrangements can be summarized as follows. In principle the two companies are left free to comply with the annual ceilings. However, the quotas are import/export neutral; they are set on the basis of zero net import of electricity. If net import of electricity occurs, the emissions that would have occurred if the electricity now imported had been produced in Denmark are added to the actual emissions to calculate corrected emissions. These corrected data have to be below the quotas (see *Figure 6.4*). Permanent import of electricity will lead the Ministry to consider a decrease in quotas. Permanent net export of electricity in principle will not lead to an increase in the quotas. Moreover, emission standards for new plants as laid down in the EC directive on large combustion plants and in Danish regulations for NO_x also have to be met (Sørensen, 1993). Finally, compliance with national ceilings that are set for existing large combustion plants under the same EC directive is also necessary.

Regarding monitoring and enforcement, each year the electricity companies have to submit reports containing information on past and future development of emissions as well as foreseen pollution control measures. Local authorities are responsible for controlling the actual emission levels.

6.6.2 Cost savings and environmental impacts

The power companies estimated that up to the year 2000 cumulative investments will amount to 6.3 billion Danish krone (1992 price level) to meet the current quotas. Annual operating and maintenance costs (excluding capital costs) will be around 250 million krone (1992 price level) in 1995 and will increase to around 360 million krone in the year 2000 (Elkraft, 1993). Estimates of the cost savings of the bubble concept are not available. Savings are expected, because the power companies can shift to low-sulfur fuels or other fuels (natural gas, biomass) in addition to applying flue gas desulfurization and other technical measures (Sørensen, 1993).

The structure of the sector may have some influence on the quest for cost-efficient measures for meeting the national quotas. The Danish energy law allows the inclusion of all costs in the electricity price (Ølsgaard, 1994), although prices are controlled by the electricity price committee (UNIPEDE, 1993). This suggests the existence of a so-called soft budget constraint. This means that the strict relationship between expenditure and earnings of an economic unit is relaxed, because extra expenditures are covered by others (typically the state) (Kornai, 1986). Such a constraint might reduce incentives to minimize costs. Furthermore, the shares of the formally private power companies are generally either owned by municipalities, among them the city of Copenhagen, or by consumer-owned cooperatives (UNIPEDE, 1993). They might have objectives other than pure cost minimization. The electricity distribution companies form a de facto monopoly without effective market competition (UNIPEDE, 1993). In spite of the semi-public character, the lack of competition, and the cost mark-up policy, the power companies do face incentives to keep electricity prices at low levels. Denmark's electricity supply system is connected to that of its neighbors. Currently, the electricity price is between those of Germany and Sweden. If the price were lower, less electricity would be imported. Also, some sort of price competition does seem to exist because producers do not want consumers to regard their company as being expensive (UNIPEDE, 1993). In order to minimize electricity production costs the companies operate power plants in merit order to minimize operating costs (Elkraft, 1993, Appendix B, p. 7; Ølsgaard, 1994).

Regarding environmental effectiveness, available data suggest that the actual emissions for 1992 for both sulfur and nitrogen oxides were around 10 percent below the allowed quota. In previous years emissions were also below the quota (see *Figure 6.4*), and they are expected to be lower than the quota in the future (Elsam/Elkraft, 1993). One of the goals of power companies is to ensure that the quotas are met because they want to avoid negative publicity. This, combined with uncertainties inherent in planning electricity production, implies that in practice emissions might turn out to be lower than the quotas allow (Ølsgaard, 1994).

Little is known about the administrative practicability of the quota system and its impact on innovation. The quota system might be less practical than emission standards due to the requirement of additional meetings between the environmental Ministry and the power companies (Sørensen, 1993).

6.7 Sector Covenants in the Netherlands

6.7.1 The covenant with the electricity producers

In 1990, the state of the Netherlands and the 12 provinces signed a covenant with the association of electricity producers [Samenwerkende Electriciteits Producenten (SEP)] and the four individual electricity producers on the reduction of sulfur and nitrogen oxides emissions. The covenant is an agreement between the parties to further reduce acidifying pollution up to the year 2000 (SEP, 1991). The crucial point is that the covenant establishes ceilings on the total emissions of both pollutants from the public power plants, which results in lower emissions than would have resulted from existing legislation in the form of emission and fuel standards. The motive for the ceilings is that they offer the electricity producers more possibilities to reduce emissions in a more cost-efficient way than requiring similar emission standards for classes of installations (SEP, 1991).

The objective of the covenant is to reduce SO_2 emissions to 18 ktons and NO_x emissions to 30 ktons in the year 2000. In 1989, SO_2 emissions of SEP were 41 ktons and NO_x emissions were 74 ktons (SEP, 1991). Although there are four electricity producers, decisions on the fuel use and the utilization of the power plants are made by the SEP. The costs of production are also settled by the SEP and distributed among the producers. Electricity generating companies and almost all distribution companies are joint stock companies. Distribution companies, provinces, and local authorities own the generating companies. Public authorities hold the shares of the distribution companies (UNIPEDE, 1993). In a sense the bubble implies intrafirm rather than interfirm trading, because the SEP operates as a kind of holding company.

The SEP can implement the bubbles within certain limitations:

Emission Trading for Air Pollution in Practice 161

- Existing power plants have to meet the existing emission standards.
- Relatively new plants have to meet the more stringent emission standards laid down in the covenant.

The SEP can increase the NO_x cap by 5 million kg if the SEP also supplies heat from combined heat and power generation equivalent to 1250 MW(E). If the actual heat supply is less, the allowed increase is reduced proportionally. The sulfur cap can be increased by 4 million kg if flue gas desulfurization units are out of operation due to technical problems but operate within the applicable legal framework. These corrected ceilings can be exceeded once every three years by a maximum of 3 million kg. Regarding enforcement and monitoring, the SEP has to produce a plan for reducing the emissions. The plan must be judged by an expert commission consisting of the Ministry of the Environment, the SEP, and the provinces. The SEP can alter the plan if this improves cost efficiency. The SEP has to report to the commission every two years. The individual producers have to report to the provinces and to the SEP. The provinces agree to implement the covenant, but can defer if necessary in order to meet air quality standards other than those meant for the reduction of acidification. Parties can alter the covenant in the case of unexpected environmental changes. The ceilings can be altered if electricity demand or imports depart substantially from what was planned. The parties can also dismiss the covenant if they cannot agree on the emission reduction plans or if the ceilings cannot be met by any reasonable means (SEP, 1991). As in the Danish quota system, one can observe that quotas are not allocated to individual firms, but a total bubble is fixed for the SEP.

6.7.2 Is the SEP covenant cost efficient?

According to the electricity producers the covenant is expected to cut costs by 500 million Dutch guilders compared with setting stricter uniform emission standards to meet the same ceiling. The latter would have raised costs to 1 billion guilders. This implies a 50 percent reduction in additional costs. The cost savings are achieved by applying more expensive pollution control equipment that reduces emissions below the required emission standards at those installations that have a longer remaining lifetime, that have more operating hours per year, or that have to be upgraded anyway (Lubbers, 1993). One might have some doubts, however, about the cost-minimizing behavior of the SEP. There is virtually no competition between the electricity generators. Only a limited form of competition exists because distributing companies can shop among generating companies for the one with the lowest costs. Large industrial consumers may choose any distributor (UNIPEDE, 1993).

Regarding environmental effectiveness, the covenant states that in order to reduce acidification it is especially important to reduce the total volume of emissions in the Netherlands (SEP, 1991). The environmental impacts of acidification are continental rather than local. Given the fact that every plant has to meet stringent emission standards anyway, it does not matter that much where these emissions are reduced (Lubbers, 1993). Moreover, the covenant stipulates that emission reductions have to be distributed equally over all regions in the Netherlands, accounting for the southwest, the most acidified part of the country. Finally, the provinces can still require more stringent measures at certain plants if it is deemed necessary to meet local air quality standards. This suggests that environmental impacts depend not so much on the location of the emissions but more on the total volume. If this is the case, the question arises whether the covenant can meet the total emission ceiling. On the one hand, the covenant offers several possibilities to exceed the caps (such as the special cases regarding heat supply, other electricity forecasts, and nonoperating flue gas desulfurization units) and the SEP eventually can back out of the agreement, which raises some doubts about the covenant's environmental effectiveness. On the other hand, the state could probably threaten the SEP with the more expensive alternative of setting more stringent uniform emission standards in new legislation and could also lower the caps in case of unexpected environmental changes. In view of the fact that the shares of the producing companies are owned by the regional, provincial authorities, one might expect them to do their best to meet the ceilings.

6.7.3 Bubble and averaging for refineries

No covenant exists for refineries, but an agreement was made between the Ministry of the Environment and the refinery sector that capped the total SO_2 emissions of the refinery sector at 36 ktons per year from the year 2000 onward. A type of averaging concept was introduced earlier, when emission standards for large combustion plants were set. These emission standards impose a limit of 1,000 mg SO_2/m^3 flue gas for the sum of fuel-related and process emissions from each refinery. This implies that some combustion plants within a single refinery may overcomply and others may undercomply as long as the average standard is met. The level of 36 ktons will be met exactly if each refinery meets its individual emission standard of 1,000 mg SO_2/m^3 flue gas and if oil input remains at the 1980 level. The cap thus ensures that the total emissions of all refineries will remain at their 1980 level, even if oil input increases.

The distribution of the 36 ktons is left to the five companies, which together own six refineries. Each refinery still has to meet the emission standard of 1,000

Emission Trading for Air Pollution in Practice 163

mg SO_2/m^3 flue gas. Moreover, this average emission standard will be lowered to 500 mg in the near future (Dekkers, 1993).

6.7.4 Cost savings and environmental impacts of refinery bubbles

Estimates of the cost savings of bubble and averaging concepts are not available. The average emission standard, rather than the cap, probably affords the greatest flexibility for reducing costs. The cost savings are refinery specific, because some refineries contain 30 plants whereas others contain only 2 (Dekkers, 1993). Gaasbeek (1993) doubts whether the average standards give much flexibility, because refineries have fewer installations than do electricity producers. Given the actual 1993 fuel input of 3.5 to 3.7 million tons of oil, the 36 kton cap implies an actual average standard of around 800 mg SO_2/m^3 instead of 1,000 mg SO_2/m^3 (Dekkers, 1993). This suggests that the cap would give some flexibility and would allow one refinery to cut emissions further while allowing other refineries to increase emissions up to the standard. This, however, seems not to have been the intention of the bubble, and one should recall that no distribution of emission permits to individual refineries has taken place. In practice, inter-refinery trading does not seem to occur. Some companies (e.g., Esso) simply have average emission levels far below the standard (for example, 350 mg SO_2/m^3), mainly for economic reasons. The installation of catalytic and thermal crackers, such as the Flexicoker by Esso, has led to an increase in the production of light oil fractions with higher market values. As a by-product, sulfur emissions are reduced and the cap is met. Lowering the emission standard to 500 mg SO_2/m^3 given the same fuel input and quality as in 1993 would imply that total emissions would be 22.5 kton (five-eighths 36 kton). Actual emissions in 1992 and 1993 were around 60 kton. In this case the new, average emission standard, not the cap, would be the binding constraint. Consequently, the cap would not give any incentive for interfirm trading of emission permits as each refinery must meet the average standard. The new standard would reduce the demand for emission reductions by other firms to zero, thus only allowing cost savings through intrafirm trading. As in the Danish and Dutch power plants bubble, no individual emission rights are allocated to individual firms. Therefore, there are no clear incentives for individual firms to reduce their emission levels below the average emission standard or below what would result from profit-based decisions (such as using the Flexicoker), because it is unclear which part of the overcontrol could be sold to other companies. So these types of bubbles do set a quantitative goal, but they miss the transferability of individual property rights essential for the creation of a tradable emission permit market.

The environmental effects of the bubble are not that clear. Given the increased fuel input compared with 1980, the cap would prevent emissions from increasing above their 1980 level and would lower the effective average emission standard from 1,000 to 800 mg. However, with the further reduction of the average standard and the lowering of sulfur emissions due to innovation for economic reasons, emissions might already be reduced below the cap. If this were to hold, the cap would no longer be a binding instrument and would only become an actual ceiling if fuel input were to increase considerably. Monitoring does not seem to have caused any problems, because total emissions can easily be calculated on the basis of fuel input and sulfur content. Adding process emissions and accounting for catalytic crackers gives total emissions (Dekkers, 1993).

The administrative practicability of both the average emission standard and bubble are believed to be high. One single standard for each refinery requires less administration than separate standards for every single installation, as is the case in Germany. The concept of an average standard also fits better in the existing regulatory framework than it did in the previously prescribed average sulfur content of 2 percent S (3,400 SO_2/m^3) in each refinery's fuel cocktail (Dekkers, 1993).

6.8 Offsets in Germany

6.8.1 Plant renewal clause and compensation rule

In Germany, air pollution legislation allows the transfer of emission reduction obligations. Two rules exist: the plant renewal clause and the compensation rule. Operation of these rules should be seen in light of the two-tier strategy aimed at maintaining air quality standards on the one hand, and minimizing emissions as far as technically possible on the other (Schärer, 1993).

The plant renewal clause pertains to the construction of new plants in areas where air quality standards are exceeded. In principle, the Federal Environmental Protection Law (Bundesimmissionsschutzgesetz) and the Technical Guidelines for Air Pollution Control rule out construction of such plants, even if these plants meet the state-of-the-art emission standards. The 1974 technical guidelines, however, do allow the construction of a new plant in a non-attainment area if the new plant replaces an existing plant of the same kind. These plants do not have to belong to the same firm, but they must be located in the same area. In 1983 these guidelines were extended: not only the closing of an existing plant but also its renovation can be used to offset the additional emissions of a new plant. The offsets, however, have to lead to a reduction in the annual average concentration in the area and the new plant must meet the state-of-the-art emission standards (Sprenger, 1989; Opschoor and Vos, 1989).

The compensation rule was included in the 1986 revision of the technical guidelines. As part of the revision, existing installations had to be modernized to

Emission Trading for Air Pollution in Practice

meet the stricter emission standards, usually within five years. The core of the compensation rule is that the clean-up period can be extended to eight years if emission reduction measures taken at existing installations (by the firm or by other firms) will provide more emission reductions than would otherwise result from the application of the technical guidelines for each individual plant. This compensation rule can only be used by installations within the same geographical area of impact and for the same pollutants or for pollutants with comparable impacts (Schärer, 1993).

6.8.2 Cost efficiency and environmental impacts of renewal and compensation rules

The cost efficiency of the plant renewal clause is limited because air quality standards are only exceeded in a few areas. The rules of the clause further restrict its cost efficiency. The clause can only be used if the additional impact of the plant on ambient concentrations is limited to 1 percent, if the additional emissions are offset by emission reductions of plants in the same area, and if the new plant starts operating after the improvement of the existing plant becomes effective. Furthermore, the clause does not apply to the location of new plants in attainment areas, even if this would lead to exceeding air quality standards (Sprenger, 1989).

The contribution of the compensation rule to cost savings was small. The rule was used in only 50 out of 17,000 clean-up cases (Schärer, 1993). The most important reasons for the restricted use are the following (Schärer, 1993; Sprenger, 1989):

- The short time limit for approval of renewal plans (one year);
- The fact that emission reduction requirements were very strict, because the total emission reduction had to be higher than the reduction otherwise achieved by applying the strict emission standards;
- The necessity of multiple trades, because most new firms emit more than one pollutant in the air;
- The small size of the areas in which offsets are allowed.

The environmental impact of both the renewal clause and the compensation rule is neutral to positive. The renewal clause prevents increases in emissions in non-attainment areas. The compensation rule can only be used if emissions are reduced further (Sprenger, 1989).

No data are available on the administrative requirements. Sprenger (1989) believes that the plant renewal clause stimulates innovation but does not supply empirical evidence.

6.9 Transfers under the Montreal Protocol

6.9.1 Regulating the transfer of ozone-depleting substances

In 1974 scientists were already advancing the theory that CFCs might damage the ozone layer. This damage would allow more ultraviolet radiation to reach the earth, causing increases in skin cancer and mutagenic effects and inhibiting plant growth (USEPA, 1992a). In Montreal, in 1987, a number of countries signed an agreement to control the substances that deplete the ozone layer (UNEP, 1987). This protocol established that both the production and the consumption of CFCs and Halons should be reduced to their 1986 levels. Subsequent reductions of 20 and 50 percent were agreed to for the early 1990s. Responding to new evidence, the parties met in London in 1990. They agreed to completely phase out the production and use of CFCs by the year 2000 (UNEP, 1993a). In addition, three other groups of chemicals were added to the phaseout. In 1992, in Copenhagen, phaseout schedules were adjusted and again new substances were added. Halons were phased out in 1994, and CFCs, methyl chloroform, and carbon tetrachloride are to be phased out in 1996 (UNEP, 1993a). As of August 1993, there were 124 parties to the protocol, 67 ratified the London and 6 the Copenhagen amendments. The protocol regulates both production and consumption of the substances. Consumption is defined as production plus imports minus exports.

Interestingly, the Montreal Protocol allows parties to transfer production quotas for CFCs to other parties for the purpose of industrial rationalization. Industrial rationalization means the transfer of all or part of the calculated level of production of one party to another for the purpose of achieving economic efficiency or in response to anticipated shortfalls in supply resulting from plant closures (UNEP, 1987). The production quotas are "grandfathered" to the parties (countries) on the basis of their 1986 levels of production and consumption. The so-called Article 5 parties (developing countries) have to meet deadlines 10 years later and may increase their levels of consumption up to a maximum of 0.3 kg per capita.

In the original Montreal Protocol (UNEP, 1987), the consumption and production of CFCs and Halons is cut in three steps: standstill (1992), a 20 percent cutback (1993), and a 50 percent reduction (1998). However, in order to satisfy the basic domestic needs of Article 5 parties and for the purpose of industrial rationalization, the production levels agreed to may be exceeded by up to 10 percent of the base-year (1986) production levels in 1992–1993, and by 15 percent in 1998. The following rules for the transfers of production quotas apply:

- The total combined levels of production of the parties concerned are not allowed to exceed the production limits agreed to.

Emission Trading for Air Pollution in Practice   167

- The transfers are not allowed to increase the individual production levels by more than 10 to 15 percent of the base-year (1986) level, except for parties whose level of production of CFCs was less than 25 ktons.
- The secretariat must be notified of transfers no later than the time of the transfer.

Transfers of consumption quotas are only permitted for certain parties: "Any Parties which are Member States of a regional economic integration organization ... may agree that they shall jointly fulfill their obligations respecting consumption ... provided that their total combined calculated level of consumption does not exceed the levels required" (Article 2, UNEP, 1987, p. 860). This agreement pertains to the European Community (EC). The parties to such an agreement must inform the secretariat of the terms of the agreement.

With the London amendments, which went into effect August 1992, the transfer of production quotas was extended to all pollutants and the 10 to 15 percent restriction was removed. Parties may now transfer any portion of their calculated level of production of any of the controlled substances provided that the total level of production does not exceed the limits agreed to. Furthermore, the secretariat must be notified of the transfers, their terms, and the period for which they apply (UNEP, 1993a).

Regarding monitoring, the Montreal Protocol requires parties to provide the secretariat with data on the production, imports for the base year, and exports of the controlled substances for the base year and for every year thereafter. In addition, amounts destroyed or used as feedstocks are to be reported (UNEP, 1987; UNEP, 1993a).

Furthermore, domestic laws may influence international transfers. The regulations in the USA and the EC are especially relevant as roughly 70 percent of the CFCs and Halons were produced in these regions (UNEP, 1993b). The US regulations have already been described in this chapter. The EC regulations are simple. Producers are allowed to exceed their allocated production levels either domestically, within the EC, or with any other party to the protocol provided that the sum of the calculated production levels does not increase. In the case of international transfers, both the EC and the member state have to agree. Consumption is controlled through the control of supply. Both production and consumption quotas are allocated to producers on the basis of base-year levels (1986 or 1989, depending on the substance controlled). Consumption rights are only transferable to other producers within the EC and the EC has to be notified. Producers, importers, and exporters have to report data to both the EC and the member states on production, quantities recycled or destroyed, stocks, imports, exports, and amounts placed on the market (EC, 1991). This information is audited by an independent consultant and enforcement is up to the member state (Peaple, 1993).

6.9.2 Cost savings of international transfers

Estimates of the cost savings of the transfer provisions in the Montreal Protocol are not available and evidence on the actual transfers is scarce. As of January 1994, the ozone secretariat had not received any notification of transfers of production levels (Sabogal, 1994). This does not imply that such transfers did not take place. A few international transfers of production allowances involving US companies did occur, but the volume is confidential (Stavins and Hahn, 1993; Voight, 1993). During the past two years (1992–1993) around 20 production transfers involving EC-based companies took place (Peaple, 1993). For the four transfers involving companies based in the Netherlands the average volume traded was several thousand tons (Quisthout, 1993). Price information is not available. This suggests a traded volume of 20,000 to 30,000 tons per year involving EC companies. *Table 6.5* summarizes the estimates of transfers and compares these with production data for 1986 (the base year) and 1991. Compared with production in the base year (the allowable quota), roughly 5 percent of the production has been transferred. Compared with the actual production in 1991, transferred amounts were twice as high because actual production was considerably lower than in 1986. The number of transfers has increased over the last year because allowable production levels were cut back considerably (75 to 85 percent) in 1994 and a complete phase-out is scheduled in both the USA and the EC for the year 1995 (USEPA, 1993; Peaple, 1993). The principal reason for the transfers is the cost savings involved in concentrating the remaining production (Quisthout, 1993). The production of CFCs is subject to considerable economies of scale (Dudek and Palmisano, 1988).

Transaction costs are low because there is little administration involved and the chance that transfers will be accepted is high, although some delay may occur (Peaple, 1993). Delays or denial of approval are more likely when transfers pertain to Article 5 parties (developing countries). For these countries, production limits are applicable only from 1999 onward; before 1999 no explicit production (or consumption) rights will be allocated to these countries. This has created uncertainty, both about the conditions under which Article 5 countries can transfer production rights to non-Article 5 parties and about the volumes that can be transferred to other parties in order to meet the basic domestic needs of the Article 5 parties. These issues are being examined further (UNEP, 1993c). The fact that with the exception of the EC, consumption quotas cannot be traded might have had a negative impact on cost efficiency. This is because although production can take place where needed since production quotas are fully transferable, global abatement of CFC consumption cannot take place where abatement costs are expected to be small (Bohm, 1990). Voight (1993) confirms that this has had a negative impact on market activity. On the other hand, one might question the relevance of this limitation

Table 6.5. International transfers of ozone-depleting substances.

Region	Production[a] (1,000 tons) 1986	Production[a] (1,000 tons) 1991	Transfers (1,000 tons/yr) 1992/1993	Transfers (% of production) 1986	Transfers (% of production) 1991
EC-12	415	194	20–30[b]	5–7	11–15
USA	337	131	35	10[c]	27
Other non–Article 5 parties	142	127	0	0[d]	0
Article 5 parties	19	33	0	0[d]	0
Other[e]	208	208	0	0[d]	0
Worldwide	1,121	693	55–65	5–6	8–9

[a] Production of CFCs and Halons (annex I substances). Calculated from UNEP (1993a, 1993b).
[b] Based on 20 transfers during two years (Peaple, 1993) and an average amount transferred of several thousand tons (Quisthout, 1993), giving roughly 40,000–60,000 tons during two years.
[c] Voight and Lee (1993) suggest that 10 percent of the allowances (equal to the 1986 level of production) have been traded.
[d] Sabogal (1994).
[e] Includes parties that did not supply their own data and non-signatories (only 0.5 percent of total).

because although EC companies were allowed to trade consumption rights, they did not make use of this in order to keep the contacts with their customers (Peaple, 1993). International trades by US companies have certainly been negatively affected by the US regulation that reduces the production allowed to the transferor by not only the amount transferred, but also by the difference between allowable and actual US production. This impact is believed to disappear when allowable production comes closer to actual production (Voight, 1993). The extent to which market power has affected the cost efficiency of the transfer trading is unknown, although one might expect this to happen because production is concentrated in a few companies. Du Pont de Nemours serves 25 percent of the world market (Automotive News, February 6, 1989, p. 4) and ICI's market share is 15 to 20 percent (Arkansas Gazette, September 29, 1988). In the USA there are 7 producers of CFCs and in the EC there are 15 (Hahn and McGartland, 1989; Peaple, 1993). The extent to which companies exerted market power to drive up permit prices for competitors is unknown because market power depends in part on the initial distribution of permits relative to the optimal distribution. The latter is not known and price information is also absent. One may doubt the relevance of market power as the CFC market is being phased out anyway and companies appear to be more interested in producing more profitable substitutes for CFCs and Halons (Maxwell and Weiner, 1993, p. 25). Driving up the market price for CFC permits will only accelerate this substitution.

In sum, the international market for CFC production seems to have been active: 5 percent of the permits, equaling roughly 10 percent of the actual production, seem to have been transferred. The associated costs savings are unknown. Transaction costs are low, pointing to a well-functioning market. Three factors had negative impacts on the international market and impaired cost efficiency: US restrictions on international trades and, to a lesser extent, the fact that consumption transfers are not allowed, as well as the uncertainty surrounding transfers from Article 5 countries, which have no settled production rights.

The trading provisions in the Montreal Protocol do not seem to have had a significant impact on the achievement of the Protocol's environmental goals. Problems might be that Article 5 (developing) countries can increase their production limits by 10 to 15 percent to meet basic domestic needs and that they are allowed to delay meeting compliance deadlines for 10 years (UNEP, 1987). This implies that in the absence of any domestic legislation these countries at this moment (1994) have no binding production quotas. This could allow Article 5 countries to increase their unsettled baseline production level, to transfer this to other countries, and to effectively increase worldwide production. In practice, transfers with Article 5 parties appear to bear the risk of being rejected instead of leading to an increase in production (Peaple, 1993). Moreover, actual production levels in industrialized countries (at least in the USA and probably also in the EC) are below allowable levels because producers are rapidly shifting to substitutes. Although transferability of production quotas might lead to an increase in actual production, this does not seem to have occurred.

The administrative practicability of the transfers seems to be great. Little additional information is required because data on production and imports, etc., have to be checked anyway (Quisthout, 1993).

6.10 Conclusions

This chapter examined the cost efficiency, environmental impacts, administrative practicability, and innovative impacts of a number of emission trading schemes implemented in practice in the USA (the EPA's emission trading policy, lead trading, and sulfur allowance trading), in Europe (Danish and Dutch bubbles for power plants and refineries, and the German plant renewal clause and compensation rule), and worldwide.

The general conclusion is that these emission trading schemes saved costs, did not impair environmental quality, had little or an unknown impact on innovation, and required more administration than regulation (*Table 6.6*).

In terms of both market activity and cost efficiency,

Emission Trading for Air Pollution in Practice 171

Table 6.6. Evaluation of trading schemes.

	USA				Europe			World
	EPA	Lead	CFCs	Sulfur	Denmark	Netherlands[a]	Germany	Montreal
Permits traded[b]	<1	20–60	10	>7	NA[c]	NA/0	<1	5–10
Cost savings[d]	<4	15–20	40?	40	–	50/–	–	–
(billion US$)	1–13	0.3	–	9–13	–	0.3/–	–	–
Environmental effectiveness[e]	0	0/+	0	•/0/+	0	0/0	0	0
Administrative costs[e]	•	•	0	•	•	–/+	–	0
Innovation[e]	+	–	–	+	–	–/–	–	–

[a]Data on the left-hand side refer to the bubble for the electricity producers, data on the right-hand side refer to averaging and bubble for refineries.
[b]Percentage of the allocated permits or total emission per year.
[c]NA = not applicable (only intrafirm trading).
[d]Percentage of the accumulated pollution control costs under a regulatory regime.
[e]+ = more environmental protection, less administrative burden, more innovation; • = less protection, more administration, less innovation; 0 = neutral impact compared with a regulatory (CAC type) regime; – = no data available.

- Lead trading and the sulfur allowance trading market are successful; markets were active and cost savings were high.
- Performance of the EPA emission trading program and the German offset and compensation rule has been poor, although the absolute cost savings of the EPA's policy are impressive.
- The cost efficiency of CFC trading in the USA and under the Montreal Protocol is difficult to judge, but low transaction costs suggest that the market functions well.
- The Dutch and Danish bubbles only created room for intrafirm trading, which for the Dutch power plants cut the additional costs of more stringent limits by 50 percent.

Major circumstances affecting the extent to which cost efficiency was achieved are

- Regulatory constraints
- Transaction costs
- Uncertainty on the property rights that a tradable permit represents
- Market structure

Regulatory constraints in the form of emission standards limited permit supply and demand under the EPA trading policy, in Germany, and under the Dutch and Danish bubbles. Regulatory air quality constraints may consist of air quality modeling, trading in small zones, and offset rates exceeding one. Such constraints have impaired the functioning of the permit market in the USA (EPA trading) and in Germany.

Transaction costs have been high under the EPA trading program and in Germany, mainly due to the complicated administrative approval procedure including air quality modeling. In the absence of case-by-case approval or, conversely, in the presence of quick approval and a clear, up-front allocation of rights, trading has been more successful (for example, lead and sulfur trading).

Uncertainty about the value of property rights has depressed pollution rights markets as well. Uncertainty about the initial distribution of rights (the baseline emissions) depressed market activity in the EPA program and in Germany. In these cases baseline emissions and rights had to be determined on a case-by-case basis. In the Netherlands' refinery bubble, allowable emissions were set for a group and not for individual polluters, thus reducing incentives for individual refineries to reduce emissions and cut costs. Policy uncertainty and frequent changes (among them the confiscation of banked rights) plagued trading in the case of the EPA's trading program.

Finally, the structure of the market in which the polluting firms operate is a factor. The fact that some polluters are regulated monopolies implies that incentives to minimize costs might not be as strong. The state-regulated monopoly character of the electricity sector in the USA and the regulatory treatment of the revenues of emission permit sales compared with other options have so far created a bias against emission trading. In spite of this, the market functions well. In Denmark and in the Netherlands, the soft budget constraint and especially the semi-public character appear to limit incentives for power plant companies to minimize costs because other objectives, such as meeting the emission ceiling to avoid negative publicity, are inherently relevant as well. "Trade," insofar as it exists, is "intrafirm," and a genuine market does not exist.

The environmental impacts of the emission trading schemes examined have been neutral. In some cases the promised cost savings of emission trading led to lower ceilings than would have been feasible without trading (sulfur in the USA) or accelerated compliance (lead/EPA trading). In other trading schemes not every trade achieved air quality objectives as intended (EPA trading). This neutral impact of trading on environmental quality is, to a certain extent, also due to the same factors that limit the cost efficiency of trades: air quality modeling, small zones, offset rates above one, and case-by-case approval (Germany and EPA trading). In other cases (US sulfur trading, Danish and Dutch electricity bubbles), the stringent reductions combined with emission standards throughout the country turned local air quality concerns into a relatively minor problem and allowed attention to be shifted to total emissions (rather than ambient concentrations), thus simplifying policy design. Monitoring these total emission levels can be simple if the number of sources is rather small (lead trading, bubbles in Denmark and Holland), or more complex if there is a large number of smaller sources (US sulfur trading).

The flexibility promised by these economic incentive schemes has a small price: administrative costs are higher due to emission registration schemes and enforcement problems (EPA, US sulfur and lead trading), or due to increased consultation (Dutch/Danish bubbles). An exception is the average emission standard for refineries in the Netherlands and netting under the EPA's emission trading policy, which reduced administrative costs. At times high enforcement penalties and continuous emission monitoring reduce enforcement problems (US sulfur trading). No firm data were obtained to draw conclusions on the effect emission trading schemes had on innovation, although some examples show that innovation was stimulated.

In summary, high transaction costs, uncertain property rights, the presence of regulatory constraints, and state-regulated monopolies have restricted the cost savings of emission trading. In spite of these suboptimal conditions, emission trading has saved costs without jeopardizing environmental quality. It is clear that fewer restrictions (lead, sulfur, Dutch bubble) lead to higher cost savings of trade (see *Table 6.6*).

The principal lessons that can be learned from this overview appear to be the following:

- Emission rights preferably should be allocated to individual polluters before trading starts rather than being allocated to a sector as a whole or being negotiated on a case-by-case basis. This requires a clear, quantitative environmental objective.
- Emission rights should be clearly defined and secured and should be homogeneous.
- To reduce transaction costs, case-by-case approval of each individual trade and the associated use of models to calculate the impact on air quality should be avoided.
- Banking of rights should be promoted because this gives polluters more time to find trading partners, allows more flexibility, and may lead to earlier reductions in emissions.
- The obligatory use of emission standards on top of emission trading should be avoided whenever possible to improve cost efficiency.
- To avoid locally high concentrations ("hot spots"), several possibilities exist:
  - The potential cost savings of emission trading could be used to cut emissions further.

- In particular circumstances, the use of an acquired permit to actually increase emissions could be blocked without taking away the right to sell the permit or to bank it for future use.
   - Minimum emission standards could be required at every location.
- Monitoring and enforcement policies and a system of registration of trades must be in place before trading starts and must be stringent enough to discourage abuse.
- The market should be big enough to include a sufficient number of firms that can be expected to operate as cost-minimizing units.

Clearly, it might not be possible to follow these recommendations in every context. However, the more they are adhered to the greater is the chance that tradable emission permits can be used to meet environmental goals without squandering scarce resources. Before we draw attention to how we can use these practical lessons for the application of sulfur emission trading in Europe (Chapters 8 to 11), Chapter 7 will pull together the principal conclusions from the theoretical, modeling, and practical survey described in Chapters 2 to 6.

Chapter 7

Theory, Models, and Practice: Concluding Observations

7.1 Introduction

The previous chapters examined how, and the conditions under which, economic instruments, particularly emission trading, can improve the cost efficiency of environmental policy, especially for non-uniformly dispersed pollutants. The findings of economic theory, empirical simulation models, and actual experience were reviewed in both domestic and international contexts. This chapter compares the results obtained from theory, simulation models, and practice. First, the cost efficiencies of different emission trading schemes are analyzed. Second, the conditions affecting the cost efficiencies are evaluated. The last section is devoted to the environmental effectiveness, the administrative practicability, the dynamic efficiency, and the distributional consequences of tradable emission permits.

7.2 Cost Efficiency of Emission Trading

Cost efficiency implies that environmental goals will be attained at minimum costs. If the environmental objective can be translated into a single emission ceiling (such as with global CO_2 emissions), the marginal costs of pollution control need to be equalized to minimize costs. If the environmental objective consists of a set of ambient standards in the form of concentrations or depositions at certain receptors (as is the case with SO_2), marginal pollution control costs will generally be different between sources in the least-cost solution.

In theory, when dealing with uniformly dispersed pollutants, emission charges, emission standards, or emission trading can meet the emission goal in a cost-efficient way with perfect knowledge of costs. If costs are uncertain, charge levels will have to be set in a trial-and-error way to meet the goal. Emission trading guarantees that the emission ceiling will be met at minimum costs if sources minimize costs and operate on a perfect permit market with perfect enforcement. Regulation is generally inefficient.

With non-uniformly dispersed pollutants, environmental policy becomes more complicated. Again, ambient charges or ambient permits can, in theory, meet a set of receptor-specific standards in a cost-efficient way if a set of perfect deposition markets can be installed.

Simulation models confirm that ambient permit trading is cost efficient, but only because these models assume that ambient permit trading results in the least-cost solution. Theoretical analysis also suggests that the transaction costs of ambient permits are high, because sources have to operate simultaneously on different markets. In practice, neither ambient permits nor ambient charges have been implemented.

Alternatives exist in the form of trading rules such as pollution offset trading, modified trading, and non-degradation offset trading (see *Table 7.1*). The first allows trades as long as the ambient standards are met. The second does the same, but does not allow any increase in concentrations. The non-degradation offset rule requires that the ambient standards are met and prevents increases in the total level of emissions after trade. In theory, pollution offset and modified offset trading will minimize costs, but they are also expected to have high transaction costs, especially if the number of receptors is large. Simulation models confirm the cost efficiency of these trading rules but also indicate that a large part of the potential cost savings may not be achieved if information on costs is imperfect and if trading is bilateral and sequential rather than simultaneous. In practice, neither the pollution offset nor the modified offset rule has been implemented. The non-degradation offset rule was applied as part of the EPA's emission trading policy and the German renewal clause. These applications did save costs but they were not very cost efficient because the transaction costs associated with administrative approval were too high.

As simpler counterparts, single-receptor trading, single-zone emission trading, and multiple-zone trading are possible. In theory, these schemes are not fully cost efficient. Simulation models suggest that single ambient permit trading might come close to a cost-efficient solution, but without a guarantee that the ambient standards will be met at every location.

These models also indicate that whether or not single-zone emission permit trading is more cost efficient than its CAC counterpart depends on regional characteristics such as cost differences and atmospheric dispersion. With imperfect information on costs, single-zone trading does not guarantee that ambient standards will be met at all times. Multiple-zone emission trading might come close to

Concluding Observations

Table 7.1. Cost efficiency of different emission trading schemes.

Scheme	Theory (Chapters 2 to 4)	Models (Chapter 5)	Practice (Chapter 6)
Ambient permit	Yes, if ...	Yes, but ...	Not applied
Trading rules			
Pollution offset	Yes, but ...	Yes, if ...	Not applied
Modified offset	Yes, but ...	Yes, if ...	Not applied
Non-degradation offset	No	No	No, because ...
Single receptor	No, but ...	Yes, but ...	Not applied
Single-zone emission trading	No, but ...	No, but ...	Yes, but ...
Multiple-zone trading	No	No	No evidence

Table 7.2. Conditions influencing the cost efficiency of tradable emission permits.

	Theory	Models	Practice
Uncertainty about costs	xx	xx	-
Imperfect competition	xx	xx	-
Transaction costs	x	-	xx
Discrete control options	x	x	-
Imperfect enforcement	x	-	x
Political acceptability	x	-	xx
Bilateral sequential trading	x	xx	x
Size of the emission reduction	-	xx	-
CAC reference point	x	xx	x
Region-specific characteristics	-	x	x
Regulatory constraints	x	-	xx
Uncertain property rights	-	-	x

xx = more important.
x = important.
- = irrelevant or not addressed.

the cost minimum with perfect cost information. With imperfect data, zonal trading is not cost efficient and might be more expensive than regulation. Neither single ambient permit trading nor multiple-zone trading appear to have been applied in practice. Surprisingly, the single-zone trading schemes that were applied in practice (lead trading, sulfur allowance trading, the Dutch power plants bubble) have been rather successful in promoting cost efficiency, mainly because their simple design and the absence of the need for pre-approval for each trade resulted in liquid and active markets and easy trading. Remarkably, those systems (ambient permits, pollution offset, modified offset) that were expected to be cost efficient in theory and that achieved ambient standards in simulation models did not fare well in practice. However, trading schemes (especially single-zone trading) that in theory are not cost efficient for meeting local air quality standards and that might even be more costly than CAC have been successful in saving costs in practice.

The obvious question to ask, then, is why this difference occurs. The main answer is that emission trading usually is not applied in the pure way envisaged in theory. In practice tradable emission permits are combined with regulatory instruments. Moreover, single-zone emission trading has been successful in those cases where local air quality goals were already largely met (sulfur in the USA, SO_2 bubble in the Netherlands) or could be safely ignored (lead trading in the USA). Furthermore, different conditions appear to play a more important role in determining the cost efficiency of emission trading in practice than simulation models or theoretical analysis predict (see *Table 7.2*).

7.3 Relevant Conditions

In theory, uncertainty about costs and property rights and to a smaller degree imperfect competition (market power and state-regulated industries), transaction costs, imperfect enforcement, and political acceptability influence the cost efficiency of emission trading. In an international context, the starting point is also relevant. If it is a cooperative Nash equilibrium, there is room for cost-saving reallocations of emission reductions among countries, but single-zone emission trading does not necessarily constitute a Pareto-dominant improvement. Ambient permit trading or a modified offset rule is always Pareto-dominant, but might leave little room for net cost savings.

Simulation models indicate that uncertainty about costs, bilateral trading with partial information, the percentage of the emission reduction, and the CAC reference point determine the cost savings of emission trading. With partial information on costs, bilateral sequential trading does not achieve the full potential cost savings. If the CAC reference point is cost efficient to a certain degree (e.g., by allowing intrafirm trading) or if the percentage reductions are high, cost savings of trade tend to be smaller. Neither transaction costs nor imperfect enforcement have been modeled. Although market power might have a significant impact on the permit price, the model calculations performed indicate that its influence on cost efficiency has been marginal.

The latter is confirmed by practical experience where market power has not been an issue. Far more important for the actual cost savings of emission trading were transaction costs, regulatory constraints, and uncertainty about property rights. Transaction costs were high under the EPA's emission trading program and the German offset rule, mainly due to complicated administrative approval procedures linked to the lack of political acceptability of trading. In the absence of case-by-case

Concluding Observations

approval and in the presence of a clear allocation of rights before trading and single-zone trading, trading was more successful at saving costs, as the US experience with lead and sulfur allowance trading clearly shows. Regulatory constraints (in the form of emission standards or trading in small zones) have impaired the proper functioning of the EPA's emission trading policy and the German renewal and compensation rule. The presence of emission standards at every location in the Danish and Dutch power plants sector, however, enabled the use of a simple, and therefore more cost-efficient, national bubble. Uncertainty about property rights has plagued the performance of the EPA's trading policy, the Montreal Protocol, and the Netherlands' refinery bubble. This uncertainty took the form of uncertain baseline emissions, frequent policy changes, confiscation of banked rights, allocation to emission ceilings sectors rather than to individual firms, or uncertain regulatory treatment of the revenues of permits compared with other pollution control options.

It thus appears that economic theory and simulation models emphasize market imperfections and uncertainty about costs as major circumstances affecting the cost efficiency of emission trading, although the relevance of transaction costs has recently been rediscovered. Simulation models find market imperfections (market power) to be irrelevant; they ignore them, pointing to bilateral sequential trading, the reductions required, the CAC reference point, regional characteristics, and uncertainty as relevant factors determining cost efficiency. Practice, however, shows that high transaction costs due to requirements of case-by-case approval and regulatory constraints are crucial determinants. Their presence appears to be linked to the fact that other, simpler solutions were politically unacceptable in that specific context.

7.4 Administrative Practicability, Innovation, and Political Acceptability

In theory, ambient permit trading or free trading rules always meet the environmental objectives if enforcement is perfect. This environmental effectiveness is not necessarily guaranteed by simpler schemes such as single- or multiple-zone trading because the result may depend on uncertain costs. In practice, single-zone emission trading has been successfully applied to meet environmental standards at different receptors. This was achieved by combining large reductions in overall emissions with the possibility of blocking the use of an acquired permit to emit more in a specific case (the US sulfur case), or by combining large emission reductions and emission standards at every location (the Dutch and Danish power plant bubbles and lead trading).

The administrative practicability of policy instruments is determined by the information intensity and the ease of monitoring and enforcement. Dealing with non-uniformly mixed pollutants requires information on emissions, atmospheric transport, and costs of controlling emissions, as well as information from a cost-minimization model. All instruments require an emission inventory and atmospheric dispersion information in order to be effective. Emission trading (and emission charges) require less information on costs than does regulation in order to be cost efficient. Their administrative costs could be higher or lower than CAC, depending on the costs of the emission tracking system and redistribution of tax revenues.

Empirical model studies suggest that the additional administrative costs of emission trading mainly consist of transaction costs borne by the polluters. Additional costs of keeping track of trades are small and monitoring costs are comparable with what would have been required otherwise. In practice, the flexibility of emission trading has had a small price. Administrative costs are higher due to emission tracking schemes and increased monitoring and enforcement (EPA, US sulfur trading, US lead trading). The average refinery standards in the Netherlands (an example of intrafirm trade), however, reduced administrative requirements compared with a policy requiring emission standards for every single plant.

Tradable permits as well as effluent charges tend to promote innovation more than regulation does because in theory the expected cost savings are higher. If the regulator responds to innovation, the innovative push will be higher with a charge or emission trading than with regulation. Evidence from simulation models demonstrates that innovation increases the potential cost savings of trading because polluters can respond to reductions in the pollution control costs of a specific technique by altering their optimal mix of control options. Although the evidence is not overwhelming, practical experience illustrates that emission trading leads to cost-saving innovations, such as the simplified design of flue gas desulfurization equipment.

Charges and emission trading might also have different long-term impacts than regulation, because the instruments result in different incentives on the entry/exit and location decisions of firms. Regulation is less cost efficient than charges and raises short-term average costs more. However, with charges and permit trading, industry's average costs might be higher due to the price for its unabated emissions. Locational shifts are more likely with instruments that are tailored to the location of the source, such as ambient charges, deposition permits or alternatives, and point-by-point regulation.

Regarding political acceptability, the role that economic incentives can play depends, in theory, very much on the extent to which the cost savings they promise exceed the additional costs of changing existing property rights and regulations.

Concluding Observations

Whether or not, and in which form, economic instruments are accepted depends then on the extent to which actors (industry, environmentalists, government, trade unions) can articulate their own interests. "Grandfathering" of permits or earmarking the revenues could be needed to obtain support from industry.

In an international context, countries can only be expected to accept international taxes or emission trading if the instrument constitutes a Pareto-dominant improvement. An international deposition tax schedule requires the return of tax revenues. Without full knowledge of costs and benefits, tax and reimbursement parameters might be set at levels that do not minimize costs and may turn some countries into losers. When dealing with non-uniformly dispersed pollutants, emission permit trading in one zone may or may not be Pareto-dominant compared with an initial cooperative Nash agreement. Deposition permit trading and trading according to trading rules are always Pareto-dominant improvements if the deposition goals are the deposition levels agreed to under the initial agreement. Transaction complexity and costs might, however, be high.

Simulation models show that auctioning of permits might imply that the financial burden (permit outlays plus pollution control costs) exceeds that of a regulatory alternative. "Grandfathering" of permits or holding a zero-revenues auction might therefore be needed to achieve a politically acceptable solution. In practice, debates on the distribution of costs and benefits appear to have had a significant influence on how trading schemes have been applied or if they were applied at all. As the second part of this study will illustrate, this is also the case for the trading of sulfur reduction commitments in Europe.

Part II

Application to Sulfur Emissions in Europe

"In matters of trade the fault of the Dutch is offering too little and asking too much." (Dispatch to Sir Charles Bagot, British Minister at The Hague, 31 January 1826)

Chapter 8

The Institutional Framework for Controlling Sulfur Emissions in Europe

8.1 Introduction

Whereas Chapters 1 to 7 surveyed economic theory, simulation models, and practical experience, the remaining chapters examine the application of trading sulfur emission reduction commitments in Europe. This chapter gives an overview of the institutional framework for dealing with sulfur emissions in Europe. Chapter 9 provides a theoretical analysis of international emission trading with exchange rates. Chapter 10 provides the results of a simulation model, and Chapter 11 describes a possible design for trading (or joint implementation of) sulfur emission reductions.

More than a century after the British scientist R. Smith first coined the term acid rain, negotiators of a new protocol to control SO_2 emissions in Europe discovered the potential role of economic instruments, especially emission trading. To understand the potential contribution of economic instruments, it is relevant to outline the current institutional framework for handling acidification and especially SO_2 emissions in Europe. This is because experience has shown that economic instruments are usually applied as supplements to existing regulations and the form they take is affected by the existing institutional structure (Opschoor, 1994; Hahn, 1990).

The objectives of this chapter are twofold:

- To give a broad overview of the problem of acidification in Europe (especially the role of SO_2) and its environmental impacts;
- To present an overview of the current institutional framework and policies in Europe for dealing with SO_2 emissions and acid rain.

The chapter has the following structure. Section 8.2 gives a brief overview of the acidification problem in Europe and a nutshell history of major events that led to the development of institutions and regulations for dealing with the problem. Section 8.3 focuses on the principal international institution involved in the issue, the United Nations Economic Commission for Europe (UN/ECE), and its achievements. Section 8.4 summarizes the relevant regulations on SO_2 emissions in the EC, and Section 8.5 reviews the regulations in place in the legislation of different countries in Europe, most of which are command-and-control (CAC) regulations.

8.2 Acidification in Europe

Like Demsetz (1967, p. 350), we expect that externalities such as the environmental impacts caused by acidifying emissions will tend to be internalized when the gains of doing so exceed the costs of internalization. It took some time in Europe, however, before it was recognized that one of the most serious problems caused by air pollutants is the acidification of the environment. Subsequently, appropriate international institutions were developed to deal with the problem of transboundary fluxes of acidifying pollutants in Europe.

Acidification is usually linked to the atmospheric discharge of SO_2 and NO_x and the subsequent impacts of these substance on the environment. A major source of these pollutants is the burning of fossil fuels. Sulfur and nitrogen oxides can persist in the air for up to several days. In due course, however, they are returned to the earth's surface. They are either washed out by rain (hence the terms acid rain and wet deposition) or they come down on vegetation and moist surfaces (dry deposition). These oxides remain in the air long enough to be transported hundreds, sometimes thousands, of kilometers from the original emission point. During atmospheric transport these gaseous oxides can be transformed into acids (sulfuric and nitric acid) (Persson, 1982). Both sulfur and nitrogen oxides can have direct and indirect impacts on organisms or materials. The direct impacts arise from the flow of pollutants at one specific point in time. The indirect impacts are caused by the accumulation of pollution over time. Direct impacts consist of impacts on human health and on plants (such as crops and forests), and damage to materials and cultural and historical monuments. These direct impacts chiefly depend on the concentration of the pollutants in the air (CDA, 1988). In general, the severity of these impacts declines rapidly with decreasing concentrations and increasing distances from the emitting sources (Persson, 1982). The acidification effects, usually referred to as the indirect impacts, are contingent on the level of wet and dry deposition and the natural sensitivity of soil and water bodies to acidification. Indirect impacts of acidification may consist of (Hordijk et al., 1990)

- Accelerated acidification of lakes, which endangers fish species and leads to the disappearance of fish species;
- Increased acidity of forest soils over large areas of Europe, which leads to nutrient deficiencies and high concentrations of aluminum and other toxic materials;
- Increased forest die-back due to the combined impacts of different gaseous pollutants, forest soil acidification, increased deposition of nitrogen, and other stress factors such as extreme climatic episodes.

In spite of the fact that early warning signals of the damaging impact of acid rain had already been reported by Smith in the last century, it was not until the late sixties that the Swedish scientist Svante Oden demonstrated the increasing acidity of Swedish lakes and rivers and pointed toward the increasing acidity of precipitation as a regional and temporal phenomenon in Europe (Cowling, 1982). Moreover, using trajectory analysis he found that acidification in Sweden was largely due to sulfur emissions originating from England and Central Europe. It was not until the 1972 United Nations Conference on the Human Environment, however, that the threat posed by acid rain was put on the international agenda. A Swedish case study in 1972 confirmed the damaging impact of sulfur and acid precipitation on materials and ecosystems (Cowling, 1982). The UN conference accepted the principle that states must ensure that activities within their jurisdiction of control do not cause damage to the environments of other states. Subsequently, a major interdisciplinary Norwegian study was conducted, producing a stream of reports that provided new evidence that extensive damage to aquatic ecosystems in Scandinavia was indeed occurring. Obvious sources of pollution, such as Germany and the UK, were not convinced that they were responsible for damaging the Scandinavian lakes. This was in spite of the results of an international study conducted in 1977 and 1979 by the OECD (Organisation for Economic Cooperation and Development) to estimate the local and foreign contribution to each country's deposition of sulfur. This study came to the conclusion that long-range transport of sulfur compounds was indeed occurring. It showed, among other things, that more than 50 percent of the atmospheric sulfur deposition in Austria, Finland, Norway, Sweden, and Switzerland came from foreign sources. Although not all countries agreed with the result, the OECD study provided a strong impetus to the call for international policies to control transboundary sulfur pollution.

Because some European countries, particularly those countries in Eastern Europe, are not members of the OECD, the OECD was not the appropriate institution to conduct further studies and negotiations (Hordijk et al., 1990). The UN/ECE, based in Geneva, appeared to be a better candidate to act as a platform for collaboration on transboundary air pollution because it had unique experience in dealing with environmental issues and, more important, it provided a unique forum where both West and East could meet on equal footing (Wüster, 1992).

8.3 The United Nations Economic Commission for Europe

8.3.1 The Convention on Long-range Transboundary Air Pollution

In response to the acute acidification problems, a meeting at the ministerial level was held in November 1979 in Geneva, Switzerland, within the framework of the UN/ECE. This meeting resulted in the signing of the Convention on Long-range Transboundary Air Pollution. This convention constitutes a framework within which contracting parties identify the problems posed by transboundary air pollution and accept the responsibility for taking appropriate steps. Without making any commitments to reduce emissions, the convention laid down general principles for international cooperation for air pollution abatement and set up an institutional framework to bring together research and policy (Nordberg, 1993). More specifically the convention

- Recognizes that air pollution is a major problem.
- States the parties' endeavor to limit and, as far as possible, gradually reduce and prevent long-range transboundary air pollution (Article 2).
- Expresses the intention of developing policies and strategies to serve as means of combating the discharge of air pollution (Article 3).
- States the intention to use the best available technology that is economically feasible to meet the objectives of the convention.

The Convention was signed by 35 parties, including all the countries in Europe and two republics of the Soviet Union, Belarus and Ukraine. It was put into force in March 1983 after it had been ratified by 24 parties (Hordijk et al., 1990; Wüster, 1992). The Convention does not contain any tough agreements on emission reductions (that is, it has no teeth), but it establishes an institutional framework for further cooperation.

The convention establishes an Executive Body (EB) as the supreme policy-making assembly, which represents all contracting parties and meets at least annually to review the implementation of the convention and to adopt a work plan. A bureau consisting of the EB chairman and vice chairmen deals with interim matters. The second institutional layer consists of intergovernmental working groups

Institutional Framework

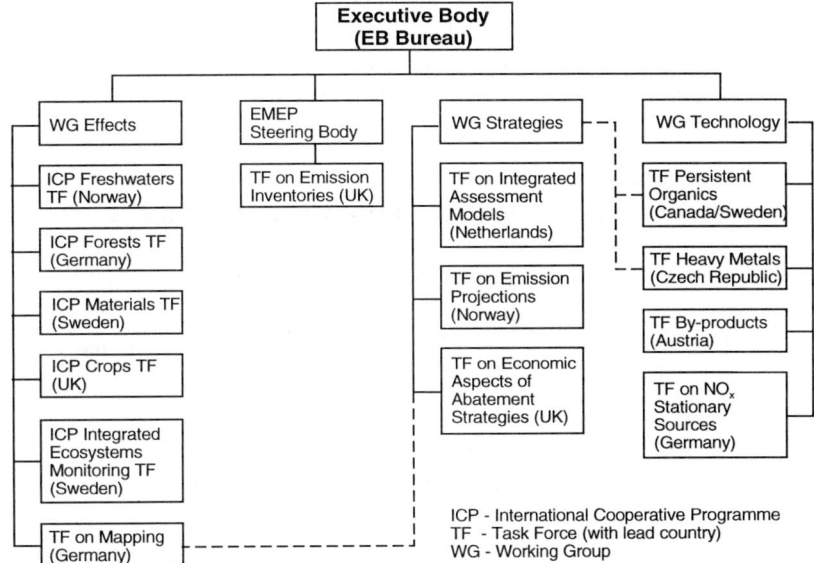

Figure 8.1. The Convention on Long-range Transboundary Air Pollution.

(effects, strategies, and technology) as well as the Steering Body to EMEP (European Monitoring and Evaluation Programme, or officially Cooperative Programme for Monitoring and Evaluation of Long-range Transmission of Air Pollutants in Europe) (see *Figure 8.1*) (Nordberg, 1993, p. 6). Under these subsidiary bodies, intergovernmental task forces can be established to deal with specific issues of the work plan. In addition, five international cooperative programs (ICP) on effect-related matters have been established.

The EMEP was originally established in 1977. In a supplementary protocol to the Convention, signed in 1984 and enacted in 1988, the 33 participating countries assumed financial responsibility through mandatory and voluntary contributions. A major objective of the EMEP is to provide information on the deposition and concentration of air pollution, and the quantity and significance of long-range transport of pollution and fluxes across borders (Nordberg, 1993). For this purpose, emission data are compiled, air and precipitation quality is monitored and measured, and atmospheric dispersion models are developed and employed (Wüster, 1992). The EMEP monitoring network consists of 101 stations in 33 countries (Nordberg, 1993). Outputs of the EMEP consist of, among other things, country-to-country blame matrices showing the extent to which one country's emissions are deposited in other countries and the contributions of countries to individual grids within Europe (Sandnes, 1993).

8.3.2 The First Sulfur Protocol

The Conference on Acidification of the Environment held in June 1982 reaffirmed the damaging impacts of sulfur and nitrogen oxides emissions and marked an important breakthrough in international air pollution control action. Germany, concerned with reports of rapidly increasing forest damage, radically changed its attitude toward acid rain (Haigh, 1989). In March 1984, ten countries agreed to reduce emissions by 30 percent: the Nordic countries (Denmark, Finland, Norway, Sweden), Austria, Canada, France, West Germany, the Netherlands, and Switzerland. The reason for selecting a 30 percent reduction was the following. According to a German report from 1979, desulfurization of renovated (replaced) installations could reduce emissions by 30 percent (Schärer, 1979, 1981, 1982, 1995). Germany therefore proposed a 30 percent reduction. Because at that time there was limited knowledge on abatement options, the 30 percent was taken as a percentage that was feasible. This 30 percent club was extended in June 1984 to a group of 18 at the Conference in Munich (Hordijk *et al.*, 1990). When so many countries opted for a reduction of a fixed percentage, it became easy to draft a protocol. This First Sulfur Protocol was opened for signatures in Helsinki in 1985 and has been signed by 20 parties: Austria, Belarus, Belgium, Bulgaria, Canada, Czech and Slovak Federal Republic, Denmark, Finland, France, Germany, Hungary, Italy, Liechtenstein, Luxembourg, the Netherlands, Norway, Sweden, Switzerland, Ukraine, and the former USSR. The basic provision of the Protocol is that parties are to reduce their annual sulfur emissions or their transboundary fluxes by at least 30 percent as soon as possible, or at the latest by 1993, using 1980 as a basis for calculating reductions (Wüster, 1992). The transboundary fluxes clause was included to convince the former USSR, in particular, to sign the Protocol. Among the countries that did not sign were Poland and the UK. The UK did not sign because it was not certain about the likely change in sulfur emissions in the next decade and because both the base year and the percentage reduction were arbitrary and did not relate deposition to environmental impacts. Poland claimed to lack control techniques and to rely too heavily on domestic coal with high sulfur contents (Hordijk *et al.*, 1990).

Table 8.1 shows the development of the SO_2 emissions over time (UN/ECE, 1993). It is based on data available in August 1993, before the signing of the Second Sulfur Protocol. The table indicates that since 1990, 1991, and 1992, sulfur emissions have dropped for all of the 25 parties to the convention that submitted data except for the former East Germany. Eight parties even achieved reductions of 50 percent or more compared with 1980; some countries (Austria and Sweden, for

Institutional Framework

Table 8.1. Development of SO$_2$ emissions in Europe.

Country	SO$_2$ emissions (ktons/yr)					
	1980	1985	1990	1991	1992	1993[a]
Austria	397	195	90	84	76	–
Belarus	740	690	–	–	–	520
Belgium	828	452	443	447	–	–
Bulgaria	2,050	–	2,020	1,600	–	1,600
Croatia[b]	59	59	69	–	–	–
Czech Republic	2,257	2,277	1,876	1,776	1,743[c]	–
Denmark	451	343	180	243	–	–
Finland	584	382	260	194	–	–
France	3,338	1,470	1,202	1,314	1,208	–
Germany[d]	3,194	2,396	1,029	–	–	–
Germany, former GDR	4,300	5,400	4,774	–	–	–
Greece	400	500	–	–	–	–
Hungary	1,632	1,404	1,010	902	–	–
Ireland	222	140	168	190[c]	–	174
Italy	3,800	2,504	–	–	–	–
Luxembourg	24	16	–	–	–	–
Netherlands	466	276	207	183	178[c]	–
Norway	142	98	54	46	40	70
Poland	4,100	4,300	3,210	2,995	2,820	–
Portugal[e]	266	198	–	–	–	270
Romania	1,800	–	–	–	–	–
Russian Federation[f]	7,161	–	4,460	4,211	–	4,961
Slovakia	700	622	539	438	–	–
Slovenia	235	240	195	180	188	–
Spain[g]	3,319	2,190	2,316	–	–	–
Sweden	503	292	130	107	–	–
Switzerland	126	96	62	62	59	–
Ukraine[h]	3,850	3,464	2,782	2,538	2,376	2,459
United Kingdom	4,898	3,724	3,780	3,565	–	3,330
Yugoslavia[h,i]	–	478	508	446	396	–

[a] Projected estimates.
[b] Only emissions from thermal power plants.
[c] Preliminary data.
[d] Figures apply to the Federal Republic of Germany prior to 1991.
[e] For the years 1990, 1993, 1995, and 2000, emissions of small combustion plants were not calculated.
[f] Figures apply to the European part of the Russian Federation within EMEP.
[g] Only emissions related to combustion for energy purposes.
[h] Only emissions from stationary sources.
[i] Refers to former Yugoslavia, excluding Croatia and Slovenia.

example) have surpassed the 30 percent goal. This has led to the suggestion that the Protocol was not very effective at gaining control beyond what countries would have done on their own anyway, although the treaty might have helped to solve the free rider problem (Ausubel and Victor, 1992). However, Chapter 4 showed that cooperative agreements on reciprocal emission reductions can be expected to lead to lower emissions and higher welfare levels than a non-cooperative Nash equilibrium (see also Nentjes, 1994).

8.3.3 The Protocol on Nitrogen Oxides and Volatile Organic Compounds

The increased recognition that NO_x contributes not only to acidification but also to the formation of ground-level ozone brought strong arguments for controlling this substance in an additional protocol. This Protocol on the Control of Emissions from Nitrogen Oxides Emissions or Their Transboundary Fluxes was signed in 1988 in Sofia by 26 parties and went into effect in 1991. The Protocol's basic requirement is that, as a first step, parties reduce or control emissions so that their national emission totals or their transboundary fluxes do not exceed their 1987 levels. Another alternative offered to the parties is to choose another base year, upon signature, provided that in addition the average annual emission level between 1987 and 1996 does not exceed the 1987 level (Nordberg, 1993; Wüster, 1992). Furthermore, the parties to the Protocol will

- Apply national emission standards to major new and substantially modified stationary sources as well as to new mobile sources based on the best available control technologies that are economically feasible;
- Introduce pollution control measures for major existing sources.

Although this Protocol does not influence sulfur emissions, it does set the stage for the important elements of the Second Sulfur Protocol. Apart from the emission standards, the Nitrogen Oxides Protocol also mentions that further steps are needed to reduce emissions, taking into account not only the best available technologies but also environmental objectives based on critical loads (Wüster, 1992), which are the levels of deposition below which no damage to sensitive ecosystems is expected to occur. This is a new element in the Protocol.

Realizing that ground-level ozone is the result of interplay between nitrogen oxides and volatile organic compounds (VOCs) in the presence of sunlight, 21 parties signed the Protocol Concerning the Control of Emissions from Volatile Organic Compounds or Their Transboundary Fluxes in 1991 in Geneva. The basic obligation here is to cut VOC emissions by 30 percent by 1999, using either 1988 levels as a basis or any other annual level between 1984 and 1990. Alternatively,

Institutional Framework 193

parties may decide to keep VOC emissions at their 1988 level in so-called Tropospheric Ozone Management Areas (TOMA), which do not generate transboundary fluxes. Or they can keep VOC emissions constant if total national emissions are below 0.5 million tons, and below 20 kg per inhabitant and 5 tons per square kilometer. Furthermore, parties have to apply best available technologies to new sources and, after five years, to existing sources in areas exceeding ozone standards or contributing to transboundary pollution (Nordberg, 1993).

8.3.4 Making a New Sulfur Protocol

Negotiations on the Second Sulfur Protocol started in 1991 (Wüster, 1992). They came to a close in June 1994, when the Protocol was signed by 26 parties in Oslo (including the EC) (UN/ECE, 1994a). At the end of 1994, 27 parties had signed. This section describes the making of the Protocol; the next section describes the Protocol's text.

A major new element in the Second Sulfur Protocol is the intention to apply an effect-oriented approach by basing the extent of emission reductions on the susceptibility of natural ecosystems to acid deposition rather than choosing an equal percentage reduction for all countries involved. Such an approach was already flagged in the Sofia Protocol on Nitrogen Oxides. The ultimate goal then is to reduce emissions so that critical loads and levels are not exceeded (Nordberg, 1993). Critical loads are defined as the maximum levels of deposition (sulfur, nitrogen, or total acidity) below which, according to current scientific knowledge, no damage to sensitive ecosystems occurs (Hettelingh *et al.*, 1991; Nordberg, 1993). Critical levels refer to the concentrations of pollutants above which adverse effects on receptors may occur. Economic, technical, and other constraints may mean that the necessary reduction cannot be achieved everywhere or immediately. Interim steps might be needed. An accepted step in this approach is the generation of target loads that not only take into account environmental sensitivity (critical loads) but also technical, social, economic, and political considerations. To apply a cost-efficient approach based on environmental sensitivity, the following information is required (Nordberg, 1993):

- Emission inventories and emission projections;
- Estimates of the potential for and costs of controlling emissions;
- Long-range transport models;
- Maps of critical loads and levels;
- Integrated assessment models that link the above elements and provide information on cost-minimizing ways to achieve environmental goals.

This type of information is available within the Convention and its subsidiary bodies.

An alternative to the receptor- or effects-oriented approach is the source- or technology-based approach. This approach is precautionary and states that emissions should be reduced as far as technically possible. Overviews of best available technologies (BAT) are required for this purpose. In the course of policy making (for example, on the Nitrogen Oxides Protocol) this BAT approach may be combined with practical, economic, and social considerations in what may be termed best available technology not entailing excessive costs (BATNEEC). Obviously, both approaches can also be merged and can serve as complements to arrive at an optimal policy mix (Nordberg, 1993).

During the making of the Second Sulfur Protocol, both the effect-oriented and the source-oriented approach were examined in order to arrive at the final Protocol, which was signed in 1994 (Amann *et al.*, 1991; Amann *et al.*, 1993). *Figure 8.2* displays the map of the five-percentile values for critical sulfur deposition as used in the negotiations. This map shows the values to which sulfur deposition must be reduced to protect 95 percent (hence five percentile) of all ecosystems in each singular grid. Obviously, the map shows that within Europe the sensitivity of soils and ecosystems are different. Highly sensitive areas are to be found in Northern Europe in particular.

Because these critical loads are so difficult to achieve in the short run, even when one uses all technical means, compromises were discussed in the course of the negotiations. The compromise that gradually emerged as a major reference point accepts that critical loads remain a long-term objective toward which the Second Sulfur Protocol makes a gradual move. As an interim target, the difference (the gap) between the sulfur deposition in 1990 and the five-percentile critical loads must be reduced by at least 60 percent (Amann *et al.*, 1993). In grids where the 1990 deposition was already below the five-percentile values (especially in parts of Portugal, Spain, and Russia), the five-percentile values were chosen as deposition targets (the long-term objective). The resulting targets for the deposition of sulfur have to be attained in a cost-efficient way, minimizing total European costs subject to the condition that countries carry out at least those reductions that they were planning to undertake anyway [their current reduction plans (CRPs)] (Amann *et al.*, 1993). In other words, scenario A5 is a design for a protocol without monetary compensation for those countries with large abatement efforts. Moreover, it is a protocol design that takes the current plans for emission reduction in each country as minimum reductions. These current reduction plans can be interpreted as a (non-) cooperative Nash equilibrium (see Chapter 4): cooperative for those countries that signed the First Sulfur Protocol, and non-cooperative for the non-signatories of the First Protocol. The resulting scenario, called A5, formed the basis for further

Institutional Framework

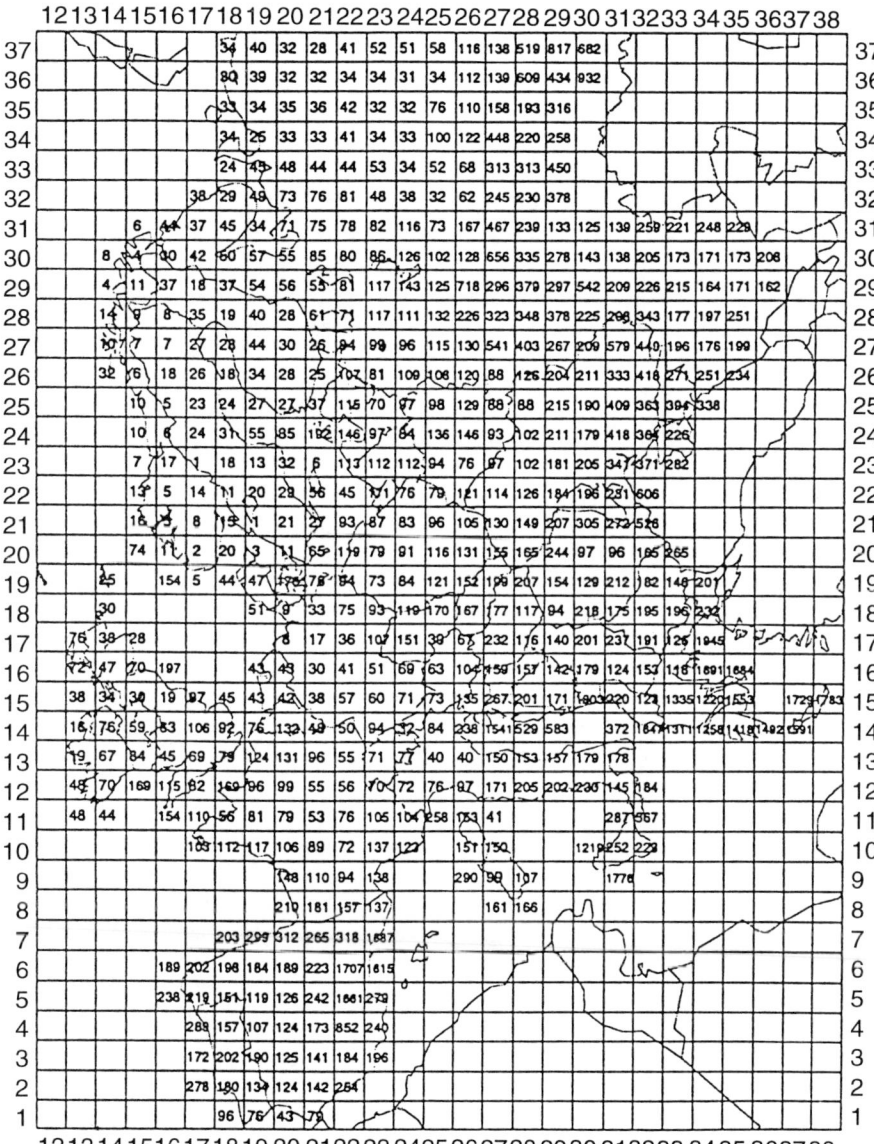

Figure 8.2. Critical sulfur deposition values (g S/m^2/yr).

Table 8.2. National ceilings in the Second Sulfur Protocol.

	1980 SO₂ (kton)	CRP	Scenario A5 RAINS Change (%)	Second Sulfur Protocol 2000 SO₂ (kton)	2005	2010	2000 Change (%) over 1980	2005	2010
Countries	(1)	(2)	(3)	(4)	(5)	(6)	(7)	(8)	(9)
Albania*	101	138	+31	–	–	–	–	–	–
Austria	390	78	–80	78	–	–	–80	–	–
Belarus*	740	456	–38	456	400	370	–38	–	–
Belgium	828	430	–77	248	232	215	–70	–72	–74
Bulgaria	2,050	520	–74	1,374	1,230	1,127	–33	–40	–45
Former CSFR									
Czech Republic	2,257	–	–72	1,128	902	632	–50	–60	–72
Slovakia	843	–	–72	337	295	240	–60	–65	–72
Denmark	448	176	–87	90	–	–	–80	–	–
Estonia,* Latvia,* Lithuania*	621	435	–86	–	–	–	–	–	–
Finland	584	116	–80	116	–	–	–80	–	–
France	3,348	1,210	–80	868	770	737	–74	–76	–78
Germany	7,494	990	–90	1,300	990	–	–83	–87	–
Greece	400	595	+49	595	580	570	+49	+45	+43
Hungary	1,632	1,094	–68	898	816	653	–45	–50	–60
Ireland	222	240	–41	155	–	–	–30	–	–
Italy	3,800	1,976	–73	1,330	1,042	–	–65	–73	–
Luxembourg	24	10	–58	10	–	–	–58	–	–
Moldavia*	330	231	–30	–	–	–	–	–	–
Netherlands	466	106	–77	106	–	–	–77	–	–
Norway	140	70	–76	34	–	–	–76	–	–
Poland	4,100	2,600	–66	2,583	2,173	1,397	–37	–47	–66
Portugal*	266	294	+11	304	294	–	+14	+11	–
Romania*	1,800	2,592	–41	–	–	–	–	–	–
Russian Federation	7,161	4,440	–38	4,440	4,297	4,297	–38	–40	–40
Spain	3,319	2,143	–55	2,143	–	–	–35	–	–
Sweden	519	100	–83	100	–	–	–80	–	–
Switzerland	126	60	–52	60	–	–	–52	–	–
Turkey*	860	2,887	1236	–	–	–	–	–	–
Ukraine	3,850	1,696	–56	2,310	2,310	2,310	–40	–40	–40
United Kingdom	4,898	2,552	–79	2,449	1,470	980	–50	–70	–80
Former Yugoslavia	1,300	1,576	–22	–	–	–	–	–	–
Croatia	150	–	–40	133	125	117	–11	–17	–22
Slovenia	230	–	–45	130	94	71	–45	–60	–70
Rep. of Yugosl.*	920	–	–13	–	–	–	–	–	–

*Indicates a non-signatory country.
Sources: Amann *et al.*, 1993; UN/ECE, 1993; UN/ECE, 1994a.

Institutional Framework

negotiations (Amann *et al.*, 1993). *Table 8.2* displays the percentage reduction in the year 2000 (over the 1980 emissions) that would have to be met according to scenario A5 (column 4) and compares it with absolute emission levels in 1980 and emission levels under CRPs for the year 2000. The second column shows that SO_2 emissions were expected to be reduced from 58,017 kton in 1980 to 31,981 kton in the year 2000 (as of the May 1993 CRP). This implies an average European cutback of 45 percent compared with 1980. The A5 scenario would require an average reduction of 59 percent over 1980 levels (column 4). Although A5 served as the reference point, further negotiations led to a slightly different schedule for emission ceilings.

8.3.5 The Second Sulfur Protocol

The reason for this change in the schedule for emission ceilings is that several countries did not agree to carry out the required reductions in the year 2000 (as scenario A5 proposed). The reductions of several countries were therefore postponed until 2005 or 2010. Apart from this postponement, several countries did not sign the Protocol at all. Again, other countries just signed up for smaller reductions. This will be explained in detail below.

The final national quotas agreed on in the final Protocol are displayed in columns 4 to 6 for the years 2000, 2005, and 2010. The resulting percentage reductions over 1980 levels are shown in the last three columns (7 to 9). *Table 8.2* shows two things. First, a comparison of column 3 (scenario A5) and columns 7 to 9 (Second Sulfur Protocol) shows the extent to which countries meet the Protocol reductions and leads to the following conclusions:

- Nine countries will meet A5 in the year 2000: Austria, Finland, Greece, Luxembourg, Netherlands, Norway, Russian Federation, Switzerland, and Slovenia (which will even surpass A5).
- Sweden will almost meet the A5 reduction in the year 2000.
- Five countries will postpone the A5 reductions until 2005 or 2010: Italy, Poland, the Czech Republic, Slovakia, and the UK.
- Three countries will almost meet the reductions in 2010: Belgium, France, and Germany.
- Seven countries signed lower reductions: Bulgaria, Denmark, Hungary, Ireland, Spain, Ukraine, and Croatia.
- Ten countries did not sign (as of December 1994): Albania, Belarus, Estonia, Latvia, Lithuania, Portugal, Moldavia, Romania, Turkey, and the rest of the former Yugoslavia (except Croatia and Slovenia).

Second, *Table 8.2* shows us the extent to which countries do more under the Protocol (columns 4 to 6) than they planned to do anyway [their (non-) cooperative Nash equilibrium, column 2]. Of the countries that signed the Protocol

- Nine countries do more than planned under CRPs: Belgium, Denmark, Hungary, Ireland, Italy, Norway, Poland, Russian Federation, and UK;
- Ten countries simply carry out what they planned to do anyway: Austria, Finland, France, Germany, Luxembourg, Netherlands, Portugal, Spain, Sweden, and Switzerland;
- Two countries do less than they planned: Bulgaria and Greece.

For several new countries a comparison cannot be made due to lack of data on CRPs. These include the Czech Republic, Slovakia, Croatia, and Slovenia.

Assuming that those countries for which no quota has been agreed on carry out their CRPs, the total SO_2 emissions would come down to 31,135 kton in the year 2000. This is a reduction of 46 percent. In 2005, SO_2 emissions would drop to 28,234 kton (49 percent cutback) and in 2010, to 25,983 kton (a 55 percent cutback).

Thus, differentiated national emission ceilings are one of the two tiers of the Second Sulfur Protocol. In principle, these ceilings were based on an effect-oriented (critical-loads-based) approach combined with political horse trading.

The second tier of the Protocol is a source-based approach consisting of emission standards for new large combustion plants (authorized after 31 December 1995) and fuel standards specifying the maximum sulfur content in gas oil. The emission standards are shown in *Table 8.3*. These standards are similar to those that were adopted by the EC at the end of 1988 in its Large Combustion Plants Directive (see Section 8.4). The fuel standards require parties to lower the sulfur content in gas oil to 0.05 percent for on-road vehicles (diesel oil) and 0.2 percent otherwise, no later than two years after the Protocol enters into force. In the case of limited supply, the lowering of the sulfur content may be postponed by 10 years.

The obvious question is why emission and fuel standards were adopted when national emission ceilings guarantee that the desired environmental effects will be achieved anyway. There are several reasons for this. First, the emission and fuel standards will assure that there will be minimum reductions beyond those expected to result from the national ceilings. Second, these standards specify a clear timing of measures, making it easier to verify the measures taken to attain the national emission ceilings. Third, several Eastern European countries were in favor of this because this enabled them to obtain funds from the World Bank, which provides funds only if the pollution control measures are legally necessary. Fourth, the same countries expected that the imposition of emission standards would improve their domestic bargaining position. Although these were the official reasons, one should

Table 8.3. Emission standards in the Second Sulfur Protocol.

Thermal power input [MW(Th)]	SO$_2$ standard (mg/m^3)		
	Solid fuels^a	Liquid fuels	Gaseous
50–100	2,000	1,700	35^b
100–300	2,400 – 4x^c	1,700	35
300–500	2,400 – 4x^c	3,600 – 6.5$x^{c,d}$	35
>500	400	400^d	35

^{a}When high sulfur contents of indigenous fuels prevent meeting the standards, minimum desulfurization rates apply depending on the size (40 to 90 percent).
^{b}For gaseous fuels in general, 5 for liquefied gas and 800 for low calorific gases of refinery residues, coke oven gas, blast furnace gas.
$^c x$ is the thermal power input, e.g., if the thermal power input is 300 MW the standard equals 2400 – 4.300 = 1200.
^{d}Minimum desulfurization rate 90 percent.

not forget the more hidden motives. The emission and fuel standards were already part of the national legislation of many countries (see Section 8.5) and part of EC legislation before the Second Protocol was signed (see Sections 8.4 and 8.5). This was the case for the EC-12 countries, Austria, Switzerland, Sweden, Finland, and to a certain degree the Czech Republic and Slovakia. Making these emission and fuel standards obligatory for all parties to the Protocol would therefore improve the competitive position of these countries.

In summary, the following are the important basic obligations of the Second Sulfur Protocol, as signed in June 1994 (UN/ECE, 1994a):

- The parties will control sulfur emissions to ensure, as far as possible and without excessive costs, that sulfur deposition does not exceed the critical loads for sulfur (*Figure 8.2*).
- As a first step, parties will, as a minimum, reduce and maintain their annual emissions of sulfur at the levels specified (see *Table 8.2*, columns 4 to 6).
- Parties will apply these emission standards, or more stringent ones, to all major, new stationary combustion sources (see *Table 8.3*).
- No later than 1 July 2004, parties must apply emission standards (set out in *Table 8.3*) for major existing combustion sources above 500 MW(Th) or take equivalent measures that achieve the national ceilings agreed on. They are also to apply emission standards or emission limitations to major existing sources between 50 and 500 MW(Th) using the values shown in *Table 8.3* for guidance.
- No later than two years after the Protocol enters into force, parties must lower the sulfur content in gas oil to 0.05 percent for on-road vehicles (diesel oil) and 0.2 percent otherwise.

Finally, a proposal was included in the Protocol on joint implementation; however, there was no firm commitment to it. The Protocol only states that the parties to this Protocol may, at a session of the Executive Body, in accordance with rules and conditions which the Executive Body shall elaborate and adopt, decide whether two or more parties may jointly implement the obligations set out in annex II (the national ceilings). These rules and conditions shall ensure the fulfillment of the obligations set out in paragraph two (the national emission ceilings) and also promote the achievement of the environmental objectives set out in paragraph one (reduce sulfur emission to protect human health and the environment from adverse effects and to ensure, as far as possible, that the deposition of sulfur remains below the critical sulfur deposition levels agreed on) (UN/ECE, 1994a, p. 6.). The proposal states that two or more parties may jointly fulfill the obligations in terms of annual emission ceilings, subject to rules and conditions to be specified. These rules are to ensure that the emission ceilings agreed on are met and to promote the achievements of the critical sulfur loads. This lack of specific conditions reflects the disagreement among parties about what conditions were necessary, and it reflects the fact that parties gave priority to reaching agreements on country-specific emission ceilings rather than to discussing joint implementation rules in detail. A more detailed assessment of the rules and conditions that have been discussed over the course of the negotiations will be given in Chapter 10. Chapter 10 will also describe how the Second Sulfur Protocol handles monitoring and enforcement and whether joint implementation requires changes in monitoring and enforcement.

8.4 The European Community

8.4.1 Introduction

The emission and fuel standards in the Second Sulfur Protocol are quite similar to the ones agreed on previously within the EC. The EC's policy to control air pollution takes two forms. First, "environment action programs" formulate the broad, midterm strategic framework. Second, specific measures are adopted as legislation. In the field of air pollution these have usually taken the form of directives (Bennett, 1991), which are items of EC legislation proposed by the European Commission and adopted by the Council of Ministers. It places binding obligations on the member states (Haigh, 1989).

The legislation regarding the control of air pollution falls into the following categories (Bennett, 1991):

- Air quality standards
- Product standards for fuels

Institutional Framework 201

- Product or emission standards for motor vehicles
- Emissions from industrial plants
- Information and monitoring
- Atmospheric change (greenhouse gases and ozone layer)

This section will be confined to product standards for fuels and emissions from industrial plants in as far as they affect SO_2 emissions in Europe. Two types of regulations influence SO_2 emissions substantially: directives regulating the sulfur content in fuels and directives on industrial, particularly large combustion, plants.

8.4.2 Fuel directives

The fuel directives regulating the sulfur content of gas oil originate in large part from the desire to prevent trade barriers caused by different national standards. It was not until the 1980s that environmental considerations came to play a more dominant role (Bennett, 1991). In 1975 a directive was adopted to arrive at reductions in SO_2 emissions caused by gas oil fuels (Johnson and Corcelle, 1989, p. 122). Gas oil includes heating oil for domestic, commercial, or industrial use, as well as diesel fuel for motor vehicles. The directive stipulates that only two types of gas oil are allowed in the internal market of the EC: type A, low-sulfur gas oil (sulfur content below 0.3 percent as of 1980), and type B, with a higher sulfur content to be used only in specific zones. In 1987 a directive was accepted to further reduce the sulfur content in gas oil (Johnson and Corcelle, 1989, p. 122). To avoid important specific air quality problems, especially in urban centers, the limit value was further reduced to 0.3 percent sulfur everywhere in each member state. Furthermore, member states were allowed to enforce the use of gas oils with a sulfur content below 0.3 percent but not below 0.2 percent. On 23 March 1993 the EC accepted a directive (93/12/EEC) to further limit the sulfur content in gas oil to 0.2 percent and, if used as diesel oil in vehicles, to not more than 0.05 percent (Dumas, 1995). In the background was the wish to further standardize the sulfur content in gas oils in order to improve air quality and fuel quality and to enable further reductions (by means of a catalyst) in exhaust gas emissions from diesel engines. The 0.2 percent standard had to be met in 1994 and the 0.05 percent standard must be met in 1996. Greece has been granted a delay; it must ultimately meet the standards on 31 December 1999.

8.4.3 The Framework Directive

The directive of 1984 on combating air pollution from industrial plants is a framework directive in that it foresees that specific emission standards will be set in subsequent directives (Bennett, 1991). The directive requires that new or significantly modified industrial plants be given prior authorization before starting operation. Before issuing authorization the following conditions have to be met:

- The best available technology not entailing excessive costs (BATNEEC) has to be applied to prevent air pollution.
- Emissions must not cause significant air pollution.
- Emission limit values have to be met and air quality limit values must be taken into account.

The plants covered by the directive are in the following sectors: energy, metal production and processing, nonmetallic mineral production, chemicals, waste disposal, and paper pulp manufacturing. Eight substances are seen as the most polluting, among which are SO_2, NO_x, asbestos, and fluorine. The Council of Ministers is empowered to fix emission limit values on the basis of BATNEEC.

8.4.4 The Large Combustion Plants Directive

In 1983 the European Commission presented a directive proposal on the limitation of emissions of pollutants into the air from large combustion plants (LCPs) (Johnson and Corcelle, 1989). Concern about air pollution and significant forest damage observed in Northern Europe triggered the proposal. The proposal accounted for the different laws already existing (Germany) or about to be accepted (the Netherlands) as well as the associated costs and the desire to harmonize national provisions in this area. Initial drafts were modeled on German legislation and included only technology-based emission standards, reflecting the German approach to pollution control called Vorsorgeprinzip, or the precautionary principle. This implies the use of the best technology to prevent pollution (Haigh, 1989). Acceptance of the directive implied that for the first time the EC fixed common emission standards for air pollution from stationary sources. It was, however, politically unacceptable for various countries (among them the UK, Spain, and Greece) to have emission standards for existing plants as is the case in Germany. Because existing plants form the bulk of the emissions, it was therefore agreed to put national ceilings on their emissions. The original proposal for a uniform 60 percent reduction in the emissions of existing plants was later turned into differentiated reductions to account for the different energy situations in countries and to make the proposal acceptable.

When it was finally adopted in November 1988 (OJ, 1988), the directive on the limitation of pollutants in the air from LCPs consisted of two main elements:

- Emission standards for new LCPs depending on the size of the plant;

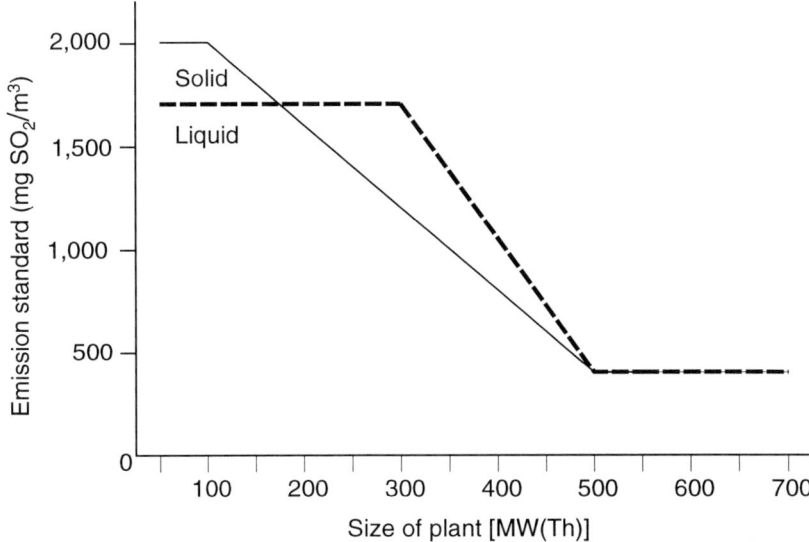

Figure 8.3. EC emission standards for SO_2 for solid and liquid fuels.

- Emission ceilings or bubbles limiting the national sulfur- and nitrogen oxides emissions from existing LCPs in three phases, 1993, 1998, and 2000.

The emission standards for SO_2 for new LCPs are shown in *Figure 8.3*. The directive applies to combustion plants with a rated heat input greater than 50 MW(Th) irrespective of the type of fuel used. A new plant is a plant for which the operating license was granted on or after 1 July 1987. *Figure 8.3* shows that the standards depend on the type of fuel and the size of the plant. The bigger the plant, the more stringent the standards. In 1994 the directive was modified and standards for small coal-fired plants [50 to 100 MW(Th)] are now fixed at 200 mg SO_2/m^3 (Dumas, 1995; see also *Figure 8.3*). If the standards cannot be met, which is sometimes the case with indigenous fuels of poor quality, a minimum degree of desulfurization has to be achieved. The standards for the biggest plants [over 500 MW(Th)] can only be achieved by flue gas desulfurization with a removal efficiency of 90 percent. For medium-sized plants [100 to 300 MW(Th)] combustion modification, partial application of flue gas desulfurization, or low-sulfur fuels (for heavy fuel oil) are sufficient. There is only one exception to these emission standards: until the end of 1999, Spain may authorize new thermal power plants greater than 500 MW(Th) that have to meet less stringent standards. In the case of imported solid fuels the emission limit value is only 800 mg/m³, and

Table 8.4. EC emission ceilings for SO_2 for existing LCPs.

Member state	Emission by LCP 1980 (kton)	Emission ceiling (kton/year)			Change over 1980 (%)		
		1993	1998	2003	1993	1998	2003
Belgium	530	318	212	159	−40	−60	−70
Denmark	323	213	141	106	−34	−56	−67
Germany (West)	2,225	1,335	890	668	−40	−60	−70
Greece	303	320	320	320	+6	+6	+6
Spain	2,290	2,290	1,730	1,440	0	−24	−37
France	1,910	1,146	764	573	−40	−60	−70
Ireland	99	124	124	124	+25	+25	+25
Italy	2,450	1,800	1,500	900	−27	−39	−63
Luxembourg	3	1.8	1.5	1.5	−40	−50	−60
Netherlands	299	180	120	90	−40	−60	−70
Portugal	115	232	270	206	+102	+135	+79
United Kingdom	3,883	3,106	2,330	1,553	−20	−40	−60

in the case of indigenous solid fuels at least a 60 percent (instead of a 90 percent) desulfurization rate is required provided, however, that the new capacity does not exceed 2,000 MW(E) for indigenous solid fuels, and 7,500 MW(E) or 50 percent of all new capacity (whichever is lower) for imported solid fuels (OJ, 1988).

Table 8.4 shows the overall national emission ceilings for SO_2 for existing LCPs. Clearly, emission reductions required from the member states differ considerably according to their respective environmental, economic, and energy situation. Belgium, France, former West Germany, and the Netherlands are aiming at 40 percent, 60 percent, and 70 percent cutbacks, and the UK is aiming at 20 percent, 40 percent, and 60 percent reductions. Greece, Ireland, and Portugal are allowed to increase emissions compared with 1980. These additional emission increases compared with 1980 might occur because plants built and authorized between 1980 and 1 July 1987 are regarded as existing plants.

8.5 National Legislation and Policies

Not all countries in Europe are members of the EC. Moreover, some countries accepted emission and fuel standards before the EC did. In addition, the EC allows countries to have standards that are more stringent than EC standards. Therefore, this section gives an overview of national legislation in European countries to reduce SO_2 emissions (UN/ECE, 1993; UN/ECE, 1991; UN/ECE, 1987) in force before the Second Sulfur Protocol was accepted. Chiefly, this legislation consists of emission standards for combustion installation, fuel standards, and the regulation of

Table 8.5. National emission standards (g SO_2/m^3).[a]

Capacity [MW(Th)]	Austria New	Austria Exist.	Belgium New	Former CSFR New/exist.	Finland New[b]	Germany New/exist.
Solid						
<50	0.4	1–2	–	2.5	–	– / 2.0
50–100	0.2–0.4	1.0	2.0	1.7	0.65	– / 2.0
100–300	0.2–0.4	0.2–1	1.2	1.7	0.4–0.65	2.0 / –
>300	0.2–0.4	0.2–0.4	0.25–0.4	0.5	0.4	– / 0.4
Liquid						
<50	1.7	1.7	–	1.7	1.7	1.7
50–100	0.35	1.1	1.7	1.7	1.7	1.7
100–300	0.35	0.35–1.1	1.7	1.7	1.7	1.7
>300	0.2	0.2	0.25–0.4	0.5	1.7	0.4

Capacity [MW(Th)]	Netherlands New	Netherlands Exist.	Switzerland New/exist.	Poland[c] New	Poland[c] Exist.	Sweden New/exist.
Solid						
<50	–	–	2.0	All sizes 0.6–1.9	All sizes 1.2–4.4	<400t S/yr = 0.57;
50–100	0.7	0.8%S	2.0			
100–300	0.7	0.8%S	0.4		After 1997 0.7–3.0	>400t S/yr = 0.29; (from 1993)
>300	0.4	0.4	0.4			
Liquid						
<50	–	–	1.7	4.5	4.5	<400t S/yr = 0.72;
50–100	1.7	1.7	1.7	0.6	0.6	
100–300	1.7	1.7	0.4	0.6	0.6	>400t S/yr = 0.36; (from 1993)
>300	0.4	1.7	0.4	0.6	0.6	

Sources: UN/ECE (1991), Federal Committee for the Environment (1991), Vernon (1988), Cofala (1991), Rentz *et al.* (1990), MOSZNIL (1990), Nowicki (1993).
[a] Conversion factors used for hard coal: 1 mg/m^3 = 0.35 g/GJ; for HFO: 1 mg/m^3 = 0.28 g/GJ.
[b] For existing > 200 MW(Th): 0.65 (230 g/GJ).
[c] Standard depends on the type of coal and fuel and the type of boiler.

industrial process emissions. In a number of countries, however, the bubble concept is used for power plants (Denmark and the Netherlands), offset rules are applied for emission trading (Germany and Austria), or sulfur taxes are used (Norway and Sweden).

Table 8.5 gives an overview of the emission standards for combustion plants in place in a number of countries in Europe. Countries that have adopted the same standards as the EC directive on LCPs are not explicitly mentioned unless

Table 8.6. National fuel standards (percentage of sulfur).

Fuel type	Austria	Former CSFR	Denmark	Finland	France	FRG	Hungary
Fuel oil							
Light	0.15–0.30	2.0	0.2	0.2	0.3	0.2	0.3
Medium	0.6	2.2	1.0	–	–	–	3.5
Heavy	1.0	3.0	1.0	–	3.0	–	5.5
Solid fuels							
Hard coal	0.4–0.6	0.6–1.5	0.9	1.0	–	–	–
Lignite	–	0.9–4.1	–	–	–	–	–

	Netherlands	Norway	Portugal	Sweden	Switzerland	USSR	Former Yugoslavia
Fuel oil							
Light	0.2	0.2	0.3–1.0	0.2	0.05	0.5–2.0	1.0–1.5
Medium	1.0	0.5	2.0	–	1.0	0.5–3.5	1.0–3.0
Heavy	1.0	1.0	3.0–3.5	–	1.0	0.5–3.5	1.0–4.0
Solid fuels							
Hard coal	1.2	0.8	–	–	0.8	–	–
Lignite	–	–	–	–	–	–	–

Sources: UN/ECE (1991), UN/ECE (1987), Federal Committee for the Environment (1991), Vernon (1988).

their standards are more stringent. The table shows that a large number of countries have accepted regulations to limit the sulfur emission from combustion. Only a few countries limit these emission standards to new installations; most countries have set emission standards for both new and existing installations. Existing installations usually have to comply with these standards within a certain period of time (Austria, Germany, former CSFR, the Netherlands) or have more lenient standards (Austria, Poland). In all but one country (Poland) the emission standards depend on the size of the installation. When comparing the emission standards of different countries, one should be aware of the fact that standards are sometimes set in different units (Sweden, Finland, Poland) and they may differ by coal type (Poland). Furthermore, conversion of standards into a common unit (g SO_2/m^3 flue gas) depends on the heat value and the flue gas volume of the fuel under consideration (Vernon, 1988).

Fuel standards are in place in a number of countries as well (*Table 8.6*). In various EC countries the sulfur content in gas oil was restricted to 0.2 percent before acceptance of the comparable EC directive. Furthermore, a number of countries limited the sulfur content in heavy fuel oil, usually to 1.0 percent, and put an upper limit on the sulfur content in hard coal (0.5–1.0 percent).

On top of this, the following countries have regulated the non-combustion emissions from industrial processes such as smelters, refineries, iron and steel plants, cement plants, paper mills, and gas plants: Austria, Finland, Germany, the Netherlands and Norway (UN/ECE, 1991). Several countries have introduced bubbles for specific sectors as well as economic instruments such as emission and fuel taxes.

Both the Netherlands and Denmark have set a cap on the total annual emissions from power plants (see Chapter 6 for details). In the Netherlands the cap is set at 18 kton SO_2, to be attained in the year 2000 (Vlieg, 1993). The cap is agreed on in a covenant between the state, provincial authorities, and the electricity production board (SEP). In Denmark the cap was set at 125 kton SO_2 to be attained in 1995 and at 100 kiloton for the year 2000 (Sørensen, 1993). In Germany the plant renewal clause allows new plants in a non-attainment area if the new plants apply state-of-the-art technologies and offset remaining emissions by renovating plants in the same area. Also, annual average concentrations must be reduced. The compensation rule is that firms can extend the clean-up period of meeting emission standards to eight years if the emission reduction measures taken at existing installations (owned by the firm or by other firms) provide more emission reductions than would otherwise result from application of the technical guidelines for each individual plant (see Chapter 6). Austria has included a similar compensation rule (or mini-bubble) in its Air Quality Law for Combustion Plants (Luftreinhaltegesetz für Kesselanlagen) (Glatz et al., 1990, p. 47).

In Norway and Sweden fuel taxes are used to promote the use of low-sulfur fuels. In Norway the sulfur content in gas oil is less than 0.15 percent as a result of the sulfur tax. In the northern parts of Norway, where the maximum allowable sulfur content in fuel oil is 0.25 percent, 0.1 percent is used to a greater extent due to the sulfur tax. In the 13 southern and southwestern counties the maximum allowable sulfur content is 1.0 percent. In Oslo and Drammen the maximum allowable level is 0.8 percent. The tax structure is as follows. There is no tax for sulfur contents below 0.05 percent. In the range between 0.05 and 0.25 percent, the tax is 0.07 Norwegian krone (0.01 ECU). For sulfur contents exceeding 0.25 percent there is an additional tax of 0.07 Norwegian krone (0.01 ECU) per liter per 0.25 percent. For example, the sulfur tax for residual oil with 0.95 percent sulfur is 0.28 Norwegian krone (0.03 ECU) per liter (Borge, 1992).

In Sweden a sulfur tax went into force in January 1991 corresponding to 30,000 Swedish krona (3,500 ECU) per ton of sulfur emitted. It is imposed on coal, peat, and oil. Technically, the sulfur tax takes the form of a fuel tax. A tax differentiation is in place for diesel fuel to stimulate the use of environmentally superior grades of diesel oil. As a result of the taxes, diesel fuel with less than 0.01 percent sulfur has a market share of 10 percent and diesel fuel with less than 0.05 percent sulfur, a market share of 60 percent. For heavy fuel oil, the average sulfur content has decreased from 0.65 to 0.4 percent (Bergman *et al.*, 1993; Lövgren, 1994).

A comparison of the emission standards in the Second Sulfur Protocol (*Table 8.3*) with national standards that existed before the Protocol (*Table 8.5*) shows that in Austria, Germany, the Netherlands, Sweden, and Switzerland national legislation was more stringent – standards are lower or they require existing plants as well as new plants to meet standards. Polish standards are sometimes more lenient, sometimes tougher. Regarding the sulfur content in liquid fuels, Austria, Denmark, the Netherlands, Norway, and Switzerland require lower sulfur contents in gas oil or they also limit the sulfur content in heavy fuel oil (see *Table 8.6*).

8.6 Concluding Remarks

This chapter gave a broad overview of the problem of acidification in Europe, especially the role of SO_2, and presented an overview of the institutional framework in Europe. The chapter showed that it was not until the early 1970s that acidification became an issue on the international political agenda. After OECD and Scandinavian studies showed the detrimental impacts of sulfur emissions and their transboundary character, the UN/ECE emerged as the principal institution for dealing with this issue. After signing a framework Convention on Long-range Transboundary Air Pollution, 20 parties signed the First Sulfur Protocol in 1985, stipulating a 30 percent uniform cutback over 1980 emissions. The Second Sulfur Protocol, signed in June 1994, substantially deviates from the first one in that it takes the cost-efficient achievement of deposition goals, more specifically critical loads for sulfur, as its explicit goal. This implies that countries' reductions are no longer uniform, but depend on their transboundary impact on sensitive areas and on their pollution control costs. The Second Sulfur Protocol merges this effect-oriented approach with a source-oriented approach. The latter is precautionary and requires that pollution be prevented as far as technically possible. For the Second Sulfur Protocol, this implies that new installations should apply emission standards based on best available technologies not entailing excessive costs and that the sulfur content of gas oils should be lowered. In addition, the Protocol opens possibilities for parties to jointly implement their national emission ceilings, but rules for this have yet to be agreed on.

The emission standards and fuel standards prescribed in the Protocol are quite similar to the EC directives on emission standards for LCPs from 1988 and on low-sulfur gas oil from 1993. A major difference is that the EC directive on LCPs puts country-specific ceilings on the emissions from existing LCPs. An overview of regulations existing in national legislation inside and outside the EC before the acceptance of the Second Sulfur Protocol shows that this legislation mainly consists of emission standards for combustion installation, fuel standards, and the regulation of industrial process emissions. Some countries have emission standards for new installations that are more stringent than those in the EC and in the Second Sulfur Protocol. Some countries also require existing plants to meet emission standards. Some countries apply the bubble concept to cap the emissions from power plants, some apply offset rules to emission trading, and other countries use sulfur taxes. The general conclusion that emerges here is that joint implementation of sulfur quotas in Europe between countries is likely to be grafted on top of an institutional framework that mainly consists of emission and fuel standards and, for EC countries, of SO_2 emission quotas for existing LCPs.

Chapter 9

Trading with Exchange Rates

9.1 Introduction

The first part of this study (Chapters 2 to 7) examined the potential cost efficiency and environmental effectiveness of emission trading in theory, in simulation models, and in practice. It was shown among other things that several emission trading schemes such as ambient permit trading or trading using rules on deposition would in theory be both cost efficient and environmentally effective. Moreover, in an international context both ambient permit trading and trading using deposition rules are Pareto-dominant improvements over initial cooperative Nash agreements. Simulation models confirm the potential cost savings of trading subject to deposition constraints (the non-degradation offset rule); however, these savings might be smaller if trade is modeled as a bilateral sequential process. The overview of practical experience shows that ambient permits have not been applied in practice. Application of the non-degradation offset rule (no increase in emissions or ambient concentrations) has not been successful. High transaction costs, such as from running an air quality model and obtaining approval for every trade, have made this form of trading unattractive in practice. The best alternative is single-zone emission trading. This system, however, is not a least-cost solution for meeting deposition standards, and simulation models show that single-zone trading might be more or less expensive than CAC types of regulations or uniform reductions. Furthermore, single-zone emission trading does not guarantee a Pareto-dominant improvement of an initial international agreement. In practice, however, single-zone trading (such as lead trading or sulfur trading) has been very successful at saving costs.

This chapter and the following chapters will discuss an alternative form of trading that tries to save costs while controlling depositions. This alternative is emission trading with a fixed exchange rate that reflects the location-specific impact of trading. The main reason for introducing this scheme in the second part of this study is that the idea of trading with fixed exchange rates has been developed in the context of discussions on how to improve the cost efficiency of the Second Sulfur Protocol. One of the possible schemes for joint implementation (see Chapter 8) under the Second Sulfur Protocol is to allow trading according to a fixed exchange rate (Klaassen and Amann, 1992). An exchange rate or offset rate is defined as "the volume of emissions that one source has to decrease if another source increases its emissions by one unit." This concept is similar to the pollution offset trading provisions in the USA in the eighties (Borowski and Ellis, 1987; Hahn, 1986). The UN/ECE also debated what the correct exchange rate should be. There are questions about whether such exchange rate trading can really achieve the cost-minimum solution when the environmental objective is the attainment of a set of region-specific targets for the deposition of sulfur in Europe. There are also questions on the environmental effects and possible violations of deposition objectives and the infringement of third-party rights.

The objective of this chapter is to analyze the cost efficiency and environmental effectiveness of exchange rate trading and to examine its merits, especially in an international context. This chapter is of a theoretical nature. Chapter 10 will show the results of a simulation model and Chapter 11 will discuss institutional design aspects.

Section 9.2 elucidates the concept of exchange rate trading. Section 9.3 examines the cost efficiency, environmental impacts, and distributional consequences of exchange rate trading.

9.2 The Concept of Exchange Rate Trading

An alternative trading system that is relatively simple and applicable in practice is emission trading subject to an exchange rate. An exchange or offset rate, as defined above, states that if one source increases emissions by one unit (it buys permits), another source must decrease its emissions by the same amount multiplied by the exchange rate (it sells permits). Therefore, an exchange rate exceeding one implies that the decrease in emissions of one source exceeds the increase in emissions of the other (buying) source. The environmental agency fixes the exchange rates exogenously. Individual sources must use these exchange rates when trading. Given their private knowledge of costs, it is left to the sources to decide what volumes they want to trade. The regulatory agency also does not put any constraints on the depositions. The system therefore differs from the pollution offset trading or non-degradation rules (see Chapter 4) in two respects:

- With exchange rate trading the exchange rates are exogenously fixed; with pollution offset trading the sources can choose the rates as long as the deposition constraints are not violated.
- With exchange rate trading there are no explicit deposition constraints.

A small model will assist in structuring the problem. The following variables are defined: E_x^0 and E_y^0 are the pre-trade emissions of sources x and y, respectively; E_x, E_y, and E_i are the post-trade emissions of sources x, y, and i, respectively; T_x and T_y are the changed emissions of sources x and y, respectively, as a result of trading (the volumes of trades); and w_{xy} is the exchange rate (the rate at which y has to decrease emissions if x increases emissions by one unit). The cost functions C_x and C_y represent costs as functions of the remaining emissions after control. The cost functions are assumed to have the following properties: marginal costs increase when emissions decrease and increase disproportionally if emissions are decreased further. Assuming that both sources involved in a trade want to minimize costs, the problem of emission trading subject to a fixed exchange rate can be stated as follows:

Minimize

$$C_x + C_y \tag{9.1}$$

subject to

$$E_x = E_x^0 + T_x , \tag{9.2}$$

$$E_y = E_y^0 - T_y , \tag{9.3}$$

$$T_y = w_{xy} * T_x . \tag{9.4}$$

All volumes are nonnegative. Conditions (9.2) to (9.4) can be reformulated as one condition:

$$E_y - E_y^0 = -w_{xy}(E_x - E_x^0) . \tag{9.5}$$

This shows that the change in emissions of source y as a result of trading depends on the exchange rate and the volume traded by source x. If the exchange rate equals one, total emissions have to be constant during trading. Shifting independent variables to the right-hand side and dependent variables to the left-hand side gives

$$E_y + w_{xy} * E_x = E_y^0 + w_{xy} * E_x^0 . \tag{9.6}$$

Trading with Exchange Rates 213

Equations (9.1) and (9.6) form a classic problem of programming subject to an equality constraint (Intriligator, 1971; Chiang, 1984). The solution can be found in formulating the Lagrangian function H:

$$H = C_x + C_y + \mu\left\{(E_y^0 + w_{xy} * E_x^0) - (E_y + w_{xy} * E_x)\right\} , \quad (9.7)$$

where μ is the Lagrange multiplier. The first-order or necessary conditions for a solution can be found by taking the partial derivatives of H with respect to E_x, E_y, and μ:

$$C'_x - \mu * w_{xy} = 0 , \quad (9.8a)$$

$$C'_y - \mu = 0 , \quad (9.8b)$$

$$E_y + w_{xy} * E_x = E_y^0 + w_{xy} * E_x^0 , \quad (9.8c)$$

in which C' indicates marginal costs. Note, however, that these first-order conditions are sufficient conditions for an absolute minimum only under certain provisions regarding the shapes of the objective and constraints functions. More specifically, if the objective function is explicitly quasi-convex and the constraint set is convex, which is the case in the example, we have an absolute constrained minimum. If the objective function is strictly quasi-convex, we have a unique constrained minimum. These first-order conditions can be combined by eliminating μ:

$$C'_x = w_{xy} * C'_y . \quad (9.9)$$

In this case the following situations are possible:

- If in the pre-trade situation $C'_x > w_{xy} * C'_y$, then country x could profit by paying country y to purify more while increasing its own emissions.
- If in the pre-trade situation $C'_x < w_{xy} * C'_y$, then it pays for country x to reduce emissions further (hence increase marginal costs) and allow country y to increase emissions.

Thus, as long as C'_x is not equal to $w_{xy} * C'_y$, there is an incentive to trade because cost savings are possible.

Condition (9.9) can be interpreted as follows: if $w_{xy} = 1$, we have the classic condition for a cost-minimum solution for meeting an emission goal stating that marginal costs of both sources have to be equal. If w_{xy} is not equal to one (e.g., $w_{xy} > 1$) and initially $C'_x > w_{xy} * C'_y$, then the exchange rate implies that source y will have to decrease its emissions more than source x is allowed to increase its emissions. Accordingly, it is more difficult for source y to enter a cost-minimizing trade (it requires more effort, hence greater costs). To compensate for the fact that emission increases for source x are smaller than the emission reduction required for source y, the marginal costs of source x (in the optimum) have to be w_{xy} times higher than the marginal costs of source y in order to achieve a cost-minimum solution.

9.3 An Exchange Rate Equal to the Ratio of the Marginal Costs in the Optimum

9.3.1 Background and rationale

One possibility for selecting the exchange rate is to base the exchange rate on the ratio of the marginal costs in the least-cost solution. As will be explained, this rule can be based on the understanding that the ratio of the marginal costs in the optimum depends on the shadow prices (relative difficulties) of attaining the binding deposition constraints. It reflects one of the conditions for a cost-minimum solution if the problem is the minimization of costs subject to constraints on deposition. This chapter will concentrate on this type of exchange rate. Chapters 10 and 11 will examine other variants, as well, and will compare them with the cost-minimum exchange rates.

In the remainder of this chapter, the conditions for a least-cost solution for attaining a set of deposition targets are analyzed (see also Chapter 3). The conditions under which exchange rate trading can achieve this cost minimum if there are only two sources are then analyzed. The remaining sections extend the analysis to three sources.

The following additional elements are defined: D_j is the level of deposition at receptor j; D_j^0 is the desired level of deposition at receptor j; and a_{ji} is a linear transfer coefficient that translates emissions of source i to deposition at receptor j. The deposition at a specific location is a function of the emissions (E_i) multiplied by their transfer coefficients. The cost-efficient solution requires the total costs (C_i) of emission reductions to be minimized subject to the constraint that the desired deposition levels are met at each receptor point. From Chapter 3 we know that the most important of the necessary Kuhn-Tucker conditions for a cost minimum are the following:

$$(E_{i,\max} - E_i)\left[C_i' - \sum_{j=1}^{J} a_{ji}\lambda_j\right] = 0 \quad \text{for every i}, \tag{9.10}$$

$$\lambda_j\left[D_j^0 - \sum_{i=1}^{I} a_j i * E_i\right] = 0 \quad \text{for every j}. \tag{9.11}$$

Trading with Exchange Rates 215

Furthermore, emission reductions and Lagrange multipliers λ_j must be nonnegative, marginal costs per ton of emissions removed have to be equal to or higher than the sum of the shadow prices for each receptor affected by that source, and the deposition at each receptor has to be equal to or less than the targets.

Equation (9.10) states that for a cost-efficient solution either the emission reduction of the source has to be zero or the marginal costs of emission reduction for each source have to equal a weighted sum of the shadow prices (λ_j) for each receptor and reflect the marginal costs of tightening the constraints. Weights are the transfer coefficients from source i to each affected receptor j. Equation (9.11) shows that either the required target load (D_j^0) is met exactly or the corresponding shadow price (λ_j) is zero. The latter means that the receptor is nonbinding. The main message is that in the cost-minimum solution the marginal costs are a function of the transfer coefficients at the binding receptors and the prevailing shadow prices unless no emission reduction is required. From equation (9.10) it follows that for two sources (or parties x and y) this first-order condition for a least-cost solution can be written as

$$\frac{C'_x}{C'_y} = \frac{\sum a_{jx}\lambda_j}{\sum a_{jy}\lambda_j} .$$

If the task is to realize the cost minimum by way of emission trading with a fixed exchange rate, one possibility for selecting an exchange rate is to base the exchange rate on the ratio of the marginal costs in the cost-minimum solution.

If the ratio of the marginal costs in the optimum is used to determine the exchange rate, the exchange rate will be [making use of equation (9.9)]

$$w_{xy} = \frac{C'_x}{C'_y} = \frac{\sum a_{jx}\lambda_j}{\sum a_{jy}\lambda_j} = \frac{a_{1x}\lambda_1 + a_{2x}\lambda_2 + \cdots + a_{jx}\lambda_j}{a_{1y}\lambda_1 + a_{2y}\lambda_2 + \cdots + a_{jx}\lambda_j} . \quad (9.12)$$

To explain how this "optimal" exchange rate governs trading, the case of one receptor is analyzed. More receptors are then included in the analysis.

9.3.2 Two countries and one receptor

To see how optimal exchange rate trading works, let us assume a transfer coefficient of 0.5 from source x to one receptor and a transfer coefficient of 1 from source y to the same receptor. In the optimum the following applies:

$$w_{xy} = \frac{C'_x}{C'_y} = \frac{a_{1x}\lambda_1}{a_{1y}\lambda_1} = \frac{a_{1x}}{a_{1y}} = 0.5 . \quad (9.13)$$

The ratio 0.5 is the exchange rate w_{xy}. This implies that if source x were to decrease its emissions by one unit, source y would be allowed to increase its

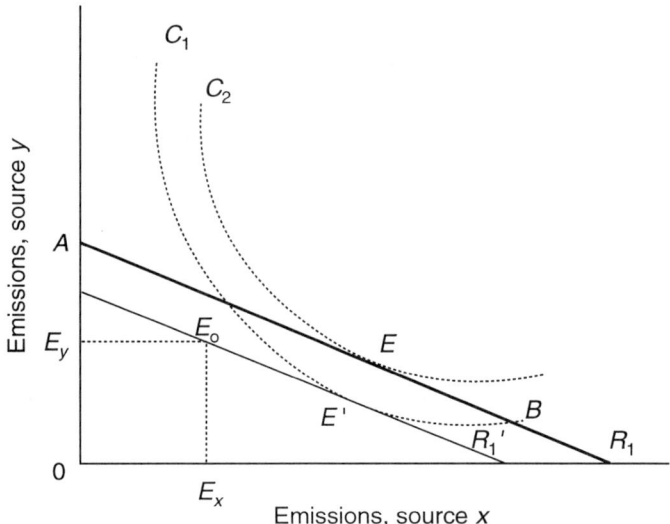

Figure 9.1. Exchange rate trading with two sources and one receptor.

emissions by only half a unit. Clearly, in this case the deposition target load will not be violated during trading, provided it was not exceeded before trading. This is because in this example the exchange rate specifies that y has to decrease its emissions by 0.5 units if x increases its emissions by 1 unit. In other words, if x wants to increase its emissions by 1 unit it must buy 0.5 emission permits from source y. If this happens, deposition at the receptor would increase by $0.5 \times 1 = 0.5$ units, because x increases by 1 unit. However, deposition would decrease by the same amount because y would have to decrease emissions by 0.5 units (thereby lowering deposition by $0.5 \times 1 = 0.5$ units). Because the exchange rate is obligatory and is based on the transfer coefficients of the receptor, deposition at the receptor will always be constant. Moreover, knowledge of the shadow prices of the receptor is not needed because knowledge of transfer coefficients would be sufficient to set the exchange rates.

Figure 9.1 illustrates the case for one receptor and two sources. The y-axis of *Figure 9.1* shows the emissions of source y and the x-axis shows the emissions of source x. Line R_1 shows combinations of emissions from both sources for which the deposition target at receptor 1 is met exactly. The curves C_1 and C_2 are iso-cost curves, combinations of emissions from both sources that lead to the same level of total costs. The closer these costs are to the origin, the higher are the costs and the lower the emissions.

Trading with Exchange Rates

Figure 9.1 shows that as long as emissions from both sources remain within area OAB the deposition targets are met. The least-cost solution is point E. At this point the ratio of the marginal costs equals the ratio of the transfer coefficients for receptor 1, which at point E are both binding. The ratio of the transfer coefficients is also the coefficient that determines the direction of line R_1. This can be seen if one rewrites condition (9.11) so that the emissions of y are a function of the emissions of x and the deposition target:

$$E_y = \frac{D_1^0}{a_{1y}} - \frac{a_{1x}}{a_{1y}} E_x \ . \tag{9.14}$$

An exchange rate based on the ratio of the optimal marginal costs allows both sources to trade along a line parallel to line R_1. Trading could, for example, take sources from point $E^0(E_y^0, E_x^0)$ to point E' (on line R_1'). From condition (9.6) for exchange rate trading, one can also express E_y as a function of other variables:

$$E_y = E_y^0 + w_{xy} E_x^0 - w_{xy} E_x \ . \tag{9.15}$$

Clearly, if both equations lead to similar results for E_y, exchange rate trading will achieve the cost-minimum solution. This is the case if the following conditions are fulfilled:

- The exchange rate has to equal the rate of the transfer coefficients: $w_{xy} = a_{1x}/a_{1y}$.
- In the pre-trade situation the target has to be met exactly: $a_{1x} * E_x^0 + a_{1y} * E_y^0 = D_1$.
- The trading partners have the objective of minimizing joint costs.

The first condition is met if the exchange rate is based on the cost-minimum solution. The second, however, is only met if initially the deposition target is met exactly. In that case the initial distribution E_x^0, E_y^0 is a point on line R_1.

A simple example can serve as an illustration. *Table 9.1* shows the data, the cost-minimum solution, the pre-trade situation (30 percent uniform cutback), and the post-trade emissions for a case with two countries and one receptor. Part I of *Table 9.1* gives the data. Marginal costs are a linear function of the emission reduction (R) in both countries. Total costs are a quadratic function of R. Currently, 5 percent of the Polish emissions and 50 percent of the Swedish emissions come down at the receptor in Sweden. The goal is a 30 percent reduction in deposition. In the least-cost solution marginal costs for Sweden have to be 10 times higher than for Poland (reflecting that the transfer coefficient of Sweden is 10 times that of Poland). In other words, the marginal costs per ton of deposition are made equal. The pre-trade situation is defined as a 30 percent flat rate reduction. Clearly, this is

Table 9.1. Two countries and one receptor.

	Marginal costs (DM/kton)	Total costs (DM)	Transfer coefficient (fraction)	Emission reduction (kton)	Unabated emission (kton)	Deposition (kton)
I. Data						
Poland	2R	R²	0.05	–	3,200.0	160.0
Sweden	20R	10R²	0.50	–	400.0	200.0
Subtotal	–	–	–	–	–	360.0
II. Pre-trade (30 percent uniform reduction in emissions)						
Poland	1,920.0	921,600	–	960.0	2,240.0	112.0
Sweden	2,400.0	144,000	–	120.0	280.0	140.0
Subtotal	–	1,065,600	–	–	–	252.0
III. Cost minimum						
Poland	392.7	38,559	–	196.4	3,003.60	150.2
Sweden	3,927.0	385,587	–	196.4	203.64	101.8
Subtotal	–	424,146	–	–	–	252.0
IV. Post-trade (Poland buys from Sweden)						
Poland	392.7	38,559	–	196.4 (–763.6)	3,003.60	150.2
Sweden	3,927.0	385,587	–	196.4 (+76.36)	203.64	101.8
Subtotal	–	424,146	–	–	–	252.0

more costly than the least-cost solution. Exchange rate trading would allow Sweden to trade with Poland using the rates of the marginal costs in the cost minimum. This implies that Poland would only have to buy 0.1 unit of reduction in Sweden to increase its emissions by 1 unit. The last part of *Table 9.1* shows that Poland would buy 76.36 kton from Sweden and could therefore increase its emissions by 763.6 kton. In this case, the countries would end up at the cost-minimum solution because they started from a situation where the constraints were binding.

The question remains whether parties that act out of self-interest will engage in trade under a regime of fixed exchange rate trading. This should be the case if given that reduction of total costs is the aim of every party total cost savings can be distributed in such a way that no single party is worse off than it was before. Let M be the monetary transfer paid to compensate the country that increases its reduction of emissions. Emission trading can then be modeled as minimizing the total costs (of emission reduction and side payments) of x (or y) under the restriction that the total costs of y (or x) do not increase. This implies minimizing the Lagrangian

$$C_x + M + \mu(C_y^0 - C_y - y) + \lambda \left\{ w_{xy}(E_x^0 - E_x) + E_y^0 - E_y \right\} , \qquad (9.16)$$

Trading with Exchange Rates

where C_y^0 are the pre-trade emission reduction costs of country y. The first-order conditions are

$$\begin{aligned} C_x' - \lambda w_{xy} &= 0, \\ -\mu C_y' - \lambda &= 0, \\ 1 - \mu &= 0. \end{aligned} \tag{9.17}$$

Combining these conditions gives

$$\frac{C_x'}{C_y'} = w_{xy}.$$

If the exchange rate is chosen *ex ante* so that $w_{xy} = a_{1x}/a_{1y}$, then $C_x'/C_y' = a_{1x}/a_{1y}$.

It follows that the joint cost minimum is also a Pareto optimum and a solution that could be acceptable for both countries x and y. In *Table 9.1* the joint cost savings are DM 640,000. This results from a cost decrease of around DM 880,000 in Poland and a cost increase in Sweden of around DM 240,000. Part of the cost decrease in Poland could be used to compensate Sweden for its cost increase.

In conclusion, with one receptor and two trading sources, the cost minimum is attained if the one receptor is binding initially. If the deposition target is not violated initially (initial point is inside the feasible area), it will not be violated after trading because trading will occur parallel to the constraint.

9.3.3 Two sources and two receptors

A pertinent question is whether this result is transferable when there are two (or more) binding receptors in the cost minimum. First, we will supply an analytical examination, then we will give a graphical illustration.

Equation (9.12) showed that the least-cost solution implies

$$\frac{C_x'}{C_y'} = \frac{a_{1x}\lambda_1 + a_{2x}\lambda_2}{a_{1y}\lambda_1 + a_{2y}\lambda_2}. \tag{9.18}$$

Equation (9.17) already showed that trade subject to the exchange rate gives

$$\frac{C_x'}{C_y'} = w_{xy}.$$

Setting the exchange rate equal to the right-hand side of equation (9.18) implies that one condition for exchange rate trading is similar to one of the conditions for

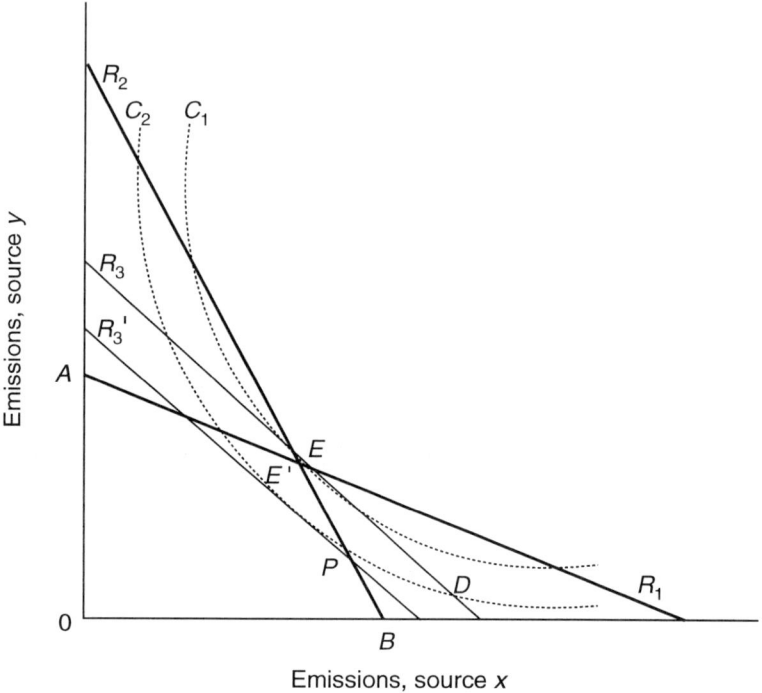

Figure 9.2. Exchange rate trading with two sources and two binding receptors.

a least-cost solution. Note that condition (9.11) for a least-cost solution is not by definition satisfied by exchange rate trading.

Figure 9.2 illustrates this; it is comparable to *Figure 9.1*, the only difference is that there are now two receptors, R_1 and R_2. These are both assumed to be binding in the least-cost solution (point E). At this point, the ratio of the marginal costs equals the weighted ratio of their transfer coefficients for R_2 and R_1 [see equation (9.18)].

This ratio is also the coefficient that determines the direction of line R_3 (weighted between R_2 and R_1). This implies that an exchange rate based on the ratio of the optimal marginal costs allows both sources to trade as long as they move (trade) along a line parallel to line R_3. Again, attainment of the optimum (E) is contingent on the initial solution. If the initial position is point P, the trading ratio prevents the optimum from being reached, although cost savings are possible. Starting from point P, sources would only be allowed to trade along line R_3' (parallel to R_3). In this case, E' would be the least-cost solution attainable from P

Trading with Exchange Rates

with the given trade ratio. Obviously E' is not identical to E, and therefore costs (C_2) are higher than in the cost minimum (C_1). It appears that trade could end up at the cost minimum if it starts from any point on line R_3. This is because given the fact that iso-cost curves cannot intersect all other points on line R_3 would be more expensive. An interesting observation is that trade could end up in a situation where both deposition constraints are met (point E) even if trade starts in a situation where only one constraint is met. This would be the case if trade were to start at point D, for example.

An important question now is whether or not exchange rate trading might result in a violation of the deposition constraints. *Figure 9.3* suggests that this might happen. Even if both deposition constraints are met initially (trade starting from inside the area $OAEB$), trade might go outside the feasible solution depending on the configuration of the iso-cost curves. If trade starts from point P, trading subject to the exchange rate could end up at point D. At point D, the deposition constraint for receptor 2 will be met, but the constraint for receptor 1 will be violated.

To simplify notation, the following matrices are defined:

$$A = \begin{pmatrix} a_{1x} & a_{1y} \\ a_{2x} & a_{2y} \end{pmatrix}, \quad D = \begin{pmatrix} D_1 \\ D_2 \end{pmatrix}, \quad E^0 = \begin{pmatrix} E_x^0 \\ E_y^0 \end{pmatrix},$$

$$S^0 = \begin{pmatrix} S_1^0 \\ S_2^0 \end{pmatrix}, \quad E = \begin{pmatrix} E_x \\ E_y \end{pmatrix}, \quad w = (w_{xy} \; 1), \quad (9.19)$$

where S^0 is a matrix of surplus variables S_j at each receptor, denoting the extent to which the actual deposition is below the deposition constraint for that receptor. The question now can be formulated in matrix notation. If initially emission levels E^0 are within the feasible region, this implies that

$$AE^0 = D^0 - S^0 . \tag{9.20}$$

The question then is whether $AE < D^0$, with D^0 being the vector of deposition targets. Recall condition (9.6) on exchange rate trading:

$$E_y + w_{xy} * E_x = E_y^0 + w_{xy} * E_x^0 ,$$

which is written in matrix form as

$$WE = WE^0 . \tag{9.21}$$

If matrix A is non-singular, one can multiply equation (9.20) by A^{-1}, the inverse matrix. This gives: $AA^{-1}E^0 = A^{-1}(D^\circ - S)$, or $E^0 = A^{-1}(D^0 - S)$. Substitution into condition (9.21) gives

$$WE = WA^{-1}(D^0 - S) .$$

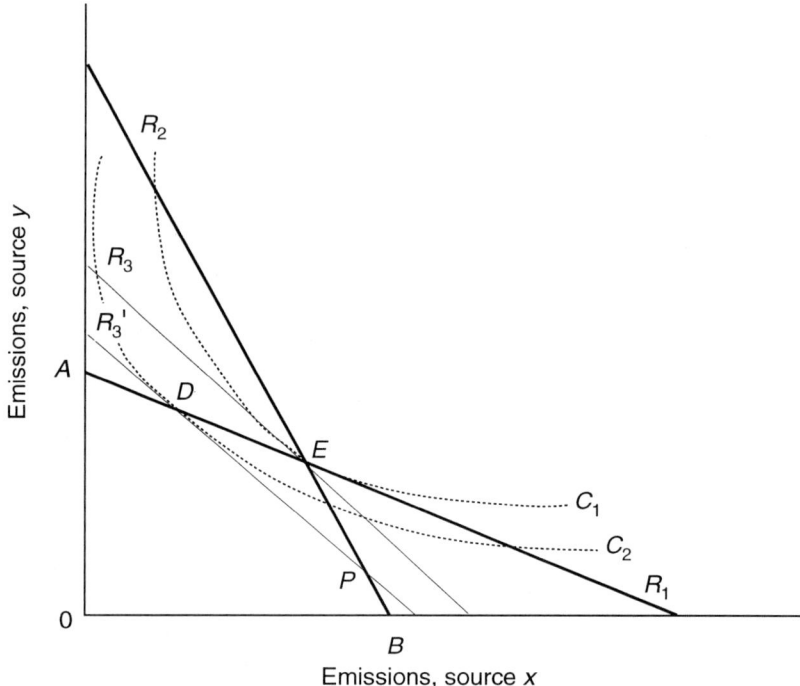

Figure 9.3. Violation with two receptors and two constraints.

Multiplying W by A and using $AE = D$, gives

$$WAE = WAA^{-1}(D^0 - S) ,$$

or

$$WD = W(D^0 - S) . \tag{9.22}$$

This does not necessarily mean that $D = D^0 - S$, because W might be a singular matrix (roughly speaking, a matrix containing a row that is a multiple of another row; Chiang, 1984, p. 81). Longhand notation of condition (9.22) may improve understanding:

$$w_{xy}D_1 + D_2 = w_{xy}(D_1^0 - S_1^0) + (D_2^0 - S_2^0) ,$$

or

$$D_2 - D_2^0 = w_{xy}(D_1^0 - D_1) - S_2^0 - w_{xy}S_1^0 . \tag{9.23}$$

If the left-hand side is greater than zero ($D_2 > D_2^0$) the deposition target at D_2 is not met. Then, however, the right-hand side must also to exceed zero. This can be the case if

$$D_1^0 - S_2^0/w_{xy} - S_1^0 > D_1 \ .$$

Because the surplus variables are positive if both deposition constraints are met initially, the deposition at receptor 1 is always met. Consequently, the deposition at one of the receptors can be exceeded even if both deposition objectives are met initially. If the constraint is exceeded at one receptor, however, deposition will always be lower than the constraint at the other receptor. This confirms what *Figure 9.3* suggests. Starting from point P and trading along the line R_3' might result in a trading equilibrium where one of the receptors is violated. If neither trading country wants this to happen, they have to run a deposition model to check whether the deposition after trade will exceed the constraints.

In conclusion, exchange rate trading with two sources and two binding receptors in the least-cost solution can only achieve the cost minimum under a special provision concerning the pre-trade emissions: these emissions have to be on a line parallel to the tangent of the iso-cost curves at the optimum going through the optimum. In all circumstances trading will improve cost efficiency (costs will be lower than in the initial situation), but the cost minimum will not be attained. A second problem is that environmental quality may deteriorate. Even if exchange rate trading starts from inside the feasible region, the analysis suggests that one deposition target might be violated even if the other is met.

9.3.4 Three countries and one receptor

This section analyzes exchange rate trading with three sources and one receptor. The cost-minimum solution is described and the optimum conditions for bilateral sequential exchange rate trading are given where two of the three countries trade bilaterally in a specific sequence.

The cost-minimum formulation is the same as before [see conditions (9.10) and (9.11)]. With three sources x, y, and z, and one receptor the first-order conditions for an optimum are

$$C_x' = a_{1x}\lambda_1 \ , \quad C_y' = a_{1y}\lambda_1 \ , \quad C_z' = a_{1z}\lambda_1 \ ,$$

$$a_{1x}E_x + a_{1y}E_y + a_{1z}E_z = D_1^0 \ . \tag{9.24}$$

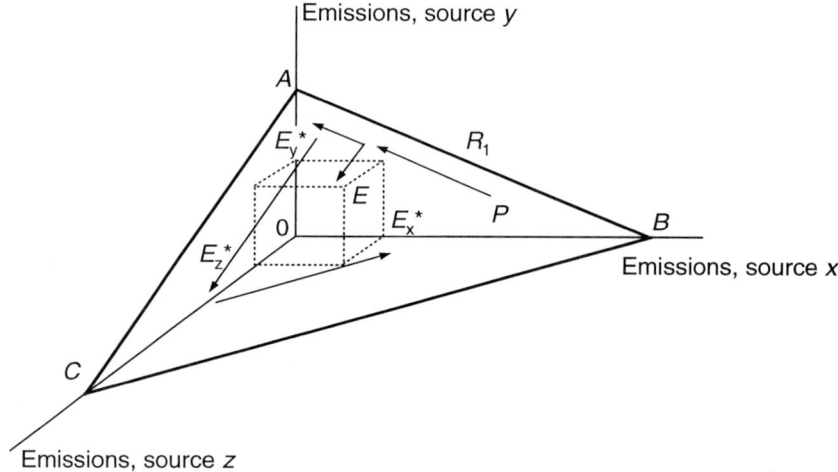

Figure 9.4. Exchange rate trading with three countries and one receptor.

In this case the ratios of the marginal costs (used for determining the "optimal" exchange rate) clearly equal the ratios of the transfer coefficients of each source at the receptor:

$$\frac{C'_x}{C'_y} = \frac{a_{1x}}{a_{1y}}, \quad \frac{C'_x}{C'_z} = \frac{a_{1x}}{a_{1z}}, \quad \frac{C'_y}{C'_z} = \frac{a_{1y}}{a_{1z}}. \tag{9.25}$$

Obviously, in this case trading with an exchange rate based on the optimum cost ratio involves trading according to each source's transfer coefficient to the receptor. Consequently, if the deposition constraint is initially met (pre-trade deposition is at or below the standard), bilateral trade cannot result in a violation of the constraint: the increase in deposition caused by the increase of one source's emissions is exactly compensated by the other source's emissions decrease. *Figure 9.4* shows this. The figure is similar to *Figure 9.1* with the exception that the constraint R_1 is now a surface rather than a line. Exchange rate trading allows trading along a surface parallel to the surface R_1.

Again, one can ask which conditions lead to attainment of the optimum using "optimal" exchange rate trading. *Figure 9.4* suggests that this is possible (but not guaranteed) only under the condition that the constraint is binding initially; that is, if we start on surface R_1. Given that we have three sources and only one receptor, it is clear that sufficient degrees of freedom exist to meet the one deposition constraint through various combinations of emissions.

In the case of simultaneous Pareto-optimal trade between the sources, it is possible to show a situation where the cost minimum is attained. As in equation (9.5), the exchange rate constraint can be written as

$$E_y - E_y^0 + w_{xy}(E_x - E_x^0) + w_{zy}(E_z - E_z^0) = 0 \ . \tag{9.26}$$

For Pareto-optimal trade, the following Lagrangian can be formulated for country x, where Y_x, Y_y, and Y_z are monetary side payments from x, y, and z, respectively:

$$\begin{aligned} C_x &+ M_x + \mu_1(C_y^0 - C_y - M_y) + \mu_2(C_z^0 - C_z - M_z) \\ &+ \rho(M_x + M_y + M_z) \\ &+ \lambda \left\{ w_{xy}(E_x - E_x^0) + (E_y - E_y^0) + w_{zy}(E_z - E_z^0) \right\} \ . \end{aligned} \tag{9.27}$$

C_y^0 and C_z^0 are the pre-trade emission control costs of y and z. The first-order conditions are

$$\begin{aligned} C_x' + \lambda w_{xy} &= 0 \ , \\ -\mu_1 C_y' + \lambda &= 0 \ , \\ -\mu_2 C_z' + \lambda w_{zy} &= 0 \ , \\ +1 + \rho &= 0 \ , \\ -\mu_1 + \rho &= 0 \ , \\ -\mu_2 + \rho &= 0 \ . \end{aligned} \tag{9.28}$$

Rewriting gives

$$\frac{C_x'}{C_y'} = w_{xy} \quad \text{and} \quad \frac{C_z'}{C_y'} = w_{zy} \ . \tag{9.29}$$

If the exchange rates are set at the cost-minimum level (that is, $w_{xy} = a_{1x}/a_{1y}$ and $w_{zy} = a_{1z}/a_{1y}$), exchange rate trading leads to one condition that is similar to one of the conditions for a cost-minimum solution.

Figure 9.4 also suggests that there might be combinations of bilateral trades that lead to the cost minimum E if one starts from surface R_1 (see arrows). There might also be paths that do not lead to the optimum. In this situation, the traded volumes depend on the cost of pollution control if both sources minimize costs. This determines the length of the arrow in *Figure 9.4*.

An analytical solution might elucidate matters. In this case we will assume that trading is a bilateral sequential process. First, two countries trade then two other countries trade, and so forth. For bilateral trading with an exchange rate for every possible trade for source y, the following conditions hold:

$$C'_x/C'_y = w_{xy}E_y = E_y^0 + w_{xy} * E_x^0 - w_{xy} * E_x ,\qquad(9.30)$$

$$C'_z/C'_y = w_{zy}E_z = E_z^0 + w_{zy} * E_y^0 - w_{zy} * E_y .\qquad(9.31)$$

From the second line in condition (9.24), the cost minimum level of E_y, analogous to equation (9.14), must be such that the deposition constraint is met. This implies

$$E_y = \frac{D_1^0}{a_{1y}} - \frac{a_{1x}}{a_{1y}}E_x - \frac{a_{1z}}{a_{1y}}E_z , \text{ or } E_z = \frac{D_1^0}{a_{1z}} - \frac{a_{1y}}{a_{1z}}E_y - \frac{a_{1x}}{a_{1z}}E_x . \qquad(9.32)$$

Assume the first trade to take place is between sources x and y. One of the exchange rate conditions for a bilateral trade is the right-hand side of condition (9.30) [see also condition (9.15)]. Exchange rate trading with one receptor has one obvious requirement if the exchange rates are based on the marginal costs in the cost-minimum solution [equations (9.25) and (9.30)]: the exchange rate must be equal to the rate of the transfer coefficients: $w_{xy} = a_{1x}/a_{1y}$. This knowledge can be used on the right-hand side of condition (9.30). Combining conditions (9.30) and (9.32) can be used for rewriting condition (9.32) to give the least-cost value of E_z:

$$E_z = \frac{D_1^0}{a_{1z}} - \frac{a_{1y}}{a_{1z}}E_y^0 - \frac{a_{1x}}{a_{1z}}E_x^0 . \qquad(9.33)$$

This gives the optimal emissions of source z as a function of the pre-trade emissions of sources x and y. Because the emissions of source z do not change when sources x and y trade, condition (9.33) should also hold for pre-trade emissions of source z. Rewriting condition (9.33) gives

$$a_{1x} * E_x^0 + a_{1y} * E_y^0 + a_{1z} * E_z^0 = D_1^0 ,$$

which simply implies, as Figure 9.4 already suggests, that one requirement for bilateral trading to attain the cost minimum is to start at a point where the deposition constraint is exactly met.

The cost-minimum value of the emissions of source y equals the optimum exchange rate value of source y if and only if the pre-trade emissions of source z are already equal to their optimal (cost-minimum) emission level. If this is not the case the emissions of sources y and x will not attain their cost-minimum values. This being the case, we can argue that further trades (where none of the emissions is at its cost-minimum value to start with) will not be able to reach the cost minimum

Trading with Exchange Rates

because the pre-trade emission value of the non-trading source (in this case, source z) will always be off the optimum value. This is, of course, tantamount to saying that for trading to be optimal all sources have to be included in the trading process. There is a way to illustrate this. After the first trade between sources x and y the marginal costs have to equal the exchange rate: $C'_x/C'_y = w_{xy} = a_{1x}/a_{1y}$. If sources x and y are both optimal, then $C'_x = \lambda a_{1x}$, $C'_y = \lambda a_{1y}$. If source x is optimal, then source y is necessarily optimal for the exchange rate condition to hold. There would be no further room for trading between sources y and z only if $C'_z/C'_y = w_{zy} = a_{1z}/a_{1y}$. Because $C'_y = \lambda a_{1y}$, this can only be the case if the marginal costs of source z are optimal: $C'_z = \lambda a_{1z}$. If C'_z is not optimal, trading between sources y and z will reduce costs, thus changing the marginal costs of source y, which then is no longer optimal.

An example can illustrate how exchange rate trading with one receptor (Sweden) and three countries would work. *Table 9.2* shows the data, the cost-minimum solution, the pre-trade emissions situation (based on a 30 percent flat rate reduction in emissions), and the cost-minimum exchange rates. Part I of *Table 9.2* shows marginal costs, total costs, and transfer coefficients (based on a single receptor, Sweden). R is the emission reduction. Part II shows the cost-minimum solution for reducing deposition at the receptor by 30 percent (down to 280 kton SO_2). Clearly, marginal costs in Sweden are the highest because it has the highest impact on the receptor. Swedish marginal costs are twice those of Norway because the transfer coefficient of Sweden is also twice as high as that of Norway. Polish marginal costs are smaller, reflecting its smaller impact on deposition in Sweden. The pre-trade emission situation is a 30 percent uniform reduction in every country. Obviously, total costs are much higher than the least-cost solution (DM 1,123,200 versus DM 480,000). *Table 9.2* also shows the exchange rates based on the least-cost solution.

The exchange rate is the volume of emissions the seller has to decrease to allow the buyer to increase emissions by one unit. Norway, for example, would have to buy five tons of emission reduction in Poland to be allowed to increase its emissions by one unit. In this single-receptor case, the exchange rate would ensure that the deposition at the Swedish receptor would remain constant, because the transfer coefficient of Norway to Sweden is five times higher than that of Poland to Sweden (0.25 versus 0.05).

The trades are shown in *Table 9.3*. In trade 1, Poland buys 763.6 kton from Sweden and increases its emissions from 2,240 to 3,003.6 kton SO_2. Sweden reduces emissions by 0.1×763.6 (0.1 is the exchange rate). Total costs are cut from DM 1,123,200 (*Table 9.2*) to DM 481,756 (*Table 9.3*), which is close to the cost minimum of DM 480,000. After this trade, Swedish marginal costs are 10 times the Polish marginal costs, reflecting the exchange rate of 10. Although the ratio of the marginal costs after trade equals the ratio of the marginal costs in the

Table 9.2. Three countries and one receptor.

I. Data

	Marginal costs (DM/kton)	Total costs (DM)	Transfer coefficient (fraction)	Unabated emission (kton)	Deposition (kton)
Poland	2R	R^2	0.05	3,200	160
Sweden	20R	10R^2	0.50	400	200
Norway	50R	25R^2	0.25	160	40
Subtotal	–	–	–	–	400

II. Pre-trade (30 percent uniform reduction)

	Marginal costs (DM/kton)	Total costs (DM)	Emission reduction (kton)	Emission (kton)	Deposition (kton)
Poland	1,920	921,600	960	2,240	112
Sweden	2,400	144,000	120	280	140
Norway	2,400	57,600	48	112	28
Subtotal	–	1,123,200	–	–	280

III. Cost minimum

	Marginal costs (DM/kton)	Total costs (DM)	Emission reduction (kton)	Emission (kton)	Deposition (kton)
Poland	400	40,000	200	3,000	150
Sweden	4,000	400,000	200	200	100
Norway	2,000	40,000	40	120	30
Subtotal	–	480,000	–	–	280

IV. Exchange rate (volume seller has to reduce to allow buyer an increase of one unit)

	Seller		
Buyer	Poland	Sweden	Norway
Poland	1.0	0.1	0.2
Sweden	10.0	1.0	2.0
Norway	5.0	0.5	1.0

least-cost solution, the absolute levels of the marginal costs differ. After trade, Sweden's marginal costs are DM 3,927; in the cost minimum they are DM 4,000. In the second trade Norway buys from Poland. Total costs again are lowered, approaching the cost minimum. Again marginal costs reflect the exchange rate (5), but are not exactly the same as in the cost minimum. In the third trade, Norway buys from Sweden, further lowering costs. Although close to the cost minimum,

Trading with Exchange Rates 229

Table 9.3. Trades with three countries and one receptor.

	Marginal costs (DM/kton)	Total costs (DM)	Emissions abated (kton)	Emissions (kton)	Change in emissions after trade (ton)	Deposition (kton)
Trade 1: Poland buys from Sweden						
Poland	392.7	38,559	196.4	3,003.6	+763.6	150.2
Sweden	3,927.0	385,597	196.4	203.6	−76.4	101.8
Norway	2,400.0	57,600	48.0	112.0	0.0	28.0
Subtotal	–	481,756	–	–	–	280.0
Trade 2: Norway buys from Poland						
Poland	436.0	47,603	218.2	2,982.0	−21.6	149.1
Sweden	3,927.0	385,597	196.4	203.6	0.0	101.8
Norway	2,180.0	47,603	43.6	116.4	+4.4	29.1
Subtotal	–	480,803	–	–	–	280.0
Trade 3: Norway buys from Sweden						
Poland	436.0	47,603	218.2	2,982.0	0.0	149.1
Sweden	3,967.0	393,416	198.4	201.7	−1.9	100.8
Norway	1,983.0	39,342	39.7	120.3	+3.9	30.1
Subtotal	–	480,361	–	–	–	280.0

the cost minimum is not yet achieved. Further trades might be able to lower costs slightly more without, however, attaining the cost minimum exactly.

In summary, the analysis suggests that with one receptor, bilateral sequential emission trading with a fixed exchange rate based on the cost-minimum solution might converge to the cost-minimum solution if the deposition constraint is binding before trade. Moreover, with every trade the deposition constraint will be met if the constraint was met before trading.

9.3.5 Three countries and two receptors

A more interesting situation occurs when there are three sources and two binding receptors. First, a graphical illustration will be given of the problem. Then the cost-minimum conditions will be compared with the exchange rate trading conditions. *Figure 9.5* shows the situation with two binding constraints (plane R_1 and plane R_2) whose intersection, line HI, is where both constraints are met. Point E is the cost minimum. The figure suggests that if trading starts at point P, bilateral trading sequences (indicated by the arrows) might or might not lead to the cost minimum.

With three countries and two binding receptors, the first-order conditions for a cost minimum are

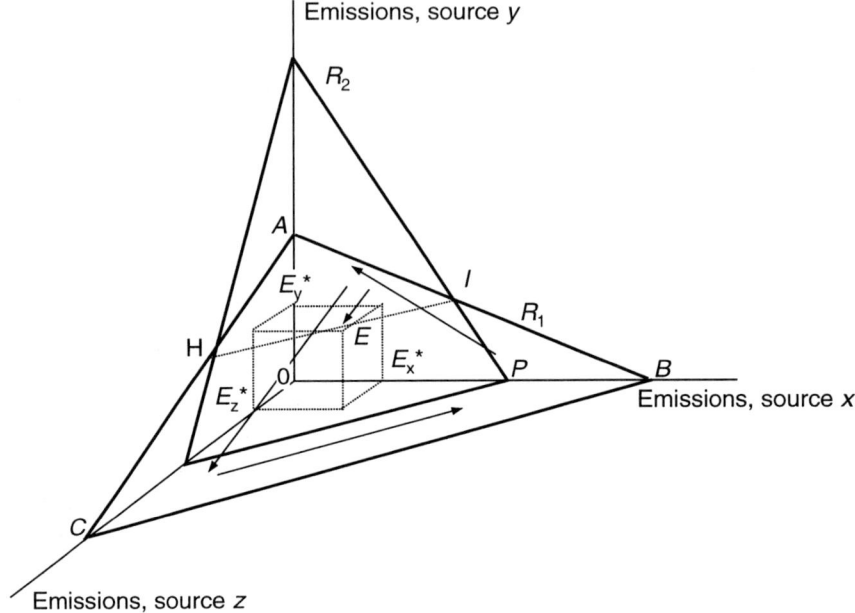

Figure 9.5. Exchange rate trading with three countries and two binding receptors.

$$\begin{aligned} C'_x &= a_{1x}\lambda_1 + a_{2x}\lambda_2 \ , \\ C'_y &= a_{1y}\lambda_1 + a_{2y}\lambda_2 \ , \\ C'_z &= a_{1z}\lambda_1 + a_{2z}\lambda_2 \ , \end{aligned} \qquad (9.34)$$

$$a_{1x}E_x + a_{1y}E_y + a_{1z}E_z = D_1^0 \ ,$$
$$a_{2x}E_x + a_{2y}E_y + a_{2z}E_z = D_2^0 \ .$$

From condition (9.34), after substituting to remove the Lagrangian multipliers and after rearranging terms, the cost-minimum value of $C_y^{\prime*}$ can be written as

$$C_y^{\prime*} = \frac{a_{2y}}{Ba_{2z} + a_{2x}} C_x^{\prime*} + \frac{Ba_{2y}}{Ba_{2z} + a_{2x}} C_z^{\prime*} \ , \qquad (9.35)$$

where $B = (a_{1x}a_{2y} - a_{2x}a_{1y})/(a_{1y}a_{1z} - a_{2y}a_{2z})$. This implies that in the optimum there is a special relation between the marginal costs of all three countries involved. This implies that if the pre-trade marginal costs of all countries differ from their

Trading with Exchange Rates

optimum values it is necessary for all three countries to change their emission levels for the above condition to be met.

The question is whether a sequence of bilateral trades using the optimal exchange rates could arrive at the cost minimum. The conditions for all the possible combinations of bilateral trades between the three sources are

$$\text{Trade 1} \quad \frac{C'_x}{C'_y} = \frac{a_{1x}\lambda_1 + a_{2x}\lambda_2}{a_{1y}\lambda_1 + a_{2y}\lambda_2} \quad E_y = E_y^0 + w_{xy}E_x^0 - w_{xy}E_x \,, \tag{9.36a}$$

$$\text{Trade 2} \quad \frac{C'_z}{C'_y} = \frac{a_{1z}\lambda_1 + a_{2z}\lambda_2}{a_{1y}\lambda_1 + a_{2y}\lambda_2} \quad E_y^2 = E_y^1 + w_{zy}E_z^1 - w_{zy}E_z^2 \,, \tag{9.36b}$$

$$\text{Trade 3} \quad \frac{C'_z}{C'_x} = \frac{a_{1z}\lambda_1 + a_{2z}\lambda_2}{a_{1x}\lambda_1 + a_{2x}\lambda_2} \quad E_x^3 = E_x^2 + w_{zx}E_z^2 - w_{zx}E_z^3 \,. \tag{9.36c}$$

Obviously, bilateral trading and the cost minimum would lead to similar results if trade were to start from the optimum, because in this case there would be no room for trading because conditions (9.36a) to (9.36c) on the left-hand side are met. A second important conclusion is that for a bilateral trade to immediately lead to the optimum the third source, which is not trading and hence not changing marginal costs, should be at its optimum. Assume that sources x and y trade; after this trade condition (9.36a) will hold. If marginal costs of both source x and source y are then optimal, there is no room for further trading because all sources are at their optimal levels. However, if country z is not at its optimal level while countries x and y are at their optimal levels, there is room for further trade because neither condition (9.36b) nor (9.36c) is met because C'_z does not equal $a_{1z}\lambda_1 + a_{2z}\lambda_2$. This implies that further trades between countries y and z or x and z will result in cost savings. Consequently, the marginal costs of either country x or country y will change after the trade with country z and will no longer equal the conditions for a cost minimum [equation (9.35)]. In conclusion, only if the non-trading country's emission level is at the cost minimum is there a chance that a bilateral trade between the other countries will end up at the cost minimum. As is the case with two trading sources, reaching the cost minimum depends on the starting point. If all three countries' emission levels diverge from the optimum before trading, a simultaneous trade between all three countries will be needed to attain the cost minimum.

Again, as in the two-country case, any bilateral trade might result in a violation of the deposition targets. The proof is similar to the proof for the two-country case, the difference being that the pre-trade emissions of country z are important now. The pre-trade emissions of country z result in the following vector of deposition at both receptors:

$$D_z^0 = \begin{pmatrix} D_{1z}^0 \\ D_{2z}^0 \end{pmatrix} \,. \tag{9.37}$$

Initially, we assume that pre-trade emission levels are within the feasible region. This gives

$$AE^0 = D^0 - S - D_z^0 .$$

Repeating the same matrix multiplications as with the two-country case leads to

$$WD = W(D^0 - S - D_z^0) .$$

Again, this does not necessarily mean that $D = D^0 + S - D_z^0$. Longhand notation clarifies this:

$$w_{xy}D_1 + D_2 = w_{xy}(D_1^0 - S_1^0 - D_{1z}^0) + (D_2^0 - S_2^0 - D_{2z}^0) ,$$

or

$$D_2 - D_2^0 = w_{xy}(D_1^0 - D_1) - S_2^0 - w_{xy}S_1^0 - D_{2z}^0 - w_{xy}D_{1z}^0 .$$

If the left-hand side is greater than zero ($D_2 > D_2^0$), the deposition target at D_2 is not met. This can be the case given the positive values of the deposition of country z and the surplus variables S_1^0 and S_1^0 if the deposition at receptor 1 is sufficiently below the desired level.

Remarkably, the post-trade emission levels of country y depend not only on the initial level, but also on the intermediary steps. In other words, the sequence is relevant. This is shown by consulting and rewriting equations (9.36a) and (9.36b). In equation (9.36b), E_y^1 is replaced by the right-hand side of equation (9.36a). This gives E_y^1 as a function of emissions before trade. Then E_z^2 is shifted to the left-hand side and E_y^2 is shifted to the right-hand side of equation (9.36b). Finally, we divide by w_{zy} (w_{zy} is by definition equal to $1/w_{yz}$, because the exchange rate is based on the cost-minimum solution). Dividing by w_{zy} is equal to multiplying by w_{yz}. This gives

$$E_z^2 = E_z^1 + w_{yz}\left\{\left[E_y^0 + w_{xy}\left(E_x^0 - E_x^1\right)\right] - E_y^2\right\} , \qquad (9.38)$$

where the superscripts 0, 1, and 2 indicate the point in time. Equation (9.38) shows that post-trade emission levels E_z^2 and E_y^2 always depend on the pre-trade emission levels E_y^0, E_x^0, and E_z^1 (which equal E_z^0 because country z does not trade in round one). After the first trade, the emissions of country y depend on the pre-trade ($t = 0$) emission levels of countries x and y. The emission level of country y is the starting point for the second trade, between countries y and z. Because of this, the emissions of country z are not only a function of the pre-trade emissions, but as equation (9.38) shows, they are also a function of the intermediary levels (E_x^1).

Table 9.4. Data for the case of three countries and two receptors.

I. Data

	Marginal costs (DM/kton)	Transfer coefficient		Deposition	
		Sweden	Poland	Sweden (kton)	Poland (kton)
Poland	2R	0.05	0.50	160	1,600
Sweden	20R	0.50	0.05	200	20
Norway	50R	0.25	0.00	40	0
Subtotal	–	–	–	400	1,620

II. Pre-trade emission (30 percent reduction in emissions)

	Marginal costs (DM/kton)	Total costs (DM)	Emission reduction (kton)	Deposition	
				Sweden	Poland
Poland	2,120	1,123,600	1,060	107.0	1,070
Sweden	2,400	144,000	120	140.0	14
Norway	1,400	19,600	28	33.0	0
Subtotal	–	1,287,200	–	280.0	1,084

III. Cost minimum

	Marginal costs (DM/kton)	Total costs (DM)	Emission reduction (kton)	Deposition	
				Sweden	Poland
Poland	1,918	919,297	958.8	112.1	1,120.6
Sweden	2,636	173,712	131.8	134.1	13.4
Norway	1,232	15,178	24.6	33.8	0.0
Subtotal	–	1,108,187	–	280.0	1,134.0

IV. Exchange rate (volume seller has to reduce to allow buyer an increase of one unit)

	Seller		
Buyer	Poland	Sweden	Norway
Poland	1.00	0.73	1.56
Sweden	1.37	1.00	2.14
Norway	0.64	0.47	1.00

An example illustrates how exchange rate trading steers trade. *Table 9.4* summarizes the data, the pre-trade situation, the least-cost solution, and the exchange rates. R is the emission reduction. We have now two receptors: Sweden and Poland. The pre-trade situation is chosen so that the receptor in Poland is not binding. For both receptors the objective is a 30 percent reduction in deposition, a level of 280 kton in Sweden and 1,134 kton in Poland.

Table 9.5. Trades with three countries and two receptors.

	Marginal costs (DM/kton)	Total costs (DM)	Emission reduction (kton)	Change in emission after trade (kton)	Deposition (kton)	
					Sweden	Poland
Trade 1: Norway buys from Poland						
Poland	2,121	1,124,618	1,060.5	−0.5	107.0	1,069.8
Sweden	2,400	144,000	120.0	0.0	140.0	14.0
Norway	1,362	18,560	27.3	+0.7	33.2	0.0
Subtotal	–	1,287,178	–	–	280.2	1,083.8
Trade 2: Norway buys from Sweden						
Poland	2,121	1,124,618	1,060.5	0.0	107.0	1,069.8
Sweden	2,442	149,035	122.1	−2.1	139.0	13.9
Norway	1,141	13,022	22.8	+4.5	34.3	0.0
Subtotal	–	1,286,675	–	–	280.3	1,083.7
Trade 3: Poland buys from Sweden						
Poland	2,066	1,067,358	1,033.0	+27.5	108.3	1,083.4
Sweden	2,840	201,612	142.0	−19.9	129.0	7.1
Norway	1,141	13,022	22.8	0.0	34.3	0.0
Subtotal	–	1,281,992	–	–	271.6	1,090.5

Table 9.5 shows three rounds of trading. Trade 1 shows that a slight violation of the deposition target for Sweden occurs immediately. The exchange rate allows Norway to increase emissions by 1 unit if Poland reduces by 0.64 units (*Table 9.4*). Given the transfer coefficients of both countries, this means an increase in deposition in Sweden of 0.2475 kton for every kton increase in Norway. Obviously, the constraint in Sweden will be violated after trade. Remarkably, the extent of the violation is small. This is because the costs of Norway and Poland steer trade, as well. Norway cannot increase its emissions excessively because this would no longer be profitable. In effect, the first trade shows that Poland decreases emissions only by 0.5 kton [the reduction is 1,060.5 instead of 1,060 (*Table 9.4*)]. *Table 9.5* shows that the second trade (where Norway buys from Sweden) further increases deposition in Sweden above the limit (280 ktons). Again, this is because the exchange rate allows Norway an increase of 2.14 units for each unit reduction bought in Sweden, whereas the transfer coefficient from Norway to Sweden is only 2 times that of Sweden to Sweden. Again, the cost functions prevent larger increases. The third trade, however, shows that trading with the cost-minimum exchange rates allows trade to be steered within the constraints again. After the

third trade, both constraints are met again (actually overfulfilled) and cost efficiency is further improved.

In conclusion, exchange rate trading with three sources and two binding receptors in the optimum can only achieve the cost minimum if the emissions of the non-trading source are already optimal and if the pre-trade emissions of the trading sources start at a special point. If this is not the case, a simultaneous trade is necessary to attain the cost minimum. The analysis shows that even if exchange rate trading starts from inside the feasible region and one deposition target is met the other might be violated, although cost-minimizing behavior tends to limit the size of the violation.

The latter conclusion is relevant in an international context. If one of the deposition constraints is met before trading and is violated after trading, there is no longer a guarantee that emission trading with this cost-minimum exchange rate is a Pareto-dominant improvement. One country might experience increases in deposition compared with the pre-trade situation (a cooperative Nash equilibrium; see Chapter 4), which constitutes a loss of environmental benefits. This loss might not be compensated for by the reduction in costs due to emission trading. One way out of this problem is to set the exchange rates so that no country experiences any increase in deposition after trade. In other words, the most binding constraints would be used to set the transfer coefficient for trading. In terms of *Figure 9.6*, this implies the following. If P is the starting point for trade, R_3 (the cost-minimum exchange rate) is no longer used to guide trade. Either R_1 or R_2 is used. If x acts as a permit seller (emission decrease) and y as a buyer, R_1 determines the exchange rate. Trade would be allowed on line R'_1, parallel to R_1. If y sells and x buys, R_2 becomes the exchange rate and trade would move along line R_2 (but only downward). In this case the deposition constraints are never violated and the trade leads to a Pareto-optimal improvement. *Figure 9.6* suggests that in this case the cost minimum will not be attained although cost savings are conceivable (trade is restricted to area $ODPB$). This type of constant deposition trading will be simulated in the next chapter.

Another solution is to use the cost-minimum exchange rates and check for violation of the deposition constraints. In the case of a violation, the volumes could be reduced so that the constraints are met.

9.3.6 Uncertainty about costs

The previous sections of this chapter assumed that the environmental agency had correct information on costs and, therefore, could determine the cost minimum and corresponding fixed exchange rates with precision.

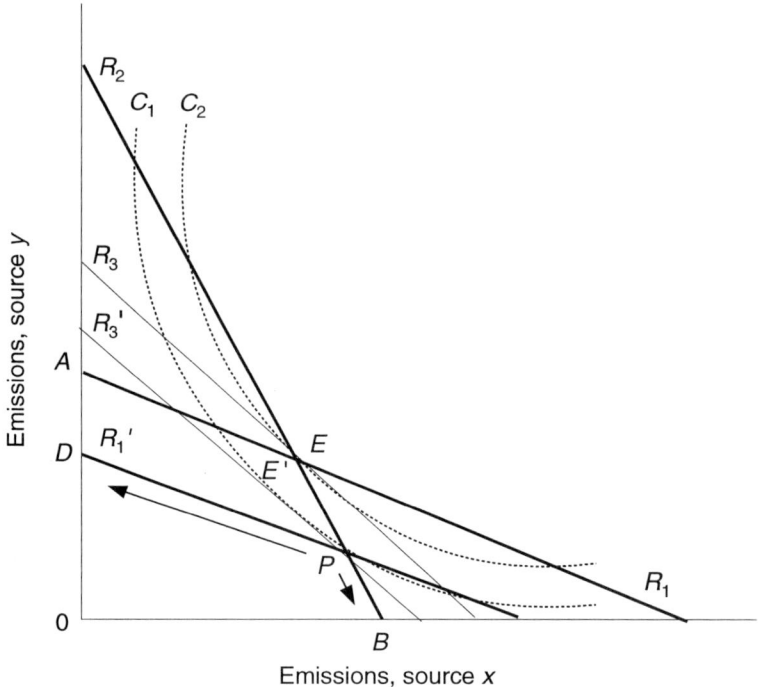

Figure 9.6. Constant-deposition trading.

If an environmental agency has incorrect information on costs, the cost minimum and the deposition targets might not be attained. However, there is a difference between the case of one receptor and the case of multiple receptors. If there is only one receptor the environmental agency does not need any information on the costs to determine the "optimal" exchange rates. This is because the exchange rates depend only on the ratios of the transfer coefficients. This is also sufficient to guarantee that the deposition target will be met if it is not exceeded before trading.

With more than one receptor, the agency cannot determine the exchange rates without knowledge of the cost functions, because without cost knowledge it does not know which receptors will be binding in the cost minimum and which will not. It needs cost information to determine the shadow prices (Lagrangian multipliers) at each receptor to arrive at the optimal marginal costs [see equation (9.12)]. The environmental agency bases its exchange rates on its own cost information. Trading partners trade using their own (the actual) costs, taking into account the exchange rates fixed by the agency.

Table 9.6. Trade with imperfect information about costs.

	Marginal costs (DM/kton)	Total costs (DM)	Emission reduction (kton)	Change in emission after trade (kton)	Deposition (kton) Sweden	Deposition (kton) Poland
Pre-trade emission						
Poland	2,120	1,123,600	1,060		107.0	1,070.0
Sweden	4,800	288,000	120		140.0	14.0
Norway	1,400	19,600	28		33.0	0.0
Subtotal	–	1,431,200	–		280.0	1,084.0
Trade 1: Sweden buys from Poland						
Poland	2,238	1,125,653	1,119	–59.0	104.0	1,040.0
Sweden	3,076	118,302	77	+43.0	161.6	16.2
Norway	1,400	19,600	28	0.0	33.8	0.0
Subtotal	–	1,263,555	–	–	299.4	1,056.2

An example can clarify this. Assume the agency faces the problem of three countries and two receptors, as in *Tables 9.4* and *9.5*. The agency uses the exchange rates from *Table 9.4*. Now assume that costs (both marginal and total) in Sweden are actually twice as high. *Table 9.6* shows what would happen if trade were to take place from the same point as in *Table 9.4* with the marginal and total costs in Sweden being twice as high (compare *Tables 9.6* and *9.4*). If Sweden buys emission permits from Poland, it will use its actual costs to minimize costs. The exchange rate says that Sweden has to buy 1.37 permits from Poland to increase by 1 unit. Given the actual costs, the Polish–Swedish trade will end up saving costs but, as *Table 9.6* indicates, deposition in Sweden will now be violated considerably (299 instead of 280 kton). Clearly, the table illustrates that with imperfect information on costs and more than one receptor, exchange rate trading based on the cost minimum might be less attractive, because violation of deposition goals is more likely to occur. This also makes it more difficult for these exchange rates to be a Pareto-dominant improvement over an international agreement.

9.4 Conclusions

The advantage of exchange rate trading is that it allows for cost savings and is simple for the trading sources. Moreover, with one receptor if the deposition constraint is initially met, deposition targets will not be violated. No knowledge of costs is needed to set the exchange rates. With one receptor, bilateral sequential trading arrives at the cost minimum only if the receptor is initially binding. With two or more binding receptors in the cost minimum, fixed exchange rate trading will reduce total abatement costs. However, there is no guarantee that the system will attain the cost minimum, nor is there certainty that deposition targets will not be violated, even if the environmental agency has complete and accurate knowledge of the costs necessary to determine the exchange rates. The system, however, tends to limit violations and moves toward the cost-minimum solution. With imperfect information on costs the exchange rates might be set at the wrong costs. This then prevents the actual cost minimum from being met and increases the probability that deposition targets will be violated. Because of the potential violation, emission trading using cost-minimum exchange rates might not constitute a Pareto-dominant improvement in an international context.

Although there is evidence that the optimum solution will only be achieved by chance, the questions of how close to the cost minimum such a trading scheme can approach and the extent to which deposition targets will be violated still remain. As usual, the magnitude depends crucially on the actual problem specification (transfer coefficients, deposition target, cost curves, initial solutions). Therefore, the following chapter describes a simulation of such trading processes.

Chapter 10

Sulfur Emission Trading in Europe: A Model Simulation

10.1 Introduction

Since Norway organized an expert meeting on emission trading in May 1991, joint implementation of emission reduction commitments has the been the subject of debate in the UN/ECE. The Second Sulfur Protocol was signed in Oslo in June 1994. The Protocol states that two or more parties may jointly fulfill the obligations in terms of annual emission ceilings subject to rules and conditions to be specified. These rules will have to ensure that the emission ceilings agreed on will be met. They should also promote the achievement of the critical sulfur loads.

The purposes of this chapter are to contribute to the discussion on possible rules and conditions for joint implementation of emission reduction commitments and to examine the extent to which rules are conceivable that will save costs without impairing the environmental objectives of the Protocol. The chapter discusses the cost efficiency, environmental effectiveness, and distributional impacts of different schemes for the joint implementation of emission ceilings under the Second Sulfur Protocol. In particular, the chapter looks at joint implementation of emission reduction commitments subject to an exchange rate, that is, the volume of emissions one country has to reduce to allow another country to increase its emissions by one unit. More specifically, the chapter will analyze joint implementation under the following alternative rules:

- Trading in one Europe-wide zone subject to the condition that the total volume of emissions remains constant;
- Trading subject to explicit deposition constraints for every EMEP (European Monitoring and Evaluation Program) grid cell;

- Trading with cost-minimum exchange rates that are based on the ratios of the marginal costs in a cost-minimum solution and hence take into account the transfer coefficients of the binding receptors;
- Trading using constant-deposition rates designed so that with every bilateral trade the average deposition does not increase in any country.

The performance of these schemes will be examined under the following conditions: without constraints, in combination with regulatory constraints in the form of emission and fuel standards, without non-signatories trading, with different sequences of trading, and with different levels of transaction costs. In general, we assume that all regions (signatories and non-signatories) trade.

Section 10.2 introduces the optimization method because the optimal solution serves as a reference point. Section 10.3 describes the bilateral sequential model used to simulate trading with a fixed exchange rate. Section 10.4 describes the data on the costs and transfer coefficients used. Section 10.5 examines several example results as well as major conditions that affect the results.

10.2 Optimization

The RAINS (Regional Acidification INformation and Simulation) model simulates the flow of acidifying pollutants (sulfur and nitrogen species) from source regions in Europe to environmental receptors (Alcamo *et al.*, 1990). The model (version 6.1) covers 38 source regions in Europe: 26 countries, 7 regions in the former Soviet Union, and 5 sea regions (for ship emissions). Analysis of deposition is performed for 547 land-based receptor sites with a regular grid size of 150 × 150 km.

The optimization mode of the RAINS model allows the user

- To identify the cost-minimum international allocation of emission reduction measures to attain a set of deposition levels for each receptor site in Europe;
- To determine the lowest costs of attaining a target level of total European emissions.

The optimization module formulates possible strategies to minimize the costs of achieving deposition targets at certain receptors as a linear optimization problem that can be solved with linear programming (LP) packages (Batterman and Amann, 1991; Amann, 1992). The cost-efficient solution requires the total costs of emission reductions to be minimized subject to the constraint that the desired depositions are met at every receptor:

$$\min C = \sum_i \sum_l C'_{i,l} R_{i,l} \; , \tag{10.1}$$

A Model Simulation

where $R_{i,l}$ is the emission reduction in region i at the lth level of emission reduction. Cost functions of emission reductions are expressed as piecewise linear curves denoting cost-minimal combinations of measures within each country to achieve certain levels of national total emissions. $C'_{i,l}$ are the marginal costs, determined as the additional cost per unit of emission reduction at level l in region i. The reduction in each of the segments is limited:

$$0 \leq R_{i,l} \leq R_{i,l,\max} \quad \text{for} \quad i = 1,\ldots,38 \quad l = 1,\ldots,L \ . \tag{10.2}$$

An identity relates emission reductions and unabated emissions (i) to calculate emissions remaining after abatement:

$$E_i = \overline{E}_i - \sum_l R_{i,l} \quad \text{for} \quad i = 1,\ldots,38 \ . \tag{10.3}$$

Total deposition (wet and dry) at each receptor j is calculated as the sum of the contributions of each source region plus the background deposition:

$$D_j = \sum_i a_{ji} E_i + N_j \quad \text{for} \quad j = 1,\ldots,J \ , \tag{10.4}$$

with a_{ji} being the linear source-receptor relationship from region i to receptor j, as based on the atmospheric transport model. N_j is the background deposition that is not attributable to specific sources in the region and is considered to be irreducible.

Furthermore, limits or targets can be set on the sulfur deposition for each receptor j ($j = 1,\ldots,J$):

$$D_j \leq D_j^* \ . \tag{10.5}$$

Alternatively, so-called policy constraints can be added on the maximum or minimum emissions remaining after abatement in each region i to reflect, for example, abatement devices already in place:

$$E_i^{\min} \leq \overline{E}_i - \sum_l R_{i,l} \leq E_i^{\max} \ . \tag{10.6}$$

The above equations form a large LP model that requires a significant amount of computer resources. The LP model can make use of an internal solver called HYBRID, developed at the International Institute for Applied Systems Analysis (IIASA), Laxenburg, Austria (Makowski and Sosnowski, 1988). For larger problems external solvers can be employed.

10.3 Simulation Method for Joint Implementation

10.3.1 Introduction

Until recently, many model studies simply assumed that the potential cost savings of emission trading schemes would equal the results of optimization procedures. In other words, a perfectly working market where emission permits are simultaneously traded was assumed. Practice (see Chapter 6; and especially Rico, 1995) as well as recent model studies (see Chapter 5) show that in reality trading tends to take place bilaterally and sequentially. With such restrictions, emission trading or joint implementation of emission reduction commitments is not expected to capture the complete cost savings possible with any LP cost-minimization procedure. The algorithm used in this chapter makes use of an adapted version of the optimization model in RAINS. The algorithm describes a process of repeated bilateral trading subject to an offset or exchange rate for every possible combination of bilateral trades. The following elements are new:

- The user can specify various exchange rates.
- The model allows transaction costs to be taken into account and permits the setting of thresholds (based on perceived transaction costs) below which joint implementation does not take place.
- The sequence for bilateral implementation can be selected because in contrast to LP optimization the final result can be path dependent.

Figure 10.1 depicts a flow diagram of the joint implementation algorithm. The diagram shows that the procedure consists of the following steps:

- The selection of an exchange rate.
- The creation of a matrix of potential cost savings from each potential bilateral joint implementation agreement.
- The calculation of transaction costs and determination of threshold level below which joint implementation will not take place.
- The selection of the sequence.
- The updating of emissions after consummation of the selected trade and the recording of cost savings.
- The updating of the matrix of possible bilateral trades or joint implementation agreements to account for the trades or agreements that have taken place (return to the second step).

A Model Simulation

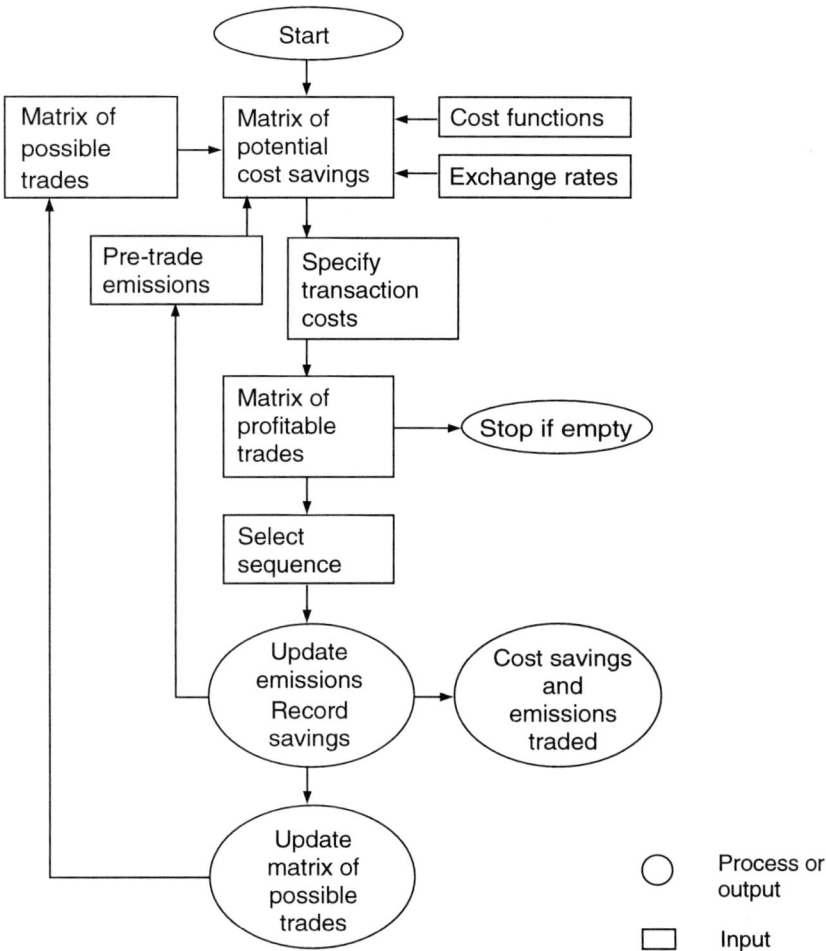

Figure 10.1. Flow diagram of joint implementation.

10.3.2 Details on the different steps

Regarding the exchange rates, the user can select from several alternatives:

- Uniform exchange rates equal to one;
- Other uniform exchange rates;
- Exchange rates based on the marginal costs in a cost-minimum solution;

- Constant-deposition rates that keep the average sulfur deposition in each country at least constant.

The third type of exchange rate takes account of the transfer coefficients of the binding receptors in a cost-minimum solution by taking the ratio of the marginal costs in the optimum solution as the exchange rate. If, for example, the marginal costs in the cost-minimum solution are 15 times higher for Norway than for Portugal, Norway would have to buy 15 emission permits from Portugal to be allowed to increase its emissions by 1 unit. Conversely, if Portugal bought only 1 emission permit from Norway it would be allowed to increase its emissions by 15 units (see Chapter 9). The piecewise linear cost functions used in RAINS affect the practical definition of the exchange rate if this is the ratio of the marginal costs of the optimum. Due to the nature of linear optimization the optimal status of most variables of an optimization problem will lie exactly at the corner points of the solution space, on the intersections of the linear cost function segments (e.g., point E in *Figure 10.2*). For these points a unique definition of marginal costs does not exist (for point E, marginal costs are either DM 3,900 or DM 5,500 per ton SO_2). The marginal costs for increasing emissions differ from the marginal costs for decreasing emissions. The algorithm uses the higher marginal costs (the additional costs of further reducing emissions represented by the next step of the cost curve) by default. This also avoids problems with marginal costs of zero, for which the definition of the exchange rate is not applicable.

Regarding the creation of the cost-savings matrix, one should realize that cost functions estimated with the RAINS model are piecewise linear functions. As a result, RAINS does not work by making marginal costs equal; it sorts and ranks elements of two or more cost functions according to their marginal costs (in ascending order). If an exchange rate is introduced, the bilateral combination of emission reductions that is cost optimal can be determined in a similar way. A cost-minimization routine has been implemented to minimize costs subject to different exchange rates and initial emission levels.

The condition for a cost-minimum solution for each bilateral trade between source x and source y is the following [compare conditions (9.1) and (9.5)]:

$$\text{Minimize } C_x + C_y \tag{10.7}$$

subject to

$$E_y - E_y^0 = -w_{xy} * (E_x - E_x^0) \; . \tag{10.8}$$

Condition (10.8) states that the emission increase by source y ($E_y - E_y^0$) should equal the emission decrease of source x ($E_x - E_x^0$) multiplied by the exchange rate w_{xy}. The exchange rate w_{xy} is defined as the volume of emissions that y can

A Model Simulation

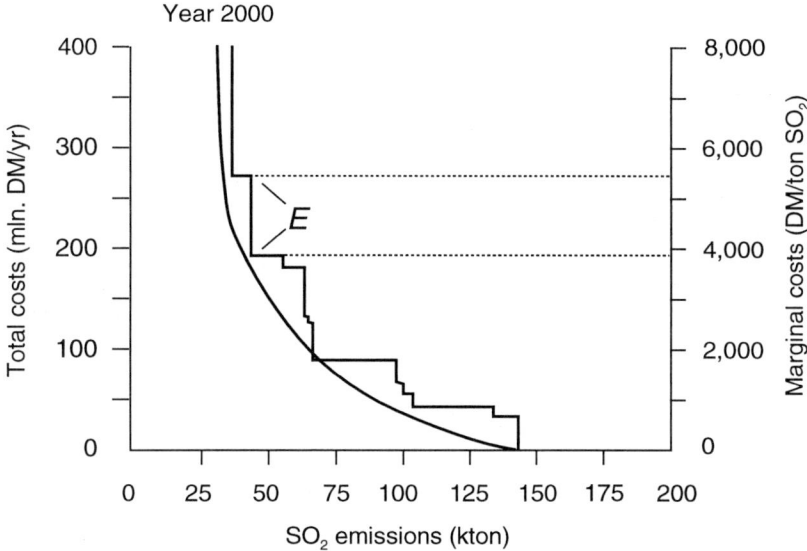

Figure 10.2. Cost function.

increase if x reduces emissions by one unit. Alternatively, the trade rule can be written as

$$E_x + E_y/w_{xy} = E_x^0 + E_y^0/w_{xy} \ . \tag{10.9}$$

On the left-hand side are the emissions after trade and on the right-hand side are the pre-trade emissions. Clearly, the pre-trade levels equal post-trade levels (in tons) only if the exchange rate is one. Equations (10.7) and (10.9) have two unknown variables in the form of the post-trade emissions. Of the 38 regions in the model, 33 are countries and 5 are sea regions. The emissions from the sea regions are emissions from ships. We do not allow sea regions to trade because there is no central authority. For each pair of the 33 countries [$n * (n-1)/2 = 528$], we can find solutions and create a matrix of cost savings of all possible bilateral trades or joint implementation agreements. Finally, the matrix of potential bilateral partners can be altered, for example, to limit joint implementation only to neighbor countries or to signatories of the Second Sulfur Protocol.

Transaction costs might be a serious obstacle to emission trading in practice if finding a trading partner and arriving at an agreement or obtaining administrative approval is difficult. Transaction costs have been estimated to be 5 to 30 percent of the value of transaction under the EPA's emission trading program and around

5 percent in the more recent sulfur allowance trading scheme in the USA (Dwyer, 1992; Rico, 1995; see also Chapter 6). Building in the transaction costs (step 2 in *Figure 10.1*) thus gives a more realistic picture of the potential cost savings of emission trading. The algorithm allows the level of fixed transaction costs for each bilateral trade to be specified exogenously. If the cost savings of a potential bilateral trade are below the threshold, the trade is not profitable and will be excluded from further consideration.

As will be shown, the final result of the bilateral sequential algorithm is path dependent. After calculating the matrix of cost savings of all possible and profitable trades, the sequence of trading is determined. Two options are available in the model:

1. The highest absolute costs savings come first in the sequence.
2. The sequence is completely random.

In case 1, the algorithm ranks all possible trades according to their absolute cost savings and always selects and implements the one with the highest cost savings first. Cost savings are defined as the difference between the total cost increase of the sources selling emission permits and the total cost decrease of the source buying emission permits. The difference pertains to the pre-trade emission level (of every round) and the post-trade emission level. This is an optimistic assumption, assuming perfect information and coordination of the selection of traders.

In case 2, the sequence of trading is completely random. This is an alternative (pessimistic) assumption based on the fact that due to the imperfect information about the costs to traders, and perhaps due to political preferences that are difficult to forecast, the sequence of trading is random.

After selecting the sequence, the next step in the algorithm is the simulation of the trades. After each trade, the following steps are implemented:

- The pre-trade emission vector is updated, accounting for the trade(s) that have taken place.
- The cost savings of the trade (compared with the pre-trade situation) are recorded in a file.

As a final step, the matrix of possible trades is updated after each bilateral trade. In order to accelerate the algorithm, sources that have already traded are allowed to trade again, but not with the same partner (as in Atkinson and Tietenberg, 1991). The cells of the cost-savings matrix that correspond to the trade between the two regions that complete a trade are skipped, and the cost savings of all the other trade relationships of these two countries are recalculated with the new emission levels.

10.4 Data on Costs and Atmospheric Transport

The RAINS model contains a sub-module to assess the potential for and costs of alternative emission abatement technologies. The evaluation is based on internationally reported performance and cost data of control devices (Alcamo *et al.*, 1990; Amann, 1992). The model extrapolates cost estimates for specific technologies to reflect country-specific conditions such as operating hours, boiler size, and fuel price. In the current version of the model, the cost evaluation of the emission reduction techniques is limited to the most relevant measures that have no impact on the underlying pattern of energy use. For the time being, energy conservation and fuel substitution are excluded from the analysis. The following technical options are implemented.

- Use of low-sulfur fuels and fuel desulfurization: this pertains to the use of fuels with a reduced sulfur content, such as fuels with a lower natural sulfur content or fuels that have undergone a desulfurization process. For low-sulfur hard coal, the sulfur content is set at 0.6 percent. Desulfurization of gas oil and diesel oil can reduce the sulfur content in two levels: to 0.3 percent and to 0.05 percent. The desulfurization of heavy fuel oil is assumed to be possible to a level of 0.6 percent.
- Desulfurization of flue gases during or after combustion: this set of measures requires investments at the plant site. Three techniques are considered: desulfurization during combustion, with removal efficiencies of 50 percent at relatively low costs; flue gas desulfurization, with a removal efficiency of 95 percent at moderate costs; and the use of advanced flue gas purification, with emission reduction of 98 percent at high costs.

Not all abatement technologies are applicable for all fuel types and energy sectors. Moreover, a distinction is made between new and existing plants to account for the additional costs of retrofitting existing plants.

To run the optimization model RAINS creates "national cost functions" for controlling emissions. National circumstances (such as sulfur content and operating hours) result in variations in the costs of applying the same technology in different countries in Europe. Other differences are the structural variations of energy systems, especially in the amount and structure of energy use, which determine the potential for application of individual control options. One way to combine these factors is to compile national cost functions (see *Figure 10.2*). These functions display the lowest costs for achieving various emission levels by applying the combination of abatement options that is cost optimal. This is done by ranking the options according to their marginal costs and their individual potential for sulfur removal and can be performed within each fuel category. The cost curves used in this paper are based on official energy use projections for the year 2000, as available in mid-1993.

If emission trading is combined with existing regulation the cost functions are slightly altered. Abatement options that have been imposed always have to be implemented, whether they are cost efficient or not. For example, new plants will have to meet emission standards requiring them to install flue gas desulfurization equipment or other devices. As a result, the cost function changes and the level of unabated emission is lower (see *Figure 10.3*). This can be easily seen by comparing *Figures 10.2* and *10.3*. *Figure 10.2* shows that without control, emissions in Norway would amount to 150 kton, whereas with existing legislation (*Figure 10.3*) they are around 80 kton. Without regulation, a ceiling of 50 kton on Norway's emissions could be met through a perfectly working national tradable permit market at an equilibrium price (marginal costs) of around DM 3,900/ton SO_2 (right-hand side *Figure 10.2*) and annual costs of approximately DM 200 million/year. If combined with existing regulations (emission and fuel standards), the permit price would only be 2,800 DM/ton SO_2 for meeting the same ceiling of 50 kton. Annual costs would only be DM 50 million. To obtain the complete costs we would, however, have to add the costs of the existing regulation (around DM 300 million/year). To simulate combinations of regulation and emission trading this chapter uses the "shrunk" cost function, like the one in *Figure 10.3*.

An overview of existing regulations is given in Chapter 8 (see also Klaassen, 1995). Because the emission standards usually depend on the size of the installation and on whether the plant is new or existing, data were collected and assumptions were made on the size distribution of the installations and the replacement of old plants by new ones. Size distribution data and the capacity of new plants were based on Klaassen (1995) and on data from the RAINS model.

Source-receptor transfer coefficients, which relate country emissions in the diffusion model to deposition at receptor points (called grids), are based on the acid deposition model developed within the EMEP (Sandnes, 1993). The RAINS model uses 547 land-based grids or receptors with a regular grid size of 150×150 km. The model calculates transboundary fluxes of oxidized sulfur and nitrogen as well as reduced nitrogen (ammonia and its product ammonium). For the trade simulations presented in this paper, EMEP model results that reflect the meteorological average of the years 1985 and 1987 to 1990 have been applied.

A Model Simulation

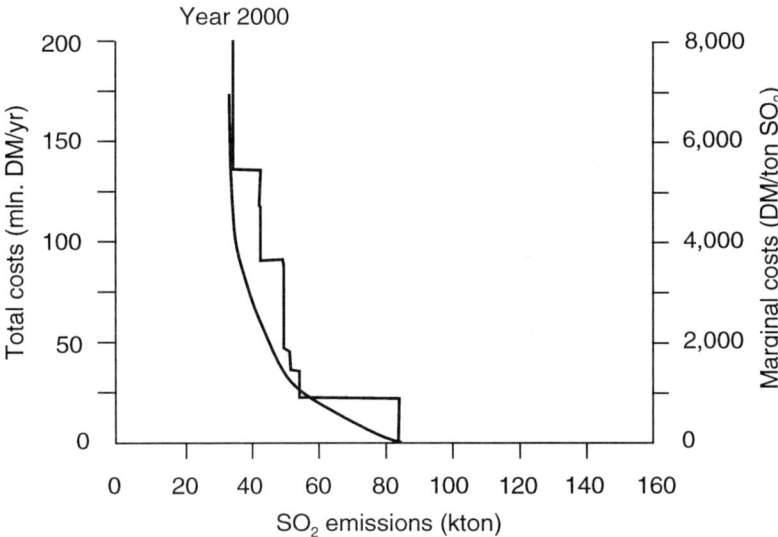

Figure 10.3. Cost function combined with regulation.

10.5 Simulation Results

10.5.1 Introduction

This section examines the cost efficiency and environmental impacts of different joint implementation schemes and compares them with a protocol without trading. The following simulations were performed:

A. The Second Sulfur Protocol without joint implementation;
B. Joint implementation with simultaneous comprehensive trading subject to the condition that the deposition is always lower than or equal to either the deposition in each grid attained by the Second Sulfur Protocol or the five-percentile critical sulfur deposition value, whichever value is higher;
C. Joint implementation with simultaneous comprehensive trading subject to the condition that the total European emissions do not increase (equivalent to emission trading in one Europe-wide zone);
D. Joint implementation with bilateral sequential trading subject to exchange rates derived from the transfer coefficients of the binding receptors in the cost minimum;

E. Joint implementation with bilateral sequential trading subject to constant-deposition rates that ensure that the average sulfur deposition in each country does not increase.

Throughout the text we will use emission trading and joint implementation as synonyms. All schemes take the emission ceilings agreed to under the Protocol as starting point. If countries did not sign the Protocol, their currently planned reductions are taken as emission ceilings.

A problem with the Second Sulfur Protocol is that the only explicit deposition constraints are the so-called five-percentile critical sulfur deposition levels – levels of deposition where 5 percent of the ecosystems in the grid are not protected. The Protocol, however, does not meet this long-term goal in all grids. Therefore, it does not make sense to take these as deposition constraints for joint implementation. Alternatively, one could take as constraints the actual deposition in each grid that results from the Protocol. This raises two problems. First of all, analysis has shown that in this case there is no room for cost savings because all 547 constraints are initially binding. Because there are only 38 instruments (regional emission levels) to meet these targets, there is an insufficient number of instruments to reach the goal of meeting the constraints at lower costs. Second, in a number of grids the deposition under the Protocol is lower than the five-percentile critical sulfur deposition loads. Calculations performed as input to the negotiations always allowed the deposition to increase to the five-percentile critical loads (Amann *et al.*, 1993). Therefore, this chapter takes the following deposition constraints as a compromise: the deposition after joint implementation should always be lower than or equal to either the deposition in each grid attained by the Second Sulfur Protocol or the five-percentile critical sulfur deposition value, whichever value is higher. These deposition constraints (option B) are depicted in *Figure 10.4*.

The simulation results for options B and C are based on the results of the LP optimization model in RAINS and thus implicitly assume a perfect market with simultaneous trading at equilibrium prices. Options D and E are based on the bilateral sequential joint implementation (emission trading) module and assume that implementation takes place so that bilateral agreements with the highest cost savings are carried out first.

More specifically, one of the Kuhn-Tucker conditions for the cost minimum [see Section 9.3, equation (9.13)] is used to determine the exchange rate of emission trading between each pair of countries. Thereby the exchange rate is set to equal the ratio of the marginal cost (C_i) of emission reduction in the cost minimum. For

A Model Simulation

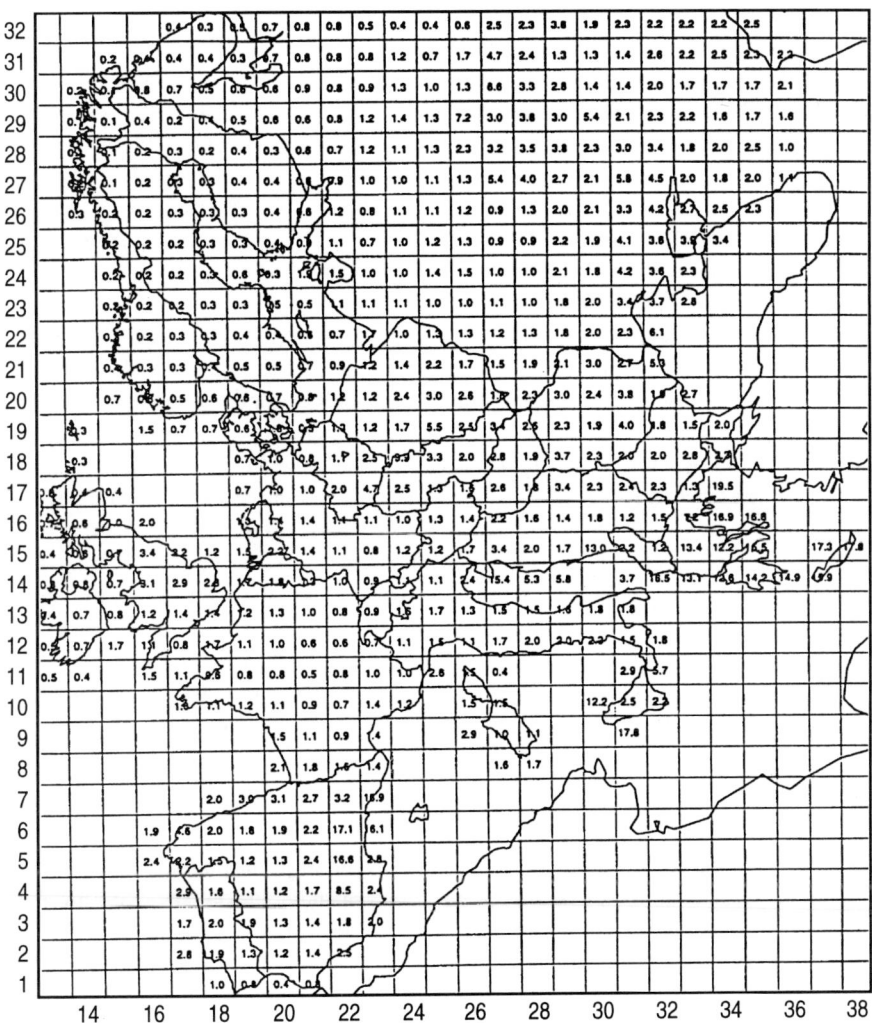

Figure 10.4. Deposition goals below Second Sulfur Protocol levels or five-percentile critical loads.

example, if in the cost minimum the marginal costs are 794 DM/ton SO_2 for Italy and are 6,002 DM/ton SO_2 for Norway, the resulting exchange rate is 7.55:

$$\frac{C'_{\text{Norway}}}{C'_{\text{Italy}}} = \frac{6,002 \text{ DM/t SO}_2}{794 \text{ DM/t SO}_2} = 7.55 \ .$$

This implies that if Norway wants to increase emissions by 1 unit, Italy would have to reduce its emissions by 7.55 units. Because marginal costs in the cost minimum generally differ among countries, these (fixed) exchange rates for bilateral trades will also be different for each trade combination. The final outcome of trading depends not only on which countries trade but also on the sequence. The sequence can be fixed so that trades with the highest absolute level of cost savings take place first, or it can be fixed so that the sequence is random. Both options are analyzed in the remainder of this chapter.

The constant-deposition rates are designed to prevent increases in average sulfur deposition in each country with every trade. These rates are derived from the five-year average country-to-country matrices developed by EMEP (Sandnes, 1993). These rates are calculated so that the average deposition in each country does not increase with every single trade. An example can illustrate this. Assume there are three countries A, B, and C. Of country A's emissions, 50 percent come down in country A, 0 percent in country B, and 50 percent in country C. Of country B's emissions, 50 percent are deposited in country A, 25 percent in country B, and 25 percent in country C. Assume country A wants to buy 1 ton from country B. As a result, deposition in country A will increase by 0.5 tons, in country B it will not change, and in country C it will increase by 0.5 tons. To keep deposition constant in country A, country B must reduce its emissions by 1 unit. To keep deposition constant in country C, however, it would have to reduce emissions by 2 units. Therefore, in this example the exchange rate for country A's buying from country B is 2. *Table 10.1* shows these constant-deposition rates.

The information in *Table 10.1* allows the following conclusions to be drawn:

- Forty-two percent of all (1,056) possible bilateral joint implementation agreements would not be possible because none of the countries deposits sulfur on all the other countries' territory (for example, Norway cannot buy emission permits from Albania because Albanian emissions do not reach Norway).
- Thirty percent of all possible agreements would not be possible because deposition in third parties would increase [for example, Remaining Russia (Russia without St. Petersburg and Kola-Karelia, RSU) cannot buy from Belgium because deposition in Greece would increase].
- In 21 percent of the cases, the exchange rates are higher than 10 (for example, 534.9 if Albania buys from Poland), putting a high penalty on bilateral trades between these parties.

Table 10.1. Constant-deposition rates.

Buyer	Seller (decreaser)										
	Alb	Aus	Bel	Bul	CSFR	Den	Fin	Fra	FRG–W	FRG–E	Gre
Alb	1.0	–	–	–	326.1	–	–	–	340.3	341.6	15.1
Aus	–	1.0	–	–	–	–	–	–	10.9	15.1	–
Bel	–	–	1.0	–	145.3	–	–	6.5	5.2	76.1	–
Bul	–	–	–	1.0	16.1	–	–	–	54.7	31.4	–
CSFR	–	–	–	–	1.0	–	–	19.3	8.1	4.0	–
Den	–	–	28.8	–	91.0	1.0	–	51.1	23.7	38.1	–
Fin	–	–	125.2	–	98.8	15.9	1.0	221.8	82.5	69.0	–
Fra	–	–	3.1	–	35.6	–	–	1.0	7.0	24.0	–
FRG-W	–	–	2.4	–	47.9	–	–	4.5	1.0	14.6	–
FRG-E	–	–	–	–	3.4	–	–	17.6	5.4	1.0	–
Gre	–	–	–	–	169.1	–	–	–	264.7	265.7	1.0
Hun	–	–	–	–	6.9	–	–	66.2	41.1	27.5	–
Irl	–	–	–	–	–	–	–	262.0	487.2	489.1	–
Ita	–	–	–	115.6	32.0	–	–	–	31.3	41.9	–
Lux	–	–	2.4	–	–	–	–	3.8	6.5	22.1	–
Net	–	–	2.4	–	166.3	–	–	15.5	3.5	58.0	–
Nor	–	–	38.2	–	120.5	9.7	28.7	90.1	50.3	72.1	–
Pol	–	–	–	–	3.3	–	–	22.7	7.9	2.8	–
Por	–	–	–	–	–	–	–	275.3	–	–	–
Rom	–	–	–	6.8	70.5	–	–	–	220.6	147.6	–
Spa	–	–	–	–	–	–	–	25.7	–	–	–
Swe	–	–	34.8	–	48.9	3.1	–	61.7	27.0	27.1	–
Swi	–	–	53.6	–	169.1	–	–	17.3	39.2	118.1	–
Tur	–	–	–	27.5	131.6	–	–	369.1	228.8	172.2	7.6
UK	–	–	20.1	–	222.2	–	–	31.2	46.4	116.3	–
Yug	–	–	–	12.9	14.7	–	–	–	34.3	28.0	–
Kol	–	–	45.0	–	35.5	7.3	2.6	79.8	29.7	24.8	–
Pet	–	–	33.0	–	18.7	7.3	2.9	60.4	29.9	17.7	–
Bal	–	–	39.4	–	35.5	–	–	69.8	37.1	26.0	–
Bye	–	–	58.7	–	25.6	–	–	94.6	42.9	25.4	–
Ukr	–	–	44.0	–	11.7	–	–	73.9	28.2	16.7	–
Mol	–	–	–	82.7	81.4	–	–	583.4	216.4	128.1	–
RSU	–	–	–	42.9	22.1	–	–	71.4	35.4	20.9	66.1

Alb – Albania; Aus – Austria; Bel – Belgium; Bul – Bulgaria; CSFR – Czechoslovakia; Den – Denmark; Fin – Finland; Fra – France; FRG-W – Germany-West; FRG-E – Germany-East; Gre – Greece; Hun – Hungary; Irl – Ireland; Ita – Italy; Lux – Luxemburg; Net – Netherlands; Nor – Norway; Pol – Poland; Por – Portugal; Rom – Romania; Spa – Spain; Swe – Sweden; Swi – Switzerland; Tur – Turkey; UK – United Kingdom; Yug – Yugoslavia; Kol – Kola-Karelia; Pet – St. Petersburg; Bal – Baltic region (Lithuania, Latvia, Estonia, and Kaliningrad); Bye – Belarus; UKR – Ukraine; Mol – Moldavia; RSU – remaining part of Russian Federation in EMEP.
Note: – implies infeasible.

Table 10.1. Continued.

	Seller (decreaser)										
Buyer	Hun	Irl	Ita	Lux	Net	Nor	Pol	Por	Rom	Spa	Swe
Alb	167.1	–	54.3	–	–	–	534.9	–	73.3	–	–
Aus	–	–	11.2	–	–	–	3.0	–	–	–	–
Bel	–	–	–	–	2.8	–	–	–	–	–	–
Bul	5.0	–	34.9	–	–	–	18.1	–	6.7	–	–
CSFR	25.8	–	–	–	–	–	4.4	–	–	–	–
Den	–	–	–	–	14.2	–	99.5	–	–	–	–
Fin	202.5	–	–	–	61.8	–	54.0	–	–	–	5.1
Fra	–	–	–	–	–	–	–	–	–	–	–
FRG-W	–	–	–	–	1.9	–	–	–	–	–	–
FRG-E	–	–	–	–	–	–	18.1	–	–	–	–
Gre	65.0	–	70.4	–	–	–	208.0	–	6.0	–	–
Hun	1.0	–	19.7	–	–	–	19.4	–	–	–	–
Irl	–	1.0	–	–	–	–	–	–	–	–	–
Ita	–	–	1.0	–	–	–	60.6	–	–	–	–
Lux	–	–	–	1.0	–	–	–	–	–	–	–
Net	–	–	–	–	1.0	–	272.7	–	–	–	–
Nor	247.0	–	–	–	25.1	1.0	98.8	–	–	–	8.6
Pol	8.5	–	–	–	–	–	1.0	–	–	–	–
Por	–	–	–	–	–	–	–	1.0	–	18.4	–
Rom	22.4	–	88.1	–	–	–	99.0	–	1.0	–	–
Spa	–	–	–	–	–	–	–	–	–	1.0	–
Swe	–	–	–	–	22.9	–	30.1	–	–	–	1.0
Swi	–	–	9.7	–	52.9	–	554.7	–	–	107.8	–
Tur	56.2	–	137.0	–	–	–	107.9	–	13.4	–	–
UK	–	–	–	–	–	–	–	–	–	–	–
Yug	4.8	–	6.5	–	–	–	29.2	–	–	–	–
Kol	72.9	–	–	–	22.0	–	19.4	–	–	–	5.9
Pet	33.6	–	–	–	24.3	–	10.3	–	–	–	6.6
Bal	42.5	–	–	–	38.9	–	13.2	–	–	–	–
Bye	25.5	–	–	–	–	–	9.1	–	–	–	–
Ukr	7.7	–	–	–	–	–	6.5	–	–	–	–
Mol	43.0	–	368.8	–	–	–	57.5	–	13.2	–	–
RSU	20.3	–	–	–	–	–	10.3	–	17.4	–	–

Alb – Albania; Aus – Austria; Bel – Belgium; Bul – Bulgaria; CSFR – Czechoslovakia; Den – Denmark; Fin – Finland; Fra – France; FRG-W – Germany-West; FRG-E – Germany-East; Gre – Greece; Hun – Hungary; Irl – Ireland; Ita – Italy; Lux – Luxembourg; Net – Netherlands; Nor – Norway; Pol – Poland; Por – Portugal; Rom – Romania; Spa – Spain; Swe – Sweden; Swi – Switzerland; Tur – Turkey; UK – United Kingdom; Yug – Yugoslavia; Kol – Kola-Karelia; Pet – St. Petersburg; Bal – Baltic region (Lithuania, Latvia, Estonia, and Kaliningrad); Bye – Belarus; UKR – Ukraine; Mol – Moldavia; RSU – remaining part of Russian Federation in EMEP.
Note: – implies infeasible.

Table 10.1. Continued.

Buyer	Seller (decreaser)										
	Swi	Tur	UK	Yug	Kol	Pet	Bal	Bye	Ukr	Mol	RSU
Alb	–	–	–	39.5	–	–	–	–	–	–	–
Aus	–	–	44.8	12.5	–	–	–	–	–	–	–
Bel	–	–	–	–	–	–	–	–	–	–	–
Bul	–	–	–	6.7	–	–	–	–	14.4		
CSFR	–	–	33.7	–	–	–	–	–	–	–	–
Den	–	–	17.8	–	–	–	–	–	–	–	–
Fin	–	–	77.4	–	9.0	6.0	8.7	26.4	135.5	–	–
Fra	–	–	–	–	–	–	–	–	–	–	–
FRG-W	–	–	18.2	–	–	–	–	–	–	–	–
FRG-E	–	–	98.6	–	–	–	–	–	–	–	–
Gre	–	–	–	43.9	–	–	–	–	149.1	–	–
Hun	–	–	115.6	–	–	–	–	–	–	–	–
Irl	–	–	32.7	–	–	–	–	–	–	–	–
Ita	–	–	–	21.1	–	–	–	–	–	–	–
Lux	–	–	–	–	–	–	–	–	–	–	–
Net	–	–	15.6	–	–	–	–	–	–	–	–
Nor	–	–	16.9	–	11.5	73.5	64.0	128.6	–	–	364.2
Pol	–	–	19.8	–	–	–	–	–	–	–	–
Por	–	–	480.3	–	–	–	–	–	–	–	–
Rom	–	–	–	13.5	–	–	–	–	62.1	–	–
Spa	–	–	64.9	–	–	–	–	–	–	–	–
Swe	–	–	20.5	–	–	–	–	–	–	–	–
Swi	1.0	–	110.4	–	–	–	–	–	–	–	–
Tur	–	1.0	–	66.4	–	–	–	–	33.0	–	–
UK	–	–	1.0	–	–	–	–	–	–	–	–
Yug	–	–	–	1.0	–	–	–	–	–	–	–
Kol	–	–	31.1	–	1.0	6.4	–	–	–	–	–
Pet	–	–	35.0	–	–	1.0	–	–	–	–	–
Bal	–	–	40.6	–	–	–	–	–	–	–	–
Bye	–	–	76.9	–	–	–	5.5	9.5	48.8	–	31.6
Ukr	–	–	–	21.4	–	–	2.3	4.4	22.5	–	17.4
Mol	–	–	479.3	108.9	–	–	1.0	7.4	64.0	–	93.9
RSU	–	–	41.4	–	–	–	4.2	1.0	22.3	–	–

Alb – Albania; Aus – Austria; Bel – Belgium; Bul – Bulgaria; CSFR – Czechoslovakia; Den – Denmark; Fin – Finland; Fra – France; FRG-W – Germany-West; FRG-E – Germany-East; Gre – Greece; Hun – Hungary; Irl – Ireland; Ita – Italy; Lux – Luxembourg; Net – Netherlands; Nor – Norway; Pol – Poland; Por – Portugal; Rom – Romania; Spa – Spain; Swe – Sweden; Swi – Switzerland; Tur – Turkey; UK – United Kingdom; Yug – Yugoslavia; Kol – Kola-Karelia; Pet – St. Petersburg; Bal – Baltic region (Lithuania, Latvia, Estonia, and Kaliningrad); Bye – Belarus; UKR – Ukraine; Mol – Moldavia; RSU – remaining part of Russian Federation in EMEP.
Note: – implies infeasible.

- In 7 percent of the cases the exchange rates are lower than 10 (for example, 2.4 if the former West Germany buys from Belgium).

The performance of the alternative joint implementation regimes will be examined under different conditions:

- As pure instruments without additional constraints (the textbook case);
- As hybrid instruments, that is, in combination with regulatory constraints in the form of emission and fuel standards;
- With non-signatories of the Second Sulfur Protocol not being allowed to participate;
- With significant transaction costs (DM 3 million/yr per trade).

Finally, the impact of the sequence of trading on the results of exchange rate trading is also analyzed.

In the remainder of this section, Section 10.5.2 gives the main results and Section 10.5.3 clarifies some of the results in more detail. Section 10.5.4 takes a look at the impact of altering the trading sequence. Section 10.5.5 discusses the impact of imperfect information on costs and behavior.

10.5.2 Main results for costs, emissions, and ecosystem protection

In this section the annual costs, the emissions, the level of ecosystem protection, and the distributional consequences of the different schemes under different conditions are discussed.

Regarding the potential costs, *Table 10.2* shows that

- Emission trading or joint implementation in one European zone will always lead to the highest cost savings – depending on the circumstances, 6 to 37 percent of the costs of the Second Sulfur Protocol could be saved;
- The deposition constraints solution leads to minor cost savings (1 to 3 percent);
- Joint implementation with cost-minimum exchange rates leads to lower cost savings (2 to 9 percent) than the Europe-wide bubble;
- Using constant-deposition rates puts a high penalty on trade. Consequently, cost savings are only 0.5 percent to 1 percent.

Another observation is worth making. The emission and fuel standards that are part of the Second Sulfur Protocol and those that exist in national and international (EC) legislation make the Protocol more expensive than necessary. Moreover, assuming that these regulations remain in place, they also limit the scope for joint implementation to save costs. This is seen from *Table 10.2* (column I versus column II).

A Model Simulation   257

Table 10.2. Annual costs for the year 2000 (mln. DM, constant 1990 values).

		Condition			
		I	II	III	IV
		No	Emission	No non-	Transaction
Instrument		constraints	standards	signatories	costs
A.	Second Sulfur Protocol	16,084 (100%)	23,941 (100%)	23,941 (100%)	23,941 (100%)
B.	Deposition constraints	15,648 (97.3%)	23,801 (99.4%)	23,821 (99.5%)	<23,821 (<99.5)
C.	European bubble	10,139 (63.0%)	22,225 (92.8%)	22,572 (94.3%)	22,580 (94.3%)
D.	Cost-minimum exchange rates	14,638 (91.0%)	23,315 (97.4%)	23,347 (97.5%)	23,349 (97.5%)
E.	Constant-deposition rates	15,954 (99.2%)	23,833 (99.5%)	23,840 (99.6%)	23,840 (99.6%)

Table 10.2 suggests that existing emission and fuel standards increase the costs of the Second Sulfur Protocol by 50 percent compared with a Protocol with only national ceilings (from DM 16 to DM 24 billion per year). This leads to three important observations:

- Allowing more flexibility in how countries meet their national bubbles would offer considerable scope for cost savings.
- The establishment of a European bubble while maintaining regulation (alternative *C*) would only reduce costs from DM 16 to DM 15.5 billion per year.
- A European bubble without regulatory impediments would be able to cut costs from DM 16 to DM 10 billion per year.

Assuming that emission standards are in place, excluding non-signatories of the Second Sulfur Protocol from trade does not have significant impacts on the potential cost savings of the emission trading schemes (compare columns II and III of *Table 10.3*). This is because only a few countries did not, and are not expected to, sign this Protocol: Albania, Romania, Belarus, Turkey, the Baltic States (Estonia, Latvia, and Lithuania), and Moldovia.

Similar remarks can be made for the impact of raising transaction costs to DM 3 million/yr per trade (5 percent of the average cost savings per trade with a European bubble). This will not have a significant impact on trading. Only a few trades will not take place under the European bubble or under the exchange rate regimes (cost-minimum rates and average deposition constant rates). Consequently, the loss of cost savings is insignificant. This is subject to one qualification, however.

Table 10.3. Emissions (kton SO$_2$/year).

Instrument		Condition			
		I No constraints	II Emission standards	III No non- signatories	IV Transaction costs
A.	Second Sulfur Protocol	31,744	31,744	31,744	31,744
B.	Deposition constraints	30,677	31,022	31,070	–
C.	European bubble	31,744	31,774	31,744	31,744
D.	Cost-minimum exchange rates	32,015	31,753	31,858	31,868
E.	Constant-deposition rates	30,898	31,086	31,142	31,142

The deposition constraints regime could, in principle, be achieved by a system of tradable deposition (or ambient) permits or by pollution offset trading. In the first case, countries would have to operate on different deposition permit markets simultaneously in order to increase or decrease emissions. This might raise the transaction costs in terms of finding partners and making agreements with them. The pollution offset trading scheme might be able to meet the cost minimum in theory (see Chapter 3), but in reality it might be difficult to find a sequence of bilateral trades to get there (see Atkinson and Tietenberg, 1991). Moreover, this solution requires the use of deposition models that show the impact on air quality before and after trading. In practice this raises transaction costs because approval becomes costly and uncertain. Almost no trades requiring deposition models took place under the EPA's emission trading program in the past (Tietenberg, 1990). For both the exchange rates regime and the European bubble, these problems are less likely to occur and transaction costs are not expected to have a significant impact on the potential cost savings.

The impacts on emissions are summarized in *Table 10.3*. Apart from the fact that the impact of all trading schemes on the total level of emissions is small (a 1 percent increase versus a 3 percent decrease) the following points can be made:

- The deposition constraints solution always reduces emissions compared with the Protocol. This implies that cost savings are not achieved because emissions are increased but because emission reductions are shifted to those countries where abatement is cheaper.

Table 10.4. Ecosystem protection (percentage protected).

Instrument		Condition			
		I No constraints	II Emission standards	III No non-signatories	IV Transaction costs
A.	Second Sulfur Protocol	82.7	82.7	82.7	82.7
B.	Deposition constraints	83.8	83.5	83.4	–
C.	European bubble	81.1	82.3	82.9	82.7
D.	Cost-minimum exchange rates	82.9	82.8	82.6	82.5
E.	Constant-deposition rates	83.5	83.3	83.3	83.3

- The cost-minimum exchange rates regime saves costs in part because it increases emissions (slightly) compared with the Protocol.
- The European bubble by definition keeps the total emission constant.
- The constant-deposition rates always imply a reduction in emissions because these rates are always higher than one, but the reduction is small because the high offset rates imply that only a few profitable trades take place.

The different joint implementation regimes have slightly different impacts on ecosystem protection. Ecosystem protection is defined here as the average percentage of ecosystems in each country that are exposed to sulfur deposition levels below the five-percentile critical sulfur deposition loads; in other words, the percentage of ecosystems not expected to be damaged. *Table 10.4* shows that

- The deposition constraints regime leads to an increase in overall Europe-wide ecosystem protection;
- The European bubble leads to a decrease in ecosystem protection unless trade is superimposed on existing regulations and is limited to signatories;
- The cost-minimum rates sometimes imply an increase in overall ecosystem protection (case I) and sometimes imply a reduction in protection (case III) compared with the Protocol;
- Joint implementation with constant-deposition rates always improves overall ecosystem protection, because it lowers the emissions while accounting for the location.

Table 10.5. Distributional impacts (number of countries with less protection).

		Condition			
		I No constraints	II Emission standards	III No non- signatories	IV Transaction costs
Instrument					
A.	Second Sulfur Protocol	0	0	0	0
B.	Deposition constraints	1(0)*	1(0)	1(0)	–
C.	European bubble	20(0)	18(1)	14(3)	14(2)
D.	Cost-minimum exchange rates	5(1)	8(3)	9(2)	9(4)
E.	Constant-deposition rates	0	0	0	0

*Figures in brackets indicate the number of countries that have less environmental protection without saving costs.

It is important to know why the one-zone regime leads to a decrease in Europe-wide ecosystem protection. This is because it shifts emission reductions toward countries in Eastern Europe with low marginal costs (especially Bulgaria, Romania, Turkey, and Ukraine), enabling increases in emissions in Scandinavia and Western Europe. Combining this European bubble strategy with regulation restricts, but does not fully eliminate, this tendency. Because the predominant winds are western and because ecosystems in these Eastern European countries are less sensitive than ecosystems in Scandinavia, the Alpine regions, or the Netherlands, emission reductions would tend to take place there, where they are less necessary to protect the environment. This contrasts with the US sulfur allowance trading program where reductions are expected to take place where they are needed to protect sensitive natural environments (see Chapter 6).

The aggregate changes in ecosystem protection are not more than a few percentage points. They result, however, from improvement in a number of countries and sometimes considerable deterioration in others. The distribution of environmental benefits and costs is important for the political acceptability of the trading or joint implementation schemes. This distribution is probably a highly problematic aspect of the one-zone trading schedule. *Table 10.5* shows that

- One-zone joint implementation implies that no fewer than 14 to 20 (out of 32) countries might be confronted with less ecosystem protection. Even though

A Model Simulation

nearly all countries also can expect to receive partial compensation for this because they trade and cut their costs, this might nonetheless be problematic;
- The deposition constraints regime saves costs while decreasing ecosystem protection only in Denmark (ecosystem protection would drop slightly, from 79 to 77 percent), but that country would also save costs;
- The cost-minimum exchange rates would decrease ecosystem protection in 5 to 9 countries (depending on the conditions) and some of these countries would not reduce costs because they would not trade;
- The constant-deposition rates either improve ecosystem protection or keep it constant in each country.

In view of the above, it seems questionable whether a European bubble is an acceptable political solution. Countries with high preferences for ecosystem protection would be confronted with considerably less protection, such as Austria (46 instead of 74 percent protected), Finland (77 versus 92 percent protected), Germany (23 instead of 46 percent protection), the Netherlands (17 instead of 23 percent protected), Sweden (70 instead of 85 percent protected), and Switzerland (58 versus 77 percent protected). This makes this one-zone solution not a viable option. Before drawing conclusions, we will examine some of the results in more detail.

10.5.3 Detailed analysis of joint implementation with exchange rates

This section examines some of the results more closely to explain what happens with both exchange rates trading options – cost-minimum exchange rates and constant-deposition rates – if they are combined with regulations and if only signatories trade (case IIID and IIIE). The following will be discussed:

- Results for cost-minimum rates, including an examination of why they do not achieve the cost-minimum (*Table 10.6*);
- Constant-deposition rates (*Table 10.7*);
- Country results for the case of emission standards and no trading by non-signatories (*Table 10.8*).

Joint implementation subject to a cost-minimum exchange rate would lead to seven bilateral agreements (see *Table 10.6*) with exchange rates varying from a low of 0.80 to a high of 7.55. It is noteworthy that the emissions under the exchange rate regime (scenario D) are higher than they are under the cost-minimum solution (scenario B), and consequently costs are lower. Obviously, the costs can be lower only because the deposition constraints of option B are violated, as we already expected from the theoretical analysis in Chapter 9.

Table 10.6. Joint implementation with cost-minimum exchange rates (case III).

Trade	Buyer	Amount (kton SO$_2$)	(%)*	Seller	Amount (kton SO$_2$)	(%)*	Exchange rate	Cost savings (mln. DM/yr)
1	Ukraine	794	34.0	Bulgaria	637	43.0	0.80	505
2	Norway	8	24.0	Italy	58	4.0	7.55	64
3	Denmark	7	8.0	Former East Germany	20	3.4	2.82	15
4	Greece	81	14.0	Russia	81	2.1	0.99	7
5	Belgium	1	0.4	Russia	1	0.0	1.85	1
6	Austria	2	2.5	Russia	7	0.0	3.04	1
7	Finland	2	2.0	Ukraine	7	0.3	3.12	0
Total	All	895	39.0	All	811	2.5	0.91	593

*As percentage of the Protocol emissions of each country.

Table 10.7. Potential buyers and sellers with constant-deposition rates (case III).

	Buyers			
Sellers	Norway	Former Yugoslavia	Russian Federation	Italy
Bulgaria		x	x	x
Kola-Karelia	x			
United Kingdom	x			
Former West Germany	x			
Former East Germany	x			
Netherlands	x			

Several factors are responsible for the fact that the cost-minimum solution is not attained. First, the exchange rate is based only on one Kuhn-Tucker condition for the cost-minimum solution, the condition that marginal costs per ton of emissions abated have to equal the weighted sum of the transfer coefficient and the shadow prices of the binding receptors (see also Chapter 3). The exchange rate ignores the condition that depositions have to be below the the deposition standards. The introduction of the cost-minimum exchange rates is a simple way to internalize the deposition constraints in the trade process and they are clearly not perfect for realizing the goal of ecosystem protection.

Second, the concept of marginal costs has no unique definition for piecewise linear cost functions related to technological indivisibilities. The cost-minimum

Table 10.8. Emissions and ecosystem protection without non-signatories and with emission and fuel standards (case III).

Scenario	Emissions (kton SO$_2$)			Ecosystem protection (%)		
	A	D	E	A	D	E
Country/region	Second[a] Sulfur Protocol	Cost-min. exchange rates	Constant-deposition rates	Second Sulfur Protocol	Cost-min. exchange rates	Constant-deposition rates
Albania	138	138	138	100	100	100
Austria	78	80	78	74	74	74
Belgium	248	249	248	8	8	8
Bulgaria	1,473	737	737	54	80	92+
CSFR (former)	1,465	1,465	1,465	28	28	28
Denmark	90	97	90	79	77−	79
Finland	116	118	116	92	92	92
France	868	868	868	94	94	94
Germany-West (former)	710	710	710	43	43	43
Germany-East (former)	590	570	590	43	43	43
Greece	595	676	595	98	99	99+
Hungary	899	898	899	53	53	53
Ireland	155	155	155	94	94	94
Italy	1,330	1,272	1,330	86	87	86
Luxembourg	10	10	10	30	30	30
Netherlands	106	106	106	22	22	22
Norway	34	42	35	70	70	70
Poland	2,583	2,583	2,583	27	26	27
Portugal	304	304	304	98	98	98
Romania	2,592	2,592	2,592	38	39	40+
Spain	2,143	2,143	2,143	98	98	98
Sweden	100	100	100	84	84	84
Switzerland	60	60	60	77	77	77
Turkey	2,887	2,887	2,887	98	98	99+
UK	2,449	2,449	2,449	79	79	79
Yugoslavia (former)	1,415	1,415	1,464	88	89	89+
Kola-Karelia	320	320	304	95	94	95
Petersburg	248	248	248	95	94	95
Russia[b]	3,872	3,783	3,872	95	94	95
Baltic[c]	435	435	435	93	92	93
Belarus	456	456	456	95	94	95
Ukraine	2,310	3,097	2,310	95	94	95
Moldavia	231	231	231	95	94	95
Europe[d]	31,774	31,858	31,142	82.7	82.6	83.3

[a] Non-signatories are assumed to carry out current reduction plans.
[b] Russian Federation in EMEP area; Kola-Karelia, St. Petersburg, and Kaliningrad regions excluded.
[c] Includes Estonia, Latvia, Lithuania, and Kaliningrad.
[d] Includes emissions in Baltic and North Seas and Atlantic Ocean (74+174+317 kton SO$_2$).

solution for these stepwise functions sometimes allows for two cost-minimum marginal costs (in the case of corner solutions). An example is the trade between Norway and Italy. In the cost-minimum solution the marginal costs of Norway are DM 6,002/ton SO_2. The marginal costs of Italy in the optimum are DM 794/ton SO_2. The Norway-Italy exchange rate is therefore 7.55 (6,002 divided by 794). Norway has to buy 7.55 emission permits from Italy to increase its emission by 1 unit. Before trade Norway's marginal costs are DM 53,412/ton SO_2 and Italy's marginal costs are DM 714/ton SO_2. After trade both Norway and Italy have marginal costs that correspond to their cost-minimum marginal costs. So the ratio of the marginal costs coincides with the exchange rate; this is precisely the condition for a cost minimum. Their emission levels, however, deviate from the cost-minimum levels. Norway emits 42 kton instead of 39 kton SO_2 (the cost-minimum amount), and Italy emits 1,272 instead of 1,298 kton SO_2. Due to the piecewise linear cost functions the same level of marginal costs (the same permit price) corresponds to different levels of emissions, the cost-minimum and post-trade levels. This is not a new observation; Rose-Ackermann (1973) noted that an emission charge set at a certain level of marginal costs might not determine a unique treatment level in the case of technological indivisibilities. In summary, due to the piecewise linear functions, one exchange rate based on marginal cost ratios does not define unique emission levels. So although the marginal costs correspond to the cost-minimum levels, the emissions do not. Post-trade emission levels depend not only on the marginal costs but also on the initial, pre-trade emission levels because the exchange rates apply to all intramarginal trades and the final trade equilibrium thus depends on the initial distribution of emission permits.

Joint implementation with the constant-deposition rates would only enable two cost-saving trades if it were combined with emission and fuel standards. *Table 10.7* shows which countries could act as buyers and sellers. It should be noted that when Norway trades first with the Netherlands it can always trade afterward with one of the other four countries or regions (Kola-Karelia, UK, Former West Germany, or Former East Germany). The cost savings are DM 40 million/year.

If those trades that result in the highest absolute cost savings are implemented, only two trades occur. Norway buys approximately 16 kton SO_2 emission permits from Kola-Karelia (part of Russia) and increases its emissions by about 1.4 kton (exchange rate 11.5). The annual cost savings are DM 61 million. Bulgaria sells 637 emission permits (each equal to 1 kton SO_2) to the former Yugoslavia. This allows the former Yugoslavia to increase its emissions by 49 ktons (exchange rate 12.9). The annual cost savings are around DM 40 million. Norway has eight possible combinations for trading: four single trades with Kola-Karelia, UK, former West Germany, or former East Germany, and four combinations where Norway first buys from the Netherlands and then buys from Kola-Karelia, UK, former West

A Model Simulation 265

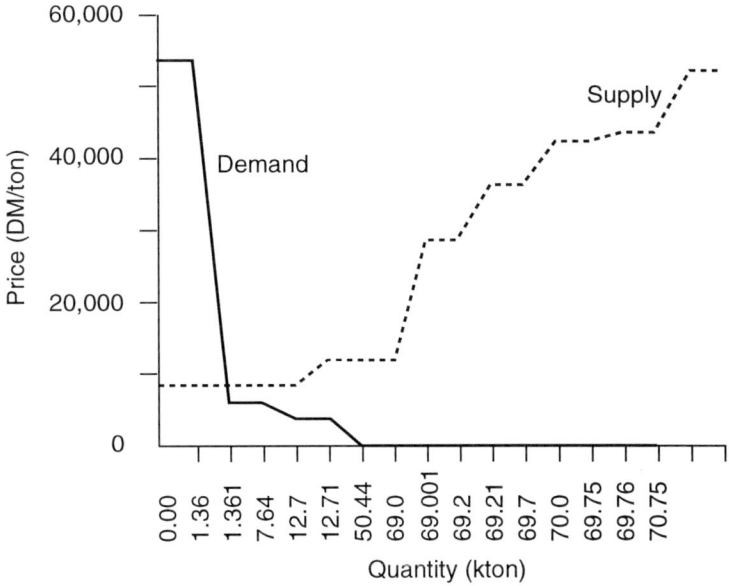

Figure 10.5. Demand and supply on Norway's market (case III*E*).

Germany, or former East Germany. Bulgaria can only trade with three countries. This gives a total of 8 × 3 possible outcomes of bilateral sequential trading.

Trading with constant-deposition rates results in two separate markets: one in the northwest with Norway apparently acting as a monopsonist facing five sellers, and one in the east where Bulgaria seems to act as a monopolist facing three buyers. It looks as if this gives both Norway and Bulgaria the opportunity for strategic behavior.

Norway might consider reducing its demand in order to lower the price it has to pay to obtain permits and thus raise its cost savings above a competitive market equilibrium. In this way the trade would be less cost efficient but Norway's profits might be higher. *Figure 10.5* shows the demand and supply curve on the Norwegian market. All quantities are expressed in ktons bought by Norway.

With perfect competition the permit price would be DM 8,287/ton SO_2. In this special case Norway could not lower the price below this level, because no country would be willing to supply below this level. There is some room for strategic behavior for Kola-Karelia, however, which could limit supply in order to drive up the price. Kola-Karelia, however, is faced with a limit price of DM 12,067/ton

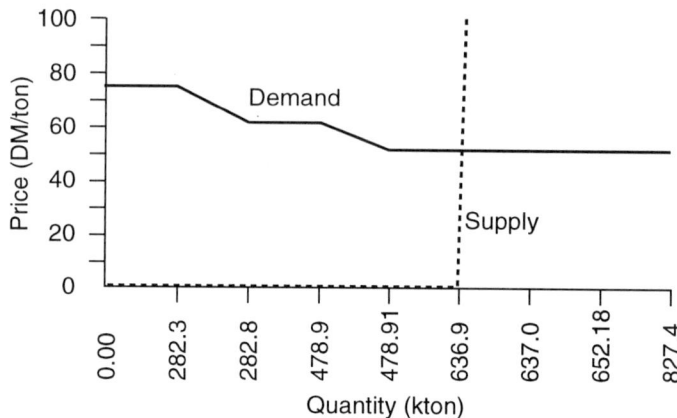

Figure 10.6. Demand for and supply of Bulgarian emission reductions (case IIIE).

SO_2. Above this price, it would become attractive for the UK to act as a seller. In conclusion, there appear to be possibilities for strategic behavior for the supplier, Kola-Karelia, rather than for the buyer, Norway.

Being in a monopoly position, Bulgaria might limit permit supply in order to increase the price it would get from selling permits and thus raise its profits. The extent to which it can do this depends on the price elasticity of the buyers' demand and the elasticity of its own marginal cost curve. The demand for and supply of Bulgarian emission reductions are shown in *Figure 10.6*. Under a competitive equilibrium Bulgaria would sell 637 ktons to the former Yugoslavia. The equilibrium price would be DM 51/ton (the former Yugoslavia's demand price, i.e., its marginal costs saved per ton bought from Bulgaria).

It can be shown that Bulgaria has no real market power. Its marginal costs are very low (DM 0.2/ton) up to a supply of 637 kton and then they rise steeply. At the intersection with the demand curve, supply is nearly perfectly inelastic. On the other hand, Bulgaria is faced with a highly elastic demand; for a price increase from DM 51 to DM 61/ton the price elasticity of demand is –2.5, and for a price increase from DM 61 to DM 75 this elasticity is –1.8. This implies that limiting supply to less than 637 kton will decrease the profits Bulgaria can make from trade. However, the buyer, the former Yugoslavia, has market power. The former Yugoslavia is the only country willing to buy at a maximum price of DM 51/ton. It could buy less to lower the price to nearly DM 16.7/ton, which is the price at which Italy would enter the market as a buyer. If both Bulgaria and the former Yugoslavia want to minimize

A Model Simulation

costs it is likely that they will agree on the competitive equilibrium amount of 637 kton. The price will be between DM 0.2 and DM 51 per ton.

In conclusion, the high penalty that constant-deposition rates trading puts on trading in principle may result in thin markets where price taking may no longer be guaranteed. For the specific demand and supply curves examined, however, the elasticities of supply and demand are such that price-setting behavior will not affect quantities. Therefore, one might expect that the volumes traded will equal volumes under a competitive market equilibrium. Therefore, the monopoly and monopsony situations will not affect the cost efficiency in this particular case.

To finalize this section, *Table 10.8* shows the country-specific results for both the constant-deposition rates and the cost-minimum exchange rates and compares them with the Second Sulfur Protocol. It is assumed that only signatories trade and that trades take place in compliance with existing regulations (case III). *Table 10.8* indicates that with the cost-minimum rates several countries will slightly alter their emissions and will trade. *Table 10.6* provides a listing of countries that will trade. In most cases the change in emissions is small. There are two exceptions. Bulgaria will sell rights and reduce emissions from 1,473 kton to 737 kton, thus enabling Ukraine to increase emissions from 2,310 kton to 3,097 kton. Bulgaria is able to sell this amount of rights because the marginal costs of reducing these emissions are, to a large extent, zero. This is because without any control Bulgaria's emissions are already expected to go down to 815 kton in the year 2000. Bulgaria can thus make windfall profits. This results from the fact that Bulgaria originally was to reduce emissions to 50 percent of its 1980 level, bringing emissions down to 520 kton in the year 2000 (Amann *et al.*, 1993). Before the Protocol was signed, however, the official emissions of Bulgaria for 1980 increased from 1,034 to 2,050 kton. This implies that a 50 percent reduction in allowed emissions would be 1,025 kton (instead of 520 kton). On top of this, Bulgaria, along with other countries, postponed and slightly relaxed its commitment to 45 percent (to be met in 2010) and made a commitment to reduce emissions by 23 percent (1,473 kton) in 2000. *Table 10.8* also shows that the cost-minimum exchange rate trading has only small impacts on ecosystem protection. A number of countries would face slightly less protection: Denmark, Poland, Russia, the Baltic States, Ukraine, Belarus, and Moldavia. Some countries stand to gain, most notably Bulgaria, which would see its level of ecosystem protection increase from 54 to 80 percent because it would sell a large amount of permits. As a result, ecosystem protection would also slightly increase in neighboring countries (Romania and Greece). In conclusion, the constant-deposition rate exchanges have a small but positive impact on emissions and ecosystem protection.

For countries or regions that trade (see *Table 10.8*) the percentage of cost savings (compared with the Second Sulfur Protocol) under cost-minimum exchange

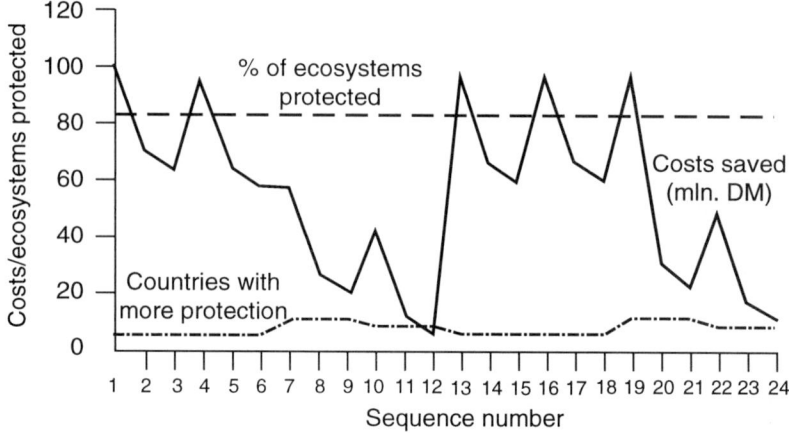

Figure 10.7. Constant-deposition rates and sequence.

rates trading are as follows: Austria and Belgium hardly save any costs (0 percent), Bulgaria saves more than 100 percent, Denmark 1 percent, Finland 0 percent (rounded), former East Germany and Greece 1 percent, Italy 2 percent, Norway 9 percent, Russia hardly anything (0.2 percent), and Ukraine 18 percent. With constant-deposition rates, only four countries or regions trade. The cost savings of Bulgaria are 12 percent; Norway, 9 percent; the former Yugoslavia, 8 percent; and Kola-Karelia, 9 percent.

10.5.4 The impact of the sequence

The impact of the trading sequence has been examined for both exchange rate schemes. As previously explained (*Table 10.7*), Norway can buy from five countries and Bulgaria from three countries. However, if Norway buys first from the Netherlands it can still trade with one of the four other countries and save costs. This implies that Norway has eight different possibilities for trade: four single trades (with Kola-Karelia, UK, former West Germany, or former East Germany), or first with the Netherlands and then with Kola-Karelia, the UK, former West Germany, or former East Germany. Because Bulgaria has three trade options, a total of 24 combinations of bilateral sequential trades exist for a situation with constant-deposition rates. The results of each possible combination are depicted in *Figure 10.7*. With completely random trading each combination would have an equal chance of being selected.

A Model Simulation

Figure 10.7 shows that the cost savings are highly dependent on the sequence. The average cost savings are DM 25.3 million/year and the standard deviation is DM 14.6 million/year. The costs saved depend on the sequence because if one trade takes place (e.g., Norway buys from former West Germany) it precludes other trades (e.g., Norway buying from another country) because the number of profitable trades is small. The percentage of ecosystems protected does not depend on the sequence, because the volumes traded are small and emissions always decrease. Consequently, ecosystem protection varies from 83.3 to 83.4 percent and is always better than under the Protocol. Finally, irrespective of the sequence, there is no country that has less ecosystem protection after trading. By contrast, 5 to 11 countries have more ecosystem protection. Five countries always have higher levels of protection: Bulgaria, Romania, Greece, Turkey, and the former Yugoslavia, because Bulgarian emissions are reduced after trade. Whether protection in Belgium, the Netherlands, France, Germany, Luxembourg, and Sweden improves depends on which country trades with Norway.

Surprisingly, the cost savings of the cost-minimum exchange rate trading appear to be quite sequence independent. The average cost savings are DM 596 million/year and the standard deviation is only DM 4.8 million/year. The average cost savings are exactly the same amount as when trading is implemented so that the trade with the highest cost savings is implemented first. This is because the number of profitable trades is fairly large and trading can continue until the cost savings are depleted. In the random-order case, the number of trades necessary to achieve the cost savings is much larger than when the trades with the highest cost savings are implemented first. The average number of trades needed is 43.5 for the former case and only 7 trades for the latter case. Quite independent of the sequence, the cost-minimum regime always increases emissions and results in a small expected loss of overall ecosystem protection compared with the Protocol (82.3 percent versus 82.7 percent protection). The exact loss depends on the sequence and ranges between 81.5 and 82.3 percent protection. The number of countries that are faced with less protection varies from a low of 1 to a high of 22, the average being 13 (standard deviation 6.7). In conclusion, the cost savings of trading with cost-minimum rates do not appear to be very sequence dependent. However, the overall level of ecosystem protection and the number of countries that face less protection is path dependent. Of the countries facing less environmental protection, the reduction varies from below 5 percent for 16 countries, to between 5 and 10 percent for 7 countries, to over 10 percent (up to 50 percent) for two countries, Belgium and Luxembourg.

10.5.5 Imperfect information about costs and behavior

The results discussed above depend on the specific situation assumed for these scenarios and they may be sensitive to other conditions than those quantitatively examined. Several assumptions deserve qualitative discussion.

First, to set the cost-minimum exchange rates it is assumed that the environmental agency has perfect information about abatement costs when determining the optimal solution. It might be more realistic to assume that this information is less than perfect, not least because cost functions depend on the underlying prediction of future energy consumption and energy use patterns. Consequently, even if the agency sets exchange rates on centrally perceived cost curves, it might well be that countries trade on the basis of what they individually see as their own realistic cost functions. The results of emission trading will therefore differ from the expectations of the agency. As a result, the emission pattern (which determines environmental impacts) will also be uncertain.

Second, if costs are uncertain the cost savings that emission trading can achieve are likely to be higher. This is because the emissions agreed to under the Protocol have been based to a certain degree on a cost-minimum solution (Amann *et al.*, 1993). This limits the room for additional cost savings. If, however, the actual costs differ from the ones that have been used to calculate the cost-minimum solution and determine the emissions of the Protocol, the emissions agreed to are more likely to deviate from the cost minimum that would result from the actual costs. To show this, we have examined the potential for cost savings if the costs are different. More specifically, we have assumed that due to structural changes in Eastern Europe (Amann *et al.*, 1993) energy efficiency in these countries has increased and it has become cheaper to reduce emissions in CEE countries. If no other constraints apply (no regulations, all countries trade: case I), calculations suggest that the costs of meeting the no-trade Protocol will also be lower. However, the potential cost savings will be much higher with deposition constraint trading (8 percent instead of 3 percent) as well as with the European bubble (56.5 percent instead of 37 percent). Similar results can be expected for both cost-minimum exchange rate trading and trading with constant-deposition rates. In conclusion, when the costs deviate from the cost estimates used to reach an agreement on the Protocol, joint implementation can be expected to lead to higher cost savings.

Finally, given initial emission and exchange rates, it is assumed that trading partners will try to minimize costs. Although this is a realistic assumption for firms operating on a competitive market, one may doubt whether this is an adequate assumption for the behavior of countries in an international context. Countries might place more emphasis on their environmental targets than on cost efficiency. Moreover, aspects of administrative practicability and political considerations are likely to play important roles in shaping environmental policies (see Opschoor and Vos, 1989).

10.6 Conclusions and Discussion

This chapter has examined the cost efficiency, the environmental impacts, and the distribution of costs and environmental benefits of the following trading or joint implementation schemes:

- Trade in one Europe-wide zone subject to the condition that the total emissions remain constant;
- Trade subject to explicit deposition constraints for every EMEP grid cell;
- Trade using cost-minimum exchange rates (and taking into account the transfer coefficients of the binding receptors);
- Trade using constant-deposition rates designed so that with every bilateral trade the average deposition does not increase in any country.

The performance of these schemes was examined under different conditions: without constraints, in combination with regulatory constraints in the form of emission and fuel standards, excluding non-signatories from trading, with different sequences of trading, and with different levels of transaction costs.

Table 10.9 summarizes the results and shows that depending on the circumstances

- Emission trading in one single European zone may not be politically acceptable, in spite of the fact that this scheme generates the highest cost savings, because too many countries would be confronted with losses in ecosystem protection.
- Joint implementation subject to deposition constraints reduces ecosystem protection in only one country, which saves costs. The scheme offers room for cost savings but high transaction costs may preclude actual achievement of these savings.
- Emission trading on the basis of cost-minimum exchange rates does not have significant impacts on overall levels of ecosystem protection and it saves considerably more costs than deposition constraint trading. However, it leads to small decreases in ecosystem protection in some countries without reducing their costs.
- International trading with constant-deposition rates ensures that every country is at least as well off as under the Protocol; limited cost savings (0.5 to 1 percent) are combined with improvements in ecosystem protection.

Regarding the importance of the different conditions that constrain trading, it appears that combining the national ceilings in the Protocol with existing emission

Table 10.9. Summary of emission trading schemes.

Scenario	A Second Sulfur Protocol	B Joint implementation Deposition constraints	C Joint implementation Europe-wide bubble	D Cost-minimum exchange rates	E Constant-deposition rates
Annual cost savings (as % of SSP)	0	0.5 to 3	5.7 to 37	2.5 to 9	0.4 to 0.8
Change in emissions (as % of SSP)	0	−2.2 to −3	0	0 to 1	−2 to −3
Change in ecosystem protection (% of ecosystems)	0	0.7 to 1.1	−1.6 to 0.2	−0.2 to 0.2	0.6 to 0.8
Countries with lower protection	0	1	14 to 20	5 to 9	0

SSP = Second Sulfur Protocol.

and fuel standards is the most relevant. This makes the Protocol 50 percent more expensive than national ceilings. Surprisingly, the best chance for cost savings thus appears to be when existing regulations are circumvented or altered so that countries have more freedom in meeting their own national bubble in a cost-efficient manner. This can be done, for example, by organizing national permit markets. International reallocation can do little to increase cost savings if existing regulations remain in place.

Excluding non-signatories from trade or increasing transaction costs would not lead to significant losses in cost savings. If the trading sequence were random the expected cost savings of constant-deposition rates trading would be much lower, but in any sequence ecosystem protection would be better than under the Protocol. The constant-deposition rates lead to thin markets where price-setting behavior might occur. This, however, is not expected to affect the volumes of emissions traded. Random trading does not alter the expected cost savings of cost-minimum exchange rate trading, nor does it change the number of countries confronted with less ecosystem protection than under the Protocol.

In closing, it appears to be possible to design several rules for the joint implementation of sulfur emission reduction commitments in Europe that allow cost savings while meeting the environmental objectives of the Protocol.

Chapter 11

An Institutional Design for Joint Implementation

11.1 Introduction

Joint implementation, generally speaking, implies that two or more actors cooperate to complete specific tasks or to fulfill specific commitments or obligations. When dealing with international pollution, joint implementation by two or more parties (usually countries) to an agreement usually pertains to the joint fulfillment of emission reduction commitments. The Montreal Protocol on ozone-depleting substances was the first international protocol that allowed this. In the context of the Montreal Protocol, it is perfectly clear that joint implementation means the joint fulfillment of agreed (and quantified) production and consumption ceilings. These ceilings are clear for all parties to the Protocol, apart from several developing countries whose ceilings depend on uncertain future levels of development (see Chapter 6 for details). Joint implementation is also foreseen under the Framework Convention on Climate Change (FCCC). Article 4, paragraph 2 mentions that parties may implement policies and measures (to limit emissions of greenhouse gases) jointly with other parties (Ministry of Housing, Physical Planning, and the Environment, 1992, p. 7). The FCCC, however, cannot be implemented as the Montreal Protocol, because it does not specify precise limits on emissions for individual countries. It only states that parties shall undertake policies and measures to return to their 1990 levels of man-made emissions of CO_2 and other greenhouse gases (Barrett, 1993). What is lacking is a clear allocation of property rights to individual countries.

Given these international precedents it should come as no surprise that joint implementation has also become a subject of debate in the negotiations on the Second Sulfur Protocol. At a meeting of experts in Oslo in May 1991, it was concluded that the use of economic instruments could play an important role in reducing costs and that different possibilities for trading of emission commitments exist. The experts agreed that further analytical work was needed before recommendations could be made regarding trading schemes (Ministry of the Environment, 1991). In spite of this early Norwegian initiative, the Second Sulfur Protocol was signed in June 1994 in Oslo without any firm commitment to emission trading or joint implementation. Article 2, paragraph 7 of the Protocol merely states, "The Parties to this Protocol may, in accordance with rules and conditions which the Executive Body shall elaborate and adopt, decide whether two or more Parties may jointly implement the obligations set out in Annex II. These rules and conditions shall ensure the fulfillment of the obligations set out in paragraph 2 above and also promote the achievement of the environmental objectives set out in paragraph 1 above" (UN/ECE, 1994a, p. 6). The obligations meant are the national emission ceilings for the years 2000, 2005, and 2010. The objectives of paragraph 1 are to protect human health and the environment and to contribute to meeting the critical loads in the long run. The procedure to adopt these rules is as follows. Possible rules are discussed in the Task Force on Economic Aspects of Abatement Strategies, which reports to the Working Group on Strategies. This Working Group proposes a set of rules to be accepted by the Executive Body at one of its annual sessions. Three years of discussions at the Task Force led to a proposal to the Working Group in March 1995 without specific results, let alone acceptance of specific proposals (UN/ECE, 1994c).

In contrast to the FCCC, many countries have signed the Second Sulfur Protocol and, therefore, have agreed on clearly specified quantitative emission ceilings. The main reason no agreement has yet been reached on specific rules for joint implementation is the fact that sulfur is a non-uniformly dispersed pollutant; its environmental impacts depend not only on the total volume emitted but also on the location of the emission source. Because of this, countries fear that joint implementation between two or more other parties will result in sulfur deposition in their country.

Given this background, this chapter explores the possibilities for an institutional design of joint implementation under the Second Sulfur Protocol that takes due account of the atmospheric dispersion characteristics of sulfur. In doing so, Section 11.2 elaborates on the background and objectives of joint implementation and Section 11.3 explores the costs and environmental benefits of several schemes. Section 11.4 outlines major rules and conditions for joint implementation. Section 11.5 sketches how joint implementation among countries could be translated into national policies and measures. Section 11.6 discusses how international transactions could be organized. Finally, Section 11.7 explores whether joint implementation requires improvements in the existing monitoring and enforcement practices under the Protocol.

11.2 Background and Objectives

Joint implementation implies that one party to the Second Sulfur Protocol will accept additional emission reduction obligations to enable another party to reduce emissions less than is specified under the Protocol. The basic advantage of this approach is that it allows countries to meet their (joint) commitments in a more cost-efficient way. This is especially true when the country that increases its level of emissions compared with its Protocol commitment faces higher marginal emission control costs than the country that reduces emissions below its original commitments. Joint implementation thus enhances the cost efficiency of the Protocol (UN/ECE, 1994b). It should be recalled that the emission reductions for each country under the Second Sulfur Protocol were to a certain degree based on cost minimization using the RAINS model (Amann *et al.*, 1993) (see Chapter 8). Europe-wide costs of meeting a set of targets for the deposition of sulfur were minimized subject to the condition that countries would at least carry out their current reduction plans (their non-cooperative Nash equilibrium, so to speak). In principle, this limits the potential cost savings of joint implementation. There is, however, still room for cost savings for three reasons.

First, the Protocol that has been signed differs from the cost-minimum solution. Some countries will not reduce emissions as much as the model calculations suggested (e.g., Bulgaria, Hungary, Greece, and Spain; see Chapter 8, *Table 8.3*), and other countries (e.g., former CSFR, Belgium, Ukraine, UK) will only meet their modeled reductions in the year 2010, not in the year 2000 as modeled. Because of these differences, and because costs in the year 2010 will differ from those in the year 2000, the Protocol still offers room for cost savings.

Second, the cost estimates used in the RAINS model are uncertain. Although they are based on many data, the cost estimates are particularly dependent on future energy consumption patterns. This makes both energy and cost estimates dependent on assumptions about future economic growth, energy demand, oil and coal prices, and specific national policies (such as nuclear expansion or protection of domestic coal production). Analysis has shown that the cost-minimum, country-specific, emission reduction commitments depend to a certain degree on the expected energy demand and supply (Amann and Schöpp, 1993; Altman *et al.*, 1994). The differences in Europe-wide annual costs between the different energy scenarios are around 20 percent. These differences in cost-minimum individual emission reductions vary from country to country; in some cases the estimates agree, in others one cost-minimum reduction estimate is four times that of the other energy scenario. Chapter 10 has shown that when costs differ substantially from the expected costs used for the Protocol, the potential cost savings of joint implementation can be considerably higher than in the case of perfect information about costs.

Finally, technological development, which is not anticipated by the model, may lead to cost reductions.

In conclusion, joint implementation offers room for cost savings because the emission commitments agreed to in the Protocol differ from the cost-minimum model calculations and because costs are uncertain.

Improving cost efficiency is the main reason for including a provision for joint implementation in the Second Sulfur Protocol. Chapters 2 and 3 showed that cost efficiency may mean the minimum costs of meeting an emission level or of meeting sulfur deposition targets. Article 2, paragraph 1 of the Protocol makes clear that it is the latter definition that applies because the Article states that joint implementation shall promote the achievement of the environmental objectives of the Protocol (UN/ECE, 1994a, p. 4). These objectives are to protect human health and the environment from adverse effects and to ensure that deposition of sulfur does not exceed the critical loads for sulfur.

Furthermore, although it is clear that cost efficiency is the prime objective, it is also clear from our analysis in Chapter 4 that political acceptability demands that joint implementation constitute a Pareto-dominant improvement to the Protocol. The net benefits (the additional environmental benefits of reduced sulfur depositions minus the costs of additional pollution control plus the net transfer payment due to joint implementation) should be zero or positive for each party. If this is not the case, parties cannot be expected to agree on the joint implementation provisions. This is not merely a theoretical condition (cf. Chapter 4) but is also defended in practice by several parties (e.g., Sweden, Switzerland, the Netherlands). These parties stress that negative environmental impacts on third parties (parties not carrying out a specific joint implementation agreement) should be avoided. An exact delineation and measurement of environmental benefits (or negative third-party impacts) appears to be difficult, as later sections of this chapter show.

Finally, joint implementation and emission trading may also promote technological innovation (Mason and Swanson, 1994; see also Chapter 3 and Chapter 6).

11.3 The Potential Costs and Benefits of Different Schemes

In the past, the UN/ECE's Task Force on Economic Aspects of Abatement Strategies concluded (UN/ECE, 1994b) that schemes for joint implementation should be as simple as possible while ensuring that most of the potential cost savings are realized,

Table 11.1. Joint implementation rules.

	Exchange rates specified	Exchange rates unspecified
Without restrictions	A	B
With restrictions	C	D

that the environmental objectives are achieved, and that no conflict occurs with any other obligation of the Protocol. The task force came to the conclusion that two basic issues determine joint implementation schemes:

- Whether or not exchange rates are specified, the exchange rate being the volume of emissions one party would have to reduce if another party wanted to increase its emissions by one unit above the agreed emission commitment.
- Whether other additional restrictions apply and if so which ones.

Thus, four cases were identified (*Table 11.1*).

Because of the importance of the location of the source for the level of damage to the environment, option B was not considered relevant. It was also noted that the use of combinations of C and D would be possible for formulating exchange rates. This could, for example, take the form of indicative exchange rates to give guidance on potentially attractive trades combined with deposition modeling showing the average protection of the ecosystems (the percentage of ecosystems not exceeding the critical loads of sulfur deposition; compare Chapter 8).

Different ways to set exchange rate were discussed:

- Exchange rates of one.
- Exchange rates based on the marginal costs of the cost-minimum solution; this in effect applies exchange rates that are a weighted average of the transfer coefficients of those receptors (EMEP grids) that are binding (that is, exactly meet the deposition target) in the cost-minimum solution (see Chapters 9 and 10).
- Exchange rates that keep the average deposition of sulfur in every country constant.

In the second case the exchange rates vary from bilateral trade to bilateral trade. The rates are sometimes higher than one (decreasing emissions) and sometimes lower than one (increasing emissions). In the third case exchange rates tend to be higher than one, so emissions will be reduced.

In addition to these rates, several restrictions are conceivable:

- Sulfur deposition at each grid must be below certain levels (e.g., the deposition under the Second Sulfur Protocol, critical sulfur deposition loads).

- The total emissions after joint implementation should not exceed the total emissions agreed to under the Protocol.
- The average protection of ecosystems in each country (measured as the percentage of ecosystems where critical sulfur loads are not exceeded) should be higher than or equal to that under the Second Sulfur Protocol.
- Ecosystem protection in each grid should not be allowed to be lower that under the Second Sulfur Protocol.

These restrictions reflect various tools for meeting the environmental objectives of the Protocol with the degree of flexibility needed for joint implementation. At one extreme, joint implementation subject to the condition that deposition would not be allowed to increase in any single grid would leave hardly any room for cost savings. At the other end, allowing joint implementation subject only to the condition that the total volume of emissions would not increase would imply considerable cost savings.

Model simulations were carried out for a number of restrictions in the past (Førsund, 1992a, 1992b, 1993; Førsund and Nævdal, 1993, 1994a, 1994b, 1994c; Klaassen, 1993a, 1993b, 1994; Klaassen and Amann, 1992; Klaassen *et al.*, 1994; Ruyssenaars *et al.*, 1993). Although these studies do give an indication of the results that might be expected, their actual quantitative results are less useful because they are based on outdated drafts of the Second Sulfur Protocol. For this reason we will only make use of the recent model calculations. They were described in detail in Chapter 10. Some additional work was done by Førsund and Nævdal (1994b). These results are summarized in *Table 11.2*. The table describes the rule, the potential cost savings compared with the Second Sulfur Protocol, the impact on emissions, average European ecosystem protection (percentage of ecosystems exceeding the critical sulfur deposition values), as well as the impacts on ecosystem protection in parties to the Protocol.

In the remainder of this section we will discuss the cost efficiency, environmental effectiveness (contribution toward meeting the environmental objectives of the Protocol), and distributional consequences (the potential Pareto optimality) of the schemes and assess their administrative practicability (information requirements and administrative costs) and the potential transaction costs.

The first option is to allow trading in one European bubble as long as total emissions do not increase. *Table 11.2* shows that in this case the cost savings compared with those under the Protocol would be considerable (5.7 to 57 percent). The lower bound will occur if existing regulations and the emission and fuel standards of the Second Sulfur Protocol are taken as side constraints. Although Europe-wide emissions would remain constant, average ecosystem protection would be reduced, because reductions would move to less sensitive areas. Regarding third-party

An Institutional Design for Joint Implementation

Table 11.2. Summary of model results.

Scenario	Scenario			
	A	B	C	D
	Joint implementation			
	European bubble	Deposition constraints	Cost-minimum exchange rates	Constant-deposition rates
Annual cost savings (as % of SSP)	5.7 to 37% (57.2%)	0.5 to 3%	2.5 to 9% (17.9 to 19.2%)*	0.4 to 0.8%
Change in emissions (as % of SSP)	0% (0%)	−2.2 to 3%	0 to 1% (0 to 7%)*	−2 to −3%
Ecosystem protection (% of ecosystems)	−1.6 to 0.2% (NA)	0.7 to 1.1%	−0.2 to 0.2% (NA)	0.6 to 0.8%
Countries with less ecosystem protection	14 to 20 (NA)	1 (NA)	5 to 9 (NA)	0 (NA)
Transaction costs/ administrative costs	Low	High	Medium	Medium

Sources: Data in brackets based on Førsund and Nævdal (1994b), other data based on Chapter 10 of this study.
*The lower value applies when the total sum of the emissions is not allowed to increase.
SSP = Second Sulfur Protocol; NA = not available.

effects, a large number of countries would have less average ecosystem protection than under the Second Sulfur Protocol. Although a large number of these parties would also experience considerable cuts in costs, discussions at the UN/ECE made it clear that this is not a politically acceptable solution; it does not constitute a Pareto improvement.

A second alternative is a scheme that allows joint implementation subject to deposition constraints in each grid. In this particular scheme the deposition would have to be lower than the critical loads for sulfur (the long-term objective of the Protocol) or the actual deposition expected under the Protocol, whichever value is higher. If the deposition under the Second Sulfur Protocol were lower than the critical loads, this deposition would be allowed to increase to the critical loads. Such a scheme could be realized through deposition permit trading or, perhaps more realistically, pollution offset trading. In the latter schedule, each trade would have to be accompanied by deposition model runs showing that the proposed trade would not violate the deposition targets. *Table 11.2* shows that, assuming simultaneous trading, the cost savings would be small (0.5 to 3 percent). However, overall ecosystem protection (percentage of ecosystems not exceeding the critical loads) would be as good as under the Second Sulfur Protocol. Only one country would experience a small loss in protection, which might, however, be compensated by cost savings. A major drawback of this proposal is that although a simultaneous market might offer room for achieving the cost savings, these savings might not be reached if trading is bilateral and sequential. This is because with only two countries trading we have only 2 (instead of 33) instruments to meet the same set of deposition goals (at around 450 targets). It becomes more difficult to find feasible trades and one trade might preclude the feasibility of a second, perhaps more cost-efficient trade. The transaction costs of deposition constraint trading are likely to be higher for two reasons:

- It will be more difficult to find a partner for a trade without violating the constraints;
- A deposition model would have to be run and approval would have to be obtained from the authorities before trade would be allowed.

The question is whether these costs really are high. First, finding a partner might be difficult. Participants in an exercise at the International Institute for Applied Systems Analysis (IIASA), Laxenburg, Austria attempted to find pairs of bilateral trades that would save costs given a set of deposition constraints for each single grid (150×150 km) (Kruitwagen, 1992). Out of the 1,056 possible trades, 3 turned out to be feasible after one month of hand-made iterations. Running a deposition model would also be fairly easy because such a model is available. Including overhead, the one-time cost would be around US$ 80,000. This is comparable with other environmental impact assessments carried out under the FCCC, which cost US$ 70,000 (Barrett, 1993). Second, the cost of approving the trade would have to be incurred. As under the EPA's emission trading program, approval is now on a case-by-case basis and might take a long time. Also, this scheme does not allow an organized market, such as an auction or electronic bulletin board where trades are consummated simultaneously and can be performed more efficiently. Still, the transaction costs might not be too high. *Table 11.3* gives rough estimates of these costs for deposition constraint trading and trading with exchange rates.

The costs of finding a partner would be equal under both schemes. The cost estimate is based on one FCCC trade by a US company (Barrett, 1993). The administrative costs could be around US$ 400,000, based on recent joint implementation agreements between Norway and Mexico and Norway and Poland (Barrett, 1993). The basic cost difference between the deposition constraints and fixed exchange rates would be the deposition model costs and the costs of approval. The latter costs would not have to be incurred if it were agreed that a trade would be allowed as long as the exchange rate is used. The trade would then only have to be registered. Whether these are significant differences is difficult to judge. The cost estimates are on a project basis and a joint implementation agreement

An Institutional Design for Joint Implementation

Table 11.3. Transaction costs (in thousand US$).

	Deposition constraint	Exchange rates
1. Finding a partner and making an agreement		
Finding a partner	50	50
Administration	400	400
Deposition model	70	0
2. Obtaining approval	–	0
Total	≥ 520	450

between countries might consist of different projects. FCCC data suggest that for the trade between Norway, on the one hand, and Mexico and Poland, on the other, the budget was US$ 4.5 million. This would mean that transaction costs would be around 10 percent. Compared with the potential annual cost savings of DM 60 million/year estimated in Chapter 10 for the trade between Norway and Kola-Karelia the transaction costs would be small, even if Norway must incur the one-time transaction costs several times for different projects. In conclusion, the transaction costs of trading with deposition constraints are higher than with exchange rate trading but might be relatively small compared with the potential cost savings. The costs and the uncertainty of the approval procedure are significant cost element uncertainties.

To reduce the transaction costs, joint implementation could be allowed if fixed exchange rates are used. Two exchange rates are conceivable:

- Cost-minimum exchange rates steering the trade equilibrium toward a cost-minimum solution for meeting a set of deposition constraints;
- Rates to keep average deposition in each country at least constant.

Table 11.2 shows that the cost-minimum rates could lead to considerable cost savings. Førsund and Nævdal (1994b) estimate cost savings of nearly 19 percent. This is an unrealistically high estimate based on simultaneous trade and assuming no constraints in the form of emission and fuel standards for the year 2010. The estimate, for example, assumes that Austria doubles its emissions compared with the Second Sulfur Protocol, which is unrealistic because the actual SO_2 emissions are already at the level required under the Protocol by 1991/1992. The lower estimate of 2.5 percent assumes that all regulatory constraints (emission and fuel standards) remain in place and that trades take place bilaterally and randomly. Part of the cost savings of these schemes is obtained by increasing emissions. Førsund and Nævdal (1994b) show that if emissions are not allowed to increase, potential cost savings will drop from 19.2 to 17.9 percent. Although the overall impact of this scheme on Europe-wide ecosystem protection is fairly neutral, several countries might lose in terms of having slightly less ecosystem protection than under the Protocol. Moreover, some of these countries do not save costs, so this non-Pareto-dominant solution might not be politically acceptable. One should also realize that when costs are uncertain, it is difficult to foresee what the ultimate environmental impacts will be. This was also shown for the US sulfur allowance program, where models did not accurately predict which states would act as buyers and which would act as sellers (see Chapter 6). A major advantage, however, is that the transaction costs are relatively low. Publication of the exchange rates would make it relatively easy to work out which countries could act as potential trading partners. Approval of trades is also less costly because it would only be necessary to assess whether countries are exchanging emission volumes according to the applicable exchange rates.

Finally, allowing joint implementation with constant-deposition exchange rates also holds the promise of relatively low transaction costs. In this case, the potential cost savings are much lower because every trade must ensure that the average deposition in each country does not increase. The costs of the additional reductions in emissions and the costs of ensuring that no country is worse off than under the Protocol reduce overall cost savings. As a result, several countries are better off because they have lower costs and/or a better environment.

One important conclusion emerges from the above analysis: there is a clear trade-off between cost savings, on the one hand, and environmental effectiveness and Pareto optimality, on the other hand. Schemes with large cost savings (the European bubble and the cost-minimum exchange rates) tend to violate the environmental objectives and make several countries worse off in terms of environmental quality. Schemes that are very effective for meeting ecosystem protection goals and promoting Pareto-dominant improvements (the deposition constraints scheme and the constant-deposition rates) do this at the expense of cost efficiency. This trade-off ultimately requires a political decision. The use of exchange rates might lower transaction costs, especially because pre-approval of trades can be avoided. As the next sections will show, avoiding pre-approval might not be politically acceptable.

11.4 Rules and Conditions

11.4.1 Design issues

Before embarking on a description of rules and conditions for joint implementation it makes sense to reiterate the lessons learned from emission trading and joint implementation in practice. Experience with not always successful emission trading programs in the past has led to suggestions for developing more successful ones. On

An Institutional Design for Joint Implementation

the basis of Barakat and Chamberlin (1991), Tripp and Dudek (1989), Hourcade and Baron (1993, p. 24), Nentjes (1993), Klaassen and Nentjes (1995), and Chapter 6 of this study, one can distill the conditions for successful national schemes of tradable emission permits that meet environmental goals and promote cost efficiency. These conditions lead to the following rules:

1. Identify a single (homogeneous) good that is readily quantifiable and has a clearly defined unit and a sufficiently wide market.
2. Develop clearly defined objectives (i.e., emission ceilings and air quality constraints).
3. Allocate clearly defined and secured emission rights to individual polluters before trading starts and allow trading of these rights.
4. Allow banking of rights.
5. Aim at low transaction costs by using simple, predictable administration procedures (approval); avoid case-by-case approval and the use of air quality models.
6. Put high-level monitoring and enforcement in place before trading starts (high-quality emission inventory and a comprehensive permit system).
7. Give the institution that designs and implements the system the appropriate authority system to do so, and make sure that it possesses technical capability and can ensure that trading is evasion-proof.
8. Design the system to be simple and predictable.
9. Avoid locally high ambient concentrations, for example by
   (a) Using the potential cost saving of mission trading to cut emissions further;
   (b) Blocking the actual use of acquired emission rights in well-defined circumstances but allowing reselling or banking of rights;
   (c) Overlaying trading on emission standards that guarantee minimum levels of protection at each location.
10. Ensure that the scheme constitutes a Pareto-dominant improvement (in other words, is politically acceptable).

The question is to what extent these rules are relevant for international trade or joint implementation among countries (governments), especially under the Second Sulfur Protocol. Regarding the first three basic rules, the following apply to the Second Sulfur Protocol:

- There are clear limits on total national emissions over time and there is also a clear, quantitative, long-term objective for deposition.
- It is not clear, however, which deposition constraints should apply to guide trade for those regions (grids or receptors) where the Second Sulfur Protocol

does not meet the long-term deposition constraint (should one take the actual deposition of the Second Sulfur Protocol as constraints?).
- With respect to rule 2, the unit has been defined in the Second Sulfur Protocol (kton SO_2) but it is not necessarily a homogeneous good because the location of emission is significant.
- Regarding rule 3, there is a clear quantitative allocation of rights in the form of national ceilings to the countries that signed the Protocol, but the rules for transferability of these rights are not yet specified.

On rules 4 to 10 the following can be observed. Banking has not been discussed. The transaction costs depend on the rules adopted for trade. Monitoring and enforcement will be discussed in Section 11.7. Rule 7 applies. Rule 8 depends on the exact trading rules (see Sections 11.4.2 and 11.4.3). Rule 9a is not feasible because the Second Sulfur Protocol has been agreed on. Rule 9b is the point of discussion, and rule 9c is part of the Second Sulfur Protocol. Whether rule 10 is fulfilled depends on the rules that are adopted for joint implementation.

The remainder of this section proposes rules and conditions for joint implementation under the Second Sulfur Protocol bearing the above in mind. Section 11.4.2 describes several general rules, applicable to any scheme, whereas Section 11.4.3 describes particular rules to deal with the atmospheric dispersion of sulfur emissions and to promote Pareto-dominant improvements.

11.4.2 General rules

The following are general rules for any trading scheme.

Dimensions and baseline emissions

1. Agreements on joint implementation shall be in ktons of SO_2/year (conforming to the definition in paragraph 12 of Article 1 of the Protocol).
2. Such agreements shall specify the volumes of emissions agreed to under the Second Sulfur Protocol and after joint implementation and shall specify the duration of the agreement.

Table 11.4 gives an example where Norway strikes a deal with Russia. Norway would increase its emissions by 1 kton in the period from 2005 to 2010. To compensate for this, Russia would reduce its emissions by an additional 12 ktons.

Who is involved?

3. Only parties to the Protocol (signatory governments) may enter into joint implementation agreements. This seems sensible because the baseline emission

Table 11.4. Example of an emission trading agreement (emissions in kton SO_2).

	Year					
	2000		2005		2010	
Country	SSP	After JI	SSP	After JI	SSP	After JI
Norway	34	34	34	35	34	35
Russia	4440	4440	4297	4285	4299	4285

SSP = Second Sulfur Protocol; JI = joint implementation.

levels for non-signatories are unclear and would have to be negotiated. Experience with the Montreal Protocol, the FCCC, and the EPA's emission trading program shows that this leads to serious problems and is time consuming. Because only a few countries are not expected to sign and ratify the Second Sulfur Protocol, the loss of cost savings from excluding them is small, as was shown in Chapter 10.

4. Individual firms may play an active role in identifying and carrying out joint implementation agreements, although the responsibility for meeting the emission reduction commitments in joint implementation agreements remains with the parties (governments) that signed the Protocol.

Approval

5. Joint implementation proposals shall be submitted in written standard form to the secretariat who shall communicate them to all parties. The proposals should not be in violation of any other obligation under the Protocol, such as the agreed emission and fuel standards. Proposals must be submitted at least three months before the next session of the Executive Body.
6. All joint implementation proposals are subject to approval by consensus of the parties to the Protocol present at the first session of the Executive Body after submission of the proposal. To avoid lengthy approval procedures and associated high transaction costs, it should be made clear beforehand on which grounds parties can disapprove of a trade. Grounds for disapproval can be infringement of the positive environmental impacts of the Protocol. One could, for example, envisage a rule stating that a proposal will be accepted if the negative environmental impacts in terms of average percentage ecosystem protection are below a certain percentage. Section 11.4.3 elaborates on such rules. Such estimates of the environmental impacts would have to be based on an accepted model. Of course, such disapproval could also be based on other goals served by the Protocol, such as human health impacts due

to undesired increases in SO_2 concentrations above international or national ambient standards.
7. The secretariat shall communicate agreements on joint implementation to all parties after adoption.

Banking

8. To improve the cost efficiency of the Second Sulfur Protocol, parties will be allowed to reduce emissions below the levels agreed to under the Protocol and use the additional reductions to increase future emission ceilings by the same amount (i.e., banking of rights). This possibility could be included even without joint implementation. The Second Sulfur Protocol, however, only allows a country to take the average emissions of the three years around the required reduction before the year 2005 in case of a cold winter, dry summer, or unforeseen losses in power supply.

Practical experience with emission trading shows that those emission trading schemes that allow banking of rights (lead trading and sulfur allowance trading) are more cost efficient because they offer trading partners more flexibility in meeting commitments. Allowing this banking, or carrying forward of emission reduction commitments, would also give countries more time to find buyers for the additional emission reductions they have created. It would allow parties to use more significant reductions in the near future to meet stringent emission reductions in the more distant future. The advantage of banking is that emission reduction will take place earlier. Because the problem of acidification is one of accumulation of acidifying components, this would give endangered ecosystems more time to recover. An example might show the usefulness of banking. *Table 11.5* shows how the former CSFR could save costs by banking part of its commitments. The former CSFR would increase its emission reduction in the period from 2005 to 2009. In return, it could increase its emissions slightly in the period from 2010 to 2014, when marginal costs are expected to be higher. With banking the costs would be higher from 2005 to 2009. These extra costs would, however, be compensated for by the discounted cost savings in the period from 2010 to 2014 (given the assumed discount rate of 5 percent). The net present value of costs saved is DM 39 million (DM 8,158 million minus DM 8,119 million) or about 0.5 percent.

11.4.3 Specific rules

Because sulfur is non-uniformly dispersed, specific rules are needed to ensure that joint implementation contributes to meeting the environmental objectives of the Protocol and constitutes a politically acceptable solution for all parties to the

An Institutional Design for Joint Implementation

Table 11.5. Banking of sulfur emissions by former CSFR.

	Emissions (kton SO_2)		Net present value of costs (million DM in 2005)	
Year	Protocol	Banking	Protocol	Banking
2005	1,197	1,124	875	934
2006	1,197	1,124	834	889
2007	1,197	1,124	794	847
2008	1,197	1,124	756	807
2009	1,197	1,124	720	768
2010	872	945	919	852
2011	872	945	876	812
2012	872	945	834	773
2013	872	945	794	736
2014	872	945	756	701
Total	10,345	10,345	8,158	8,119

Protocol. The analyses in Section 11.3 suggest that either deposition constraints for each grid or exchange rates could be selected. This section chooses exchange rates, because they promise to reduce transaction costs in terms of finding trading partners, negotiating, and obtaining approval.

Two possible exchange rates are conceivable, those based on a cost-minimum solution or those that keep a country's average deposition constant. The analysis in the previous chapter showed that the cost-minimum rates may make the situations of some countries slightly worse. They also have the disadvantage that they could lead to an increase in total emissions, which as practical experience has shown has never been accepted in any trading scheme. These rates are, however, heavily promoted by at least one country (Norway). The constant-deposition rates put limits on joint implementation at the expense of cost savings but have the advantage of making no country worse off than under the Protocol (at least not in terms of the average deposition or protection of ecosystem in each country). In the remainder of this study, both options will be kept open; however, it seems that the constant-deposition rates might be more acceptable.

This leads to specification of the following specific rules:

- Joint implementation agreements will be approved provided that increases in emissions (compared with the levels agreed to in the Second Sulfur Protocol) by one party are compensated for by decreases in emissions by another party, taking into account the exchange rate for this specific bilateral agreement.
- The secretariat will publish the applicable set of exchange rates for each possible bilateral agreement as an integral part of the joint implementation rules

(see *Table 10.1*). On the basis of new information (especially EMEP country-to-country matrices) the exchange rates can be revised periodically.
- The total volume of emissions after joint implementation should not exceed the volume of emissions agreed to under the Second Sulfur Protocol.

Discussions on these proposals show that the notion of Pareto optimality is not straightforward in practice. This is because it is unclear how environmental benefits should be measured in this specific context. Do we mean benefits in terms of reductions in average levels of sulfur deposited in each country, or increases in levels of ecosystem protected? What happens to national benefits when deposition increases in one part of the national territory of a third party and decreases in another part? It is clear that in the absence of precise country-specific welfare functions this trade-off cannot be made. It has also become clear during discussions that countries are not able (or willing) to make clear statements on these trade-offs. It may be that a joint implementation agreement that keeps the average protection of an ecosystem constant for each party is still unacceptable. This is because keeping average protection constant might mean that ecosystem protection decreases in those parts of a third-party country that are considered more valuable than those areas where ecosystem protection increases. This suggests that it may be necessary to add the following conditions to ensure that third-party interests are accounted for:

- A joint implementation agreement should not lead to a decrease in the average percentage of ecosystems protected in any (third) party to the Protocol.
- Every proposed joint implementation agreement should be accompanied by an environmental impact assessment showing
    - The average degree of ecosystem protection for each party after joint implementation compared with the Second Sulfur Protocol;
    - The degree of ecosystem protection in each EMEP (European Monitoring and Evaluation Program) grid after joint implementation compared with the Second Sulfur Protocol.
- If the average ecosystem protection does not decrease in any (third) party and if the decrease in ecosystem protection in each grid is restricted to a prespecified percentage, the proposal will be accepted unless third parties make use of a right to appeal within three months after submission of the proposal.
- If the degree of ecosystem protection in any specific grid decreases by more than a specified percentage the proposal will only be accepted if the (third) parties confronted with these impacts agree in writing.

Figure 11.1 gives an example of such an environmental impact assessment. It is assumed that Norway makes an agreement with the Russian Federation (Kola-

An Institutional Design for Joint Implementation

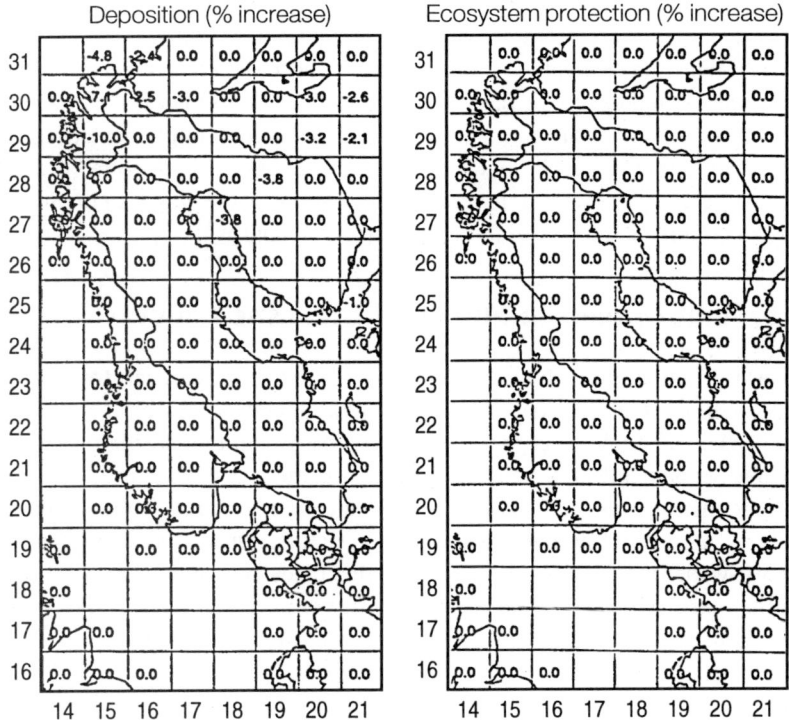

Figure 11.1. Environmental impact assessment.

Karelia). In doing so, Norway increases emissions by one kton (up to 35 ktons) and Russia reduces by 12 ktons (in Kola-Karelia). The left-hand side of *Figure 11.1* shows that deposition increases in only one grid cell in Norway-Sweden ($x - y : 18 - 21$), and is reduced in several grids in Northern Finland and Kola-Karelia (the part of the Russian Federation bordering Finland). The right-hand side of the figure shows that this will not lead to any alteration in the percentage of ecosystems protected in any grid.

Including these rules has the drawback that case-by-case decisions will have be made, which increases transaction costs and reduces cost efficiency. If possible, these rules should be avoided by using only the constant (average) deposition rates. If this is politically unfeasible (as appears to be the case), a clear, up-front statement should be made stating the conditions under which joint implementation proposals will be accepted (e.g., what percentage deviation in ecosystem protection). Otherwise, uncertainty on the approval might block the start of any agreement or might increase transaction costs so that few agreements are proposed. At least, this was the lesson learned from the EPA's emission trading policy.

In closing, the ultimate choice appears to be between a scheme based on constant-deposition exchange rates alone or a scheme combining such rates with an environmental impact assessment for each proposed joint implementation agreement. From a cost efficiency point of view, a system of exchange rates alone is preferable because it lowers transaction costs and opens the way for simultaneous trading rather than case-by-case bilateral sequential approval.

11.5 Joint Implementation and Domestic Policy

In the current situation, without joint implementation, parties are left free to decide how they will meet the national emission ceilings under the Protocol. The exception is that parties must take over the emission standards and fuel standards agreed to under the Second Sulfur Protocol. In the past, most countries used traditional regulatory instruments (fuel and emission standards) to reduce sulfur emissions, sometimes augmented by bubbles for certain sectors. Only a few countries (e.g., Norway, Sweden) use emission charges and fuel taxes as complements to emission standards (see Chapter 8). This situation will not change dramatically. First, in many countries, for example Austria, Germany, the Netherlands, Sweden, and Switzerland, existing legislation is sufficient to meet the Second Sulfur Protocol commitments. Second, countries that have to alter their policies will rely mainly on regulating instruments combined with bubbles for certain sectors (especially power plants). Only a few countries are discussing the use of tradable emission permits: Hungary, the Netherlands, Poland, and the UK.

Joint implementation under the Second Sulfur Protocol will not in itself change domestic policies. It can in principle be left to the parties to decide how they will meet their new national emission ceilings after joint implementation (Tietenberg, 1992b, p. 129), provided that they meet existing international commitments to set emission and fuel standards under EU legislation or under the Second Sulfur Protocol. Changes in domestic policies to meet new commitments after joint implementation could take the form of alterations in generic emission and fuel standards, changes in sector-specific bubbles, adaptation of emission charge levels, or changes in the initial allocation of emission rights. As in the FCCC, these changes could also take the form of an investment project at a specific plant or company. For each project the additional emission reduction compared with what otherwise would have been done to meet the Second Sulfur Protocol would have to be determined. It should be made clear that the existence of bubbles specifying the total allowable emissions for certain sectors makes it easier to agree on a joint

Table 11.6. Joint implementation and domestic policy (in kton SO_2).

Netherlands		Belgium	
Initial commitment	106	Initial commitment	248
Bought 5 kton at 1/2.5[a]	2	Sold 5 kton	5
	+		−
New commitment	108	New commitment	243
Subaccount Dutch power plants		Subaccount firm x in Belgium	
Initial commitment	18	Initial commitment	20[b]
Bought 5 kton at 1/2.5[b]	2	Sold 5 kton	5
	+		−
New commitment	20	New commitment	15

[a] Applicable exchange rate if Netherlands buys from Belgium.
[b] Would have to be determined on a case-by-case basis.

implementation contract. To give an example, the Netherlands' power plant sector could easily be allowed to buy reductions abroad because it has an agreed on ceiling of 18 kton SO_2. If the constant-deposition exchange rates are applied, this sector can buy 5 ktons of SO_2 reductions in Belgium in order to increase its emissions by 2 ktons. In this example, the problem would be on the Belgian side: the issue is whether the 5 ktons of SO_2 reductions are really additional reductions that would not have occurred under Belgium's domestic policy for meeting the Second Sulfur Protocol. The Belgian emission reductions would have to be surplus (that is, not required to meet the Protocol), enforceable, and quantifiable. It is not clear that this is the case, because Belgium, like many other countries, has not set overall emission levels for firms or sectors (it has only set emission rates). Because both the Netherlands and Belgium are parties to the Protocol, they are ultimately responsible for ensuring that the new national ceilings after such a joint implementation are met. Because they have the best knowledge on their domestic policy, only they can ensure that such reductions are additional. In this case both Belgium and the Netherlands would have to give domestic approval to the agreement before it could be forwarded to the UN/ECE for international approval. *Table 11.6* summarizes how such a transaction could be registered where we assume that firm x in Belgium is the seller.

It should be clear from the example that it is more complicated for the Belgian company to enter the agreement because, in contrast to the Dutch power sector, it has no agreed level of baseline (or pre-trade) emissions, and Belgian emission and fuel standards would have to be translated into pre-trade emission levels for this specific firm. The EPA emission trading policy showed that this can be cumbersome and time-consuming.

In summary, countries could be left free to decide how to translate joint implementation agreements into domestic policy, provided that internationally agreed emission and fuel standards are met. The existence of bubbles for certain sectors (or better, the allocation of individual emission rights to polluting firms) will facilitate both domestic policy changes and joint implementation.

11.6 Organizing International Transactions

11.6.1 Introduction

Different ways to organize transactions are possible. Barrett (1993) considers bilateral, case-by-case, joint implementation agreements to be one extreme and fully competitive markets of tradable emission entitlements to be the other extreme. Førsund and Nævdal (1994c) discuss decentralized and centralized solutions. In a decentralized solution a service center would provide indicative exchange rates and countries would undertake bilateral action. In a centralized solution the service center would act as a command center and calculate the most profitable joint implementation agreements. It would organize the trades but it would be left to the countries to actually carry them out. In an intermediate solution such a service center would collect supply and demand prices of additional emission reductions and suggest profitable trades (using exchange rates) to countries.

Joint implementation as it is formulated in the Second Sulfur Protocol, however, is not a full-fledged system of tradable sulfur permits between countries. It is, rather, a restricted attempt to introduce flexibility into a rigorously organized context of national emission ceilings and technical standards. It resembles the EPA's emission trading policy rather than the sulfur allowance trading under the 1990 Clean Air Act Amendments. This implies that the program will not start with a competitive market of tradable sulfur emission entitlements, but rather with decentralized bilateral agreements. It also implies that centralized solutions where a service center calculates the most profitable trades, to be implemented simultaneously, are less realistic (Førsund and Nævdal, 1994a, 1994b, 1994c). Such a command center solution is also less realistic because it assumes that the center has perfect information about costs. This not only conflicts with reality, it also contradicts the main reason for allowing joint implementation: information about costs is uncertain and decentralizing decisions can assist in discovering, and adapting to, new cost information. In conclusion, organizing transactions in a highly centralized way or as a complete set of tradable sulfur permit markets appears to be unrealistic in the context of the Second Sulfur Protocol.

11.6.2 Bilateral agreements

What remains are bilateral agreements. These could, however, be augmented with a clearinghouse type of arrangement to facilitate and coordinate transactions (Barrett, 1992; Roland, 1993; Hanisch et al., 1993).

The least ambitious option, and the one that would automatically evolve, is where two parties (countries) agree on a specific joint implementation project and on the associated costs and additional emission reductions compared with those agreed to under the Protocol. The parties would use the published exchange rates as guidelines. If necessary, the parties would also prepare an environmental impact assessment to evaluate third-party impacts. Finally, they would submit the agreement for approval.

Such bilateral agreements face serious drawbacks and transaction costs are likely to be high. As a first element, the costs of finding a partner would be high. First, countries would be interested in giving priority to the most cost-efficient agreements, and it would be difficult to find out which ones these are (Barrett, 1993). Of course, available models, such as the RAINS model, could assist in finding cost-efficient solutions. It might still be costly because the modeled costs might not correspond to the actual costs. Moreover, translating an international agreement into domestic measures might be costly or difficult. Second, the transaction costs for negotiating and signing the contract would be high because the net incremental costs of the agreements would have to be determined, the price and additional emission reductions would have to be agreed on, and the nature of the contract would have to be specified (see also Bohm, 1994, p. 193). Such costs, however, would have to be incurred in any form of joint implementation and need not to be significantly more difficult with bilateral actions. Third, and this might be a serious problem, the transaction costs in terms of obtaining approval might also be high, particularly if an environmental impact assessment is needed and if the conditions under which third parties would be willing to accept the agreement are not clear beforehand but are determined in the course of the approval process.

Chichilnisky and Heal (1993) remark that joint implementation as a series of bilateral bargains has other disadvantages. First, each trade will be at a price reflecting the knowledge, experience, and bargaining power of the two countries involved in the bargain between bilateral monopolies. Second, there is no price discovery and dissemination, as there is in competitive markets. This is confirmed by the model simulation in Chapter 10, which showed that random bilateral trading could have serious negative impacts on the expected cost savings of individual trades. The bilateral character itself might not necessarily be the biggest problem. The problem is that only a limited number of bilateral trades are possible because of the environmental constraints in the European sulfur case. The formation of prices is irregular. This is strengthened by the fact that the sulfur emissions traded are heterogeneous, rather than homogeneous, goods. In brief, case-by-case joint implementation agreements imply high transaction costs and low cost efficiency but are still the easiest way to begin.

11.6.3 Reducing transaction costs and a clearinghouse

To reduce transaction costs, several possibilities exist:

- Replace case-by-case approval with trading according to fixed exchange rates, with the understanding that trades are approved as long as these rates are used.
- If case-by-case approval cannot be avoided, clear rules should be adopted specifying the conditions under which agreements will be approved.
- Coordinate transactions through a clearinghouse.

Case-by-case approval could, for example, be replaced by automatic approval if the emission volumes are exchanged using the correct exchange rates and the volumes traded are not higher than say 5 percent of the countries' commitments under the Second Sulfur Protocol or do not exceed a level of 0.5 kton SO_2/year. In this case, the parties would only have to report annually the volume of actual emissions for the specific agreement. Whether or not the emission reduction (credits) bought are sufficient to cover the actual emissions allowed (the sum of the Second Sulfur Protocol–committed emissions plus the emission reductions bought times the appropriate exchange rate) would also be checked.

If such automatic approval is politically unacceptable, approval could be accelerated by creating a clear rule, stating that approval is given within a limited period of time if the volumes exchanged fit the exchange rate and an environmental impact assessment shows that ecosystem protection would not decrease by more than a certain percentage in any grid of third parties (as described in the previous section).

Finally, a clearinghouse could fulfill a useful role in facilitating transactions and reducing transaction costs (see *Figure 11.2*). A market clearinghouse can be described as an institution that organizes the dissemination of information about potential supplies of additional emission reductions, that links these offers with bids for additional emission reductions, and that assists in establishing market clearing prices (Roland, 1992, p. 27). The main benefit of such a clearinghouse would be a lowering of the search costs of finding partners. It could also bring parties together to finance emission reductions that otherwise would be too expensive for one country to finance. Furthermore, the dissemination of information about supply and demand prices would help countries to negotiate agreements (Barrett, 1993). Such

An Institutional Design for Joint Implementation 295

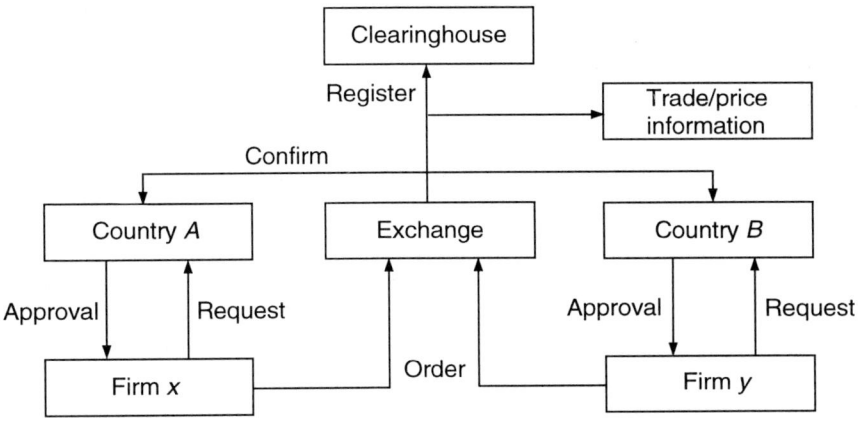

Figure 11.2. A clearinghouse.

a clearinghouse could turn bilateral agreements into more competitive, multilateral market-type operations. In the first instance, a clearinghouse could only facilitate transactions between countries. In the second instance, it could enable individual firms to sign agreements with a party to the Protocol or enable individual firms to trade emission reduction commitments. The reductions offered by these individual firms would first have to be approved by their country of origin. A clearinghouse, which can facilitate both spot and future transactions, is usually considered an essential element of a competitive market place (Sandor et al., 1993). The use of such a clearinghouse could evolve into a more organized exchange where parties to the Protocol or individual firms could place bids or make offers for sulfur emission reductions.

Obviously, such a clearinghouse can only function if transactions take place on a regular basis and if domestic and international approval of proposed joint implementation agreements is easy and not too costly. The use of such a clearinghouse is less meaningful in the case of SO_2 if only countries trade because the number of countries that can potentially trade is small. If individual firms were to be allowed to trade, the number of parties would be much bigger and a clearinghouse would be a useful institution. Allowing individual firms to participate would increase competition and could also help to reduce the market power of countries that might occur when constant-deposition rates are used. This, however, would require SO_2 emission entitlements to be allocated to individual firms before trading starts and domestic rules for using these entitlements would have to be clearly established.

Current progress in negotiating these rules suggests that joint implementation will be judged on a case-by-case basis, which limits the potential role a clearinghouse could have in lowering transaction costs.

In summary, joint implementation agreements could be organized on a bilateral, case-by-case basis, or a clearinghouse could be used to facilitate trades and reduce transaction costs. The fact that case-by-case approval is probably the only politically acceptable solution implies relatively high transaction costs and thin markets. However, one should not forget that the greatest room for cost savings is at the domestic level, where regulatory instruments make the cost of meeting national ceilings higher than necessary (see Chapter 10). Promoting the use of national bubbles for certain sectors and the use of national tradable emission permits might therefore be more important for achieving cost efficiency than using a detailed set of rules for international joint implementation. The evolution of such national tradable permit markets would also facilitate international transfers. First of all, such national permit markets would lead to the formation of national prices reflecting marginal control costs. This price information could be used to find profitable international trades. Second, the existence of national markets implies that firms have been allocated individual emission rights. At an international level there would not be any need to determine the baseline (or pre-trade) emission levels of individual firms. Third, national permit systems could be set up to avoid international transboundary complications. A few countries (Hungary, the Netherlands, Norway, Poland, the UK) are already debating the implementation of national permit markets. In brief, national permit markets could be a preparatory stage for a full-fledged international permit market with restrictions in the form of compulsory exchange rates or deposition constraints.

11.7 Monitoring and Enforcement

Does joint implementation require changes in the way monitoring and enforcement currently take place? If so, which changes are required? In answering these questions, this section first summarizes how monitoring and enforcement currently take place. Then the need for changes is discussed.

Monitoring is the process of obtaining information to facilitate implementation of an agreement and to determine whether parties are in compliance with agreed emission constraints. Determining noncompliance is sometimes called verification (Ausubel and Victor, 1992, p. 4; Russell, 1990, p. 243). Enforcement is the taking of action (sanctions and incentives) that stimulates or forces violators to comply with these constraints. Monitoring can be based on self-reporting, direct or indirect inspections, or combinations of the two (Tietenberg and Victor, 1994,

An Institutional Design for Joint Implementation 297

p. 5). Self-reporting involves the submission of reports providing details on specific data. Direct inspections involve visits to emission sources and indirect inspections consist of collecting related data (such as fuel consumption and emissions per unit of fuel) to provide indirect evidence on emission levels.

Under the Second Sulfur Protocol, monitoring is mainly based on self-reporting (UN/ECE, 1994a). Parties periodically have to give information on

- The implementation of national strategies, policies, and measures to reduce emissions;
- The national annual sulfur emission levels according to guidelines;
- The implementation of other obligations (such as the adoption of emission standards) under the Protocol.

The data on national emission levels are reported on an annual basis as national totals. Every five years more detailed reporting is required. Reporting is becoming more harmonized, following EC reporting formats. Such harmonized data were being compiled at the end of 1994 for the year 1990 within the framework of the EC-CORINAIR emission inventory. Verification takes the form of indirect auditing by independent consultants, for example, by comparing emissions per unit of fuel used with external data and submissions by other countries.

Regarding enforcement, the Second Sulfur Protocol requires the establishment of an implementation committee to review the implementation of and compliance with the Protocol (UN/ECE, 1994a, p. 30). The committee analyzes and evaluates self-reported data. The evaluation is performed by the EMEP technical centers or independent experts. Furthermore, parties that have reservations about another party's implementation of its obligations can inform the committee in writing, backed by corroborating information. In the case of possible noncompliance, the Secretariat may ask the party to supply necessary information. If the party is unable to comply, it can explain in writing the reason for its noncompliance.

In the case of a dispute between two or more parties, two options are open for settling the dispute: submission of the dispute to the International Court of Justice or arbitration according to procedures to be adopted by the parties at a session of the Executive Body. When signing the Protocol, the parties may declare which option they recognize (UN/ECE, 1994a, p. 10). No clear sanctions, not even in the form of publication of noncompliance, are mentioned in the Protocol.

Does joint implementation require any changes in the monitoring and enforcement procedures described above? The answer to this question is not straightforward and depends on who is involved in joint implementation agreements.

Hanisch *et al.* (1993) and Bohm (1994) suggest that some type of verification procedure is needed because both countries involved in a joint implementation agreement are interested in overstating the emission abatement such an agreement achieves. The buying party would be interested in overstating the emission reduction it buys in order to reduce its pollution control costs for meeting the agreed national ceilings. The selling party would be interested in overestimating the reduction in order to be paid more for seemingly doing more (Bohm, 1994, p. 194).

Tietenberg (1985) is of the opinion that emission trading between firms may create a stronger need for continuous monitoring because any cost-minimizing source would weigh the cost of compliance against the cost of noncompliance. The latter depends on the likelihood of violations being detected and on the severity of the sanctions that would then apply. If the marginal costs of continuous compliance rise, the incentive not to comply increases. Because emission trading stimulates sources to discover new means of compliance, this may mean that variable operating costs may become more relevant than under a regulatory approach. Consequently, the costs that can be saved by continuous noncompliance would be greater under emission trading than under a regulatory regime and noncompliance would increase.

Keeler (1991) finds that when tradable emission rights schemes are imperfectly enforced they can sometimes result in more pollution than would occur under a system of uniform standards. The result depends on the penalty for violation. With a constant marginal penalty, there would be less noncompliance under emission trading than under standards. This is because permit trading lowers the marginal costs of the high-cost firms and reduces their incentive not to comply without causing low-cost firms to cheat. If the permit price equals the penalty cost, the number of firms complying would be the same as or lower than under standards. The firms with high marginal costs would still face an incentive to cheat under the standards, and pollution would therefore be higher under standards than with trading. If the penalty exceeds the permit price, all firms would comply under permit trading, but firms with (very) high marginal costs might still find it profitable to cheat under regulation. With increasing marginal penalties, firms may comply or may undercomply, depending on whether the permit price is above or below the penalty price. If (marginal) penalties decline, firms that do not comply will pollute more. Keeler concludes that in some cases imperfect enforcement of a tradable permit system may lead to more pollution than a system of uniform standards. To summarize, the inclusion of joint implementation might create a certain need for stricter monitoring and enforcement, especially if individual firms become involved.

Tietenberg and Victor (1994) are of the opinion that existing institutions, especially at domestic levels, combined with self-reporting can form a sufficient basis for a viable monitoring system. They do add, however, that periodic veracity checks are needed to ensure the integrity of this self-reporting. As long as countries are the only parties that engage in joint implementation agreements, verification

An Institutional Design for Joint Implementation

of national emission reports is all that is needed to ensure that these are accurate. As soon as individual firms become active joint implementation partners, veracity checks are also needed to ensure that self-reported data by individual sources are correct. This, however, could be left to the national authorities. Recall that enabling individual firms to enter the international trading arena should only happen if individual countries make clear the extent to which the SO_2 emission reductions offered by these sources can be regarded as additional (surplus) reductions over the country's commitment under the Protocol. Such emission reductions should, therefore, be certified by the individual country as being surplus, quantifiable, and enforceable. The same applies for emission reductions bought by individual firms. Each party to the Protocol should make clear its domestic policy concerning the conditions under which sulfur reductions bought on the international market can be used to comply with domestic legislation to meet the Second Sulfur Protocol. This leads to the conclusion that in the case of joint implementation, monitoring does not need to change drastically and self-reporting can still be relied on. Because of the increased incentives to supply incorrect information, indirect veracity checks on these self-reported data may need to take place on a more regular basis.

The question that remains is whether enforcement should be tougher under joint implementation or not. It should be realized from the outset that international enforcement is based more on moral persuasion than on sanctions. It is usually done on an ad hoc basis, and potential enforcement actions usually take the form of disputes addressed through diplomatic negotiations (Tietenberg and Victor, 1994, p. 33). As long as only countries (not individual firms) trade emission reduction commitments, enforcement may not have to differ significantly from current practice. Tietenberg and Victor's (1994, pp. 44–45) suggestions for tradable entitlements to mitigate global warming can readily be adopted in the case of sulfur emissions. These suggestions consist of veracity checks, proper operation of dispute-resolution procedures and their early adoption, and transparency to ensure that noncompliance is detected and made public. If individual firms enter the international market for joint implementation agreements, qualitative changes may be needed. Although domestic enforcement by existing institutions is still heavily relied on, Tietenberg and Victor (1994, p. 44) believe that international standards for domestic enforcement must be set. More specifically, internationally agreed levels of noncompliance are necessary to ensure the stability of property rights across different national markets. If property rights are not secured or are not enforced, the confidence in emission reductions obtained internationally will be lost and thin markets may be the result. In summary, some changes in enforcement might be warranted, especially when individual firms begin to act as partners in joint implementation agreements.

11.8 Conclusions

This chapter has sought to prepare an institutional design for joint implementation under the Second Sulfur Protocol. After concluding that cost efficiency is the prime reason for joint implementation, the chapter compared several possible designs. It was concluded that schemes using exchange rates to guide bilateral agreements would probably offer sufficient room for cost efficiency while keeping transaction costs low. The chapter compared two different concepts of exchange rates: rates based on an agreed cost-minimum solution and rates that keep countries' average depositions at least constant. The former rates would save more costs but could make some countries worse off than under the Protocol. The latter rates would restrict cost savings considerably while giving a better guarantee that every country would be at least as well off. In addition to offering general rules for joint implementation, one of these exchange rates was adopted to steer joint implementation; there was a slight preference for the constant-deposition rates. It was concluded that it might be necessary to augment the use of these exchange rates (and in particular the cost-minimum exchange rates) with an environmental impact assessment of each proposed agreement to obtain political approval of joint implementation agreements. This need might arise because without a complete set of national welfare functions it is impossible to weigh the costs and benefits of joint implementation proposals. Such an assessment should show the impact on the protection of ecosystems in third parties. A joint implementation agreement would then be accepted if

- The average protection of ecosystems in third parties would be equal to or higher than under the Protocol;
- Decreases in ecosystem protection in particular parts (grids) of the territory of third parties would be either limited or explicitly approved by these third parties.

Parties could then use the non-mandatory constant deposition exchange rates to discover profitable trades that would not violate the above conditions.

The chapter also concluded that joint implementation could be translated into domestic policies in any way a party desired, but that changes in the form of adapting sectoral bubbles would be much easier to accomplish and would lower transaction costs because no determination of additional emission reductions would be needed. International transactions would mainly take the form of bilateral case-by-case agreements. They could be supplemented by a clearinghouse that would coordinate agreements and disseminate information on bids, offers, and prices, thus lowering transaction costs. This, however, would be useful only if a large number of firms were to enter trading. Monitoring and enforcement would not have to be fundamentally changed with joint implementation. The veracity of self-reported information on national emission levels would need to be checked more often to balance increased incentives for noncompliance. Enforcement could remain the same as foreseen under the Protocol. If individual firms become active joint implementation partners, international standards for domestic enforcement might have to be agreed on to ensure that property rights (the internationally acquired emission reductions) would be stable across different nations. One should recall that national emission trading schemes still have the most room for cost savings. The implementation of such national schemes should therefore also be promoted because it could be the first step toward international SO_2 trading between firms.

Chapter 12

Conclusions

12.1 Focus of the Study

Europe-wide negotiations led to the signing of the Second Sulfur Protocol in 1994. This Protocol enables the joint implementation of emission reductions, that is, the trading of agreed on national emission ceilings. Economic theory and empirical simulation models suggest that emission trading is more cost efficient than regulation, but practical experience with the instrument has been limited and not always favorable. Against this background, this study has two objectives:

- To examine the general questions of whether, when, and how the application of economic instruments (more specifically, emission trading) would be more cost efficient than regulatory instruments, especially for non-uniformly dispersed pollutants in a transboundary context, taking into account theory, empirical simulation models, and practical experience;
- To analyze the more specific questions of whether, under which conditions, and how emission trading could be implemented to improve the cost efficiency of current policies (in particular, the Second Sulfur Protocol) used to control SO_2 emissions in Europe.

12.2 Findings of Theory, Empirical Models, and Practice

A new element in this study is the combination of theory, empirical models, and practice. In this section, the main conclusions from these areas will be discussed separately. The focus will be on tradable permits. The conclusions will then be integrated to arrive at new insights.

Conclusions

The practical object of this study is to design a policy to achieve a set of exogenously fixed environmental standards (environmental effectiveness) at minimum costs (cost efficiency) accounting for administrative practicability, impacts on innovation, and political acceptability.

Theoretical analysis suggests the following conclusions.

- Regarding cost efficiency, tradable emission permits are clearly preferable to regulation.
- Market power on the permit market, high transaction costs, and imperfect enforcement negatively affect the cost efficiency of tradable emission permits.
- The environmental effectiveness of tradable permits might be at stake with imperfect enforcement, but is not affected by market power, discontinuous control options, economic growth, or inflation. The environmental effectiveness of regulation is at stake with imperfect enforcement and with continuous economic growth if regulation sets emissions per unit of output without controlling output.
- There is little reason to believe that the administrative practicability (information requirements and the implementation and enforcement costs) of tradable permits would, in general, be lower than that of regulation.
- Tradable emission permits have a clear advantage over regulations in terms of creating higher incentives for innovation.
- Whether or not tradable permits are politically acceptable depends on whether the cost savings they promise are high enough to change existing regulations. Tradable permits have political disadvantages: they put a price on pollution that was previously free, they constitute a bigger change in property rights, and they tend to raise the overall expenditures of industry. To make tradable permits acceptable, existing polluters would have to receive tradable permits for free.

Theory shows that when dealing with non-uniformly dispersed pollutants such as SO_2 the least-cost solution for meeting environmental quality standards requires that marginal costs per ton of emissions controlled generally must differ between the emission sources to reflect the impact of the location. In theory, the environmental agency could meet air quality standards at least-cost by issuing ambient or deposition permits for each receptor and allowing sources to buy and sell these deposition rights.

Without knowledge of costs regulation cannot be designed in a cost-efficient way; however, under regulation environmental goals can still be met. Theoretical analysis suggests the following conclusions on the use of instruments for non-uniformly dispersed pollutants.

- With ambient or deposition permits, the environmental goals are always met, but the least-cost solution may not be attained because transaction costs could be high.
- The following alternatives can be used to avoid the complexity of ambient permits: emission permits trading in one zone, emission trading with rules regarding ambient standards, and emission trading in multiple zones.
- Emission trading in one zone would generally not be cost efficient because it ignores the location of the source relative to the receptor. Whether single-zone emission trading or uniform regulation would be more cost efficient depends on specific regional circumstances.
- If costs are uncertain, single-zone emission trading does not guarantee that the environmental goals (ambient standards) will be met because the location of the emissions is uncertain. Uniform regulation could meet the environmental goals only with accurate knowledge of atmospheric dispersion.
- Trading rules allowing trading as long as ambient standards are met are environmentally effective, but the case-by-case approach might imply high transaction costs thus reducing cost efficiency.
- Multiple-zone trading is not cost efficient nor is it necessarily environmentally effective, especially if costs are uncertain.

As a new element, the study examined the question of the conditions under which tradable emission permits could be used to improve the cost efficiency of international agreements for non-uniformly dispersed pollutants. The theoretical analysis allows the following conclusions.

- Agreements taking the form of a cooperative Nash equilibrium of reciprocal reductions of emissions are generally not cost efficient and leave room for reallocation of obligations between countries so that total abatement costs are reduced. If countries sign an agreement based on the full equilibrium solution there is no further room for cost savings.
- Emission permit trading in one zone is not by definition a Pareto-dominant improvement over an initial agreement but can turn out to be so under specific circumstances. If individual firms instead of countries trade permits there is less of a guarantee that a Pareto-dominant change will occur, because firms ignore both domestic and transboundary externalities when trading.
- Ambient or deposition permit trading is always Pareto-dominant if deposition permits are allocated on the basis of actual deposition levels achieved under the cooperative Nash agreement regardless of whether firms or countries trade.
- Trading with rules on the deposition (the modified- and the pollution offset rule and the non-degradation offset rule) is Pareto-dominant, reduces costs, and meets environmental goals.

Conclusions

The study surveyed the cost efficiency and environmental impacts of emission trading in empirical simulation models and examined administrative practicability, innovation, and distributional implications. If the environmental objective is reduced to meeting a certain emission level, the following results are obtained.

- The potential cost savings of emission trading are greater: the lower the percentage reduction, the less the CAC alternative accounts for cost differences, the larger the number of control options, and the larger the number of sources.
- Actual cost savings could differ from modeled cost savings because the simulation models ignore that existing regulations and durable pollution control equipment are in place; they assume frictionless, regular markets with frequent and stable prices; and they ignore that trading stimulates cost-reducing innovation (they are based on accepted technologies). Also, the simulation models are based on less complete information on control options than the individual trading parties possess.
- Market power can have a substantial impact on permit prices if the market share of the price-setting firm is high without necessarily significantly affecting the cost efficiency of permit trading.

Evidence from simulation models further indicates that if ambient concentrations or deposition standards are the objective, the following conclusions emerge.

- Ambient permit trading can meet ambient air quality standards at considerably lower costs than CAC regulation if this trading equals the least-cost solution. However, the cost savings are typically achieved at the expense of increased emissions and increased local concentrations or long-range deposition.
- Trading rules restricting changes in deposition and ambient concentrations reduce the potential cost savings of emission trading (especially if such rules are modeled as bilateral sequential processes with partial information). However, they still promise cost savings compared with any initial allocation of permits without decreasing environmental quality.
- With perfect information on costs, single-zone emission permit trading may or may not be more cost efficient than CAC to meet a set of ambient standards. This depends on regional characteristics. In the case of imperfect information on pollution control costs, attainment of the environmental goals is not guaranteed.
- Given full information on costs, multiple-zone emission trading might come close to the least-cost solution. In the case of incomplete information on costs there is no guarantee that zonal emission trading is cost efficient. It might be more expensive than its CAC counterpart and might violate ambient standards, as well.

- The auction of tradable permits could imply that the financial expenditures (pollution control costs plus permit expenditures) of a system of emission trading will exceed those of the CAC alternative.

This study compiled an overview of a number of emission trading schemes implemented in practice in the USA (the EPA's emission trading policy, lead trading, CFC trading, and sulfur allowance trading), in Europe (Danish and Dutch bubbles and the German plant renewal clause and compensation rule), and worldwide (the Montreal Protocol). The overview leads to the following conclusions.

- In terms of market activity and cost efficiency, single-zone trading (lead trading and sulfur allowance trading) and bubbles (Dutch and Danish bubbles) have been successful and the performance of trading rules (EPA emission trading program and the German rules) has been poor.
- Major circumstances that negatively affect the cost efficiency are regulatory constraints, transaction costs, uncertainty about property rights, and state regulation of polluting firms.
- Regulatory constraints in the form of emission standards have limited permit supply and demand under the EPA's trading policy, in Germany, and with the Dutch and Danish bubbles. Constraints in the form of air quality modeling, trading in small zones, and offsets exceeding one have impaired proper functioning of the EPA's emission trading program and the actual use of German trading rules.
- Transaction costs have been high under the EPA's trading program and in Germany, mainly due to the complicated administrative approval procedures, among them, air quality modeling. Trading has been more successful in the absence of case-by-case approval and with a clear, up-front allocation of rights (lead and sulfur trading).
- Uncertainty about the value of property rights (the initial distribution of rights, policy uncertainty, confiscation of rights) has depressed market activity in the EPA program and in Germany, and with intrafirm trading under the Netherlands' refinery bubble.
- The fact that some polluters are state-regulated monopolies implies that incentives to minimize costs might not be very strong. This may also have created a bias against emission trading (US sulfur trading, Dutch and Danish power plant bubbles).
- The environmental impacts of the emission trading schemes examined have been neutral. In some cases this was due to the same factors that limit the cost efficiency of trades (air quality modeling, small zones, offset rates above one, and case-by-case approval). Single-zone trading has been environmentally effective because it has been combined with regulations and sufficiently high

emission reductions, and has been applied in regional settings where reductions occurred at the right locations.
- Administrative costs of emission trading are sometimes higher than regulation due to emission registration schemes, increased enforcement efforts, or increased consultation. However, sometimes they have been lower.
- No firm data were obtained to draw conclusions on the effect emission trading schemes had on innovation, although some examples show that innovation was stimulated.

The following new insights emerge from the overview of economic theory, simulation models, and practice.

- Ambient permits or emission trading subject to rules on changes in ambient concentrations are, in theory and according to most models, the least-cost solution for dealing with non-uniformly dispersed pollutants. In practice, such systems are either not applied (ambient permits) or the cost savings achieved have been limited (trading rules). The latter results are predicted by models that simulate trade as a bilateral sequential process.
- Theory and simulation models show that single-zone emission permit trading might be more or less cost efficient than regulation depending on the region under study and the quality of cost information. In practice such simple solutions proved to be highly efficient in cutting costs without impairing the environment.
- Single-zone emission trading has been environmentally effective in practice when combined with regulatory constraints and sufficiently high reductions in emissions as well as when the regional setting was such that reductions could be expected to take place at the right location.
- Economic theory points to market power, uncertainty about pollution control costs, and transaction costs as major circumstances affecting the cost efficiency of emissions trading. Simulation models suggest that bilateral sequential (rather than simultaneous) trading and uncertainty about costs affect the cost efficiency of trading. Practice shows that high transaction costs and regulatory constraints are often important in reducing cost efficiency.
- Theory, models, and practice show that there is little reason to believe that the administrative practicability of emission trading differs substantially from that of regulation.
- Evidence from simulation models and practice in support of the theory that tradable permits promote innovation more than regulation has been limited.
- The fact that tradable permits are usually "grandfathered" in practice confirms what could be expected from theory and simulation models.

12.3 Application to Sulfur Emissions in Europe

In Part II, the study examines the conditions under which the trading of sulfur emission reduction commitments could be applied to improve the cost efficiency of international agreements. From an overview of the current institutional framework for dealing with Europe-wide sulfur emissions, the following conclusions emerge.

- The national emission ceilings agreed to under the Second Sulfur Protocol are based on the transboundary impacts on sensitive areas, on pollution control costs, and on political acceptability.
- The Protocol merges this effect-oriented approach with a source-oriented approach consisting of emission standards and fuel standards, similar to EC directives and existing national legislation.
- The Protocol brings with it possibilities for parties to jointly implement their national emission ceilings, but specific rules have yet to be agreed on.
- Joint implementation of sulfur emission reduction commitments thus will have to start from an initial distribution of emission quotas set in the form of national emission ceilings, which to a certain degree are based on cost-efficiency considerations. Moreover, joint implementation has to be superimposed on existing national and international emission and fuel standards.

A new element in this study is the analysis of a new trading scheme based on fixed exchange rates simulated as a bilateral sequential process. This is also one of the options for joint implementation, or emission trading, examined in the negotiations on the Second Sulfur Protocol. A fixed exchange rate is defined as the volume of emission reduction needed in one country to allow another country to increase its emissions by one unit. Such an exchange rate could be designed to account for the location of the sources trading. One way to fix these exchange rates would be to base them on the least-cost solution for meeting a set of targets for the deposition of sulfur. In this way the exchange rates would reflect the weighted average of the transfer coefficients of those receptors that would be binding in the least-cost solution. Theoretical analysis leads to the following conclusions.

- With only one receptor, emission trading with a fixed exchange rate based on the least-cost solution will not violate the deposition target if the deposition constraint is initially met. No knowledge of costs is needed to set the exchange rates. However, even with one receptor, bilateral sequential trading achieves the cost minimum only if trading starts from a specific starting point.
- With two (or more) binding receptors there is no guarantee that exchange rate trading will attain the cost minimum nor is it certain that the deposition targets will not be violated, even if the environmental agency has accurate knowledge

Conclusions

309

of the costs. The system, however, saves costs and tends to limit violations of deposition targets.
- With imperfect information on costs the exchange rates might be set on the wrong costs. This prevents the actual cost minimum from being met and increases the probability that deposition targets will be violated.
- Because of the potential increases in deposition at some locations, emission trading using cost-minimum exchange rates might not constitute a Pareto-dominant improvement in an international context.

Although the cost-minimum solution might not be achieved, the questions of how close such a trading scheme could approach the cost minimum and the extent to which deposition targets would be violated remain. As simulation models and practice have shown for single-zone trading, the magnitude depends on the actual regional setting.

Therefore, the cost efficiency, environmental impacts, and distribution of costs and environmental benefits of emission trading with an exchange rate have been modeled in the actual context of reducing sulfur emissions in Europe. Various trading, or joint implementation, schemes have been compared:

- Trading using cost-minimum exchange rates;
- Trading using constant-deposition rates designed so that with every bilateral trade the average deposition does not increase in any country;
- Trading in one Europe-wide zone subject to the condition that the total emissions remain constant;
- Trading subject to explicit deposition constraints for every receptor.

The results of the first two schemes were based on a model that simulated trade as a bilateral sequential process. The results for the last two schemes were based on the least-cost solution using an LP model, implicitly assuming simultaneous trade on a perfect permit market. The performance of these schemes was examined under different conditions: without constraints, in combination with regulatory constraints in the form of emission and fuel standards, excluding non-signatories from trading, with different sequences of trading, and with different levels of transaction costs. The simulation results suggest the following conclusions.

- Emission trading in a single European zone would imply the highest overall cost savings (6 to 37 percent), compared with the Protocol without trade. However, it might not be politically acceptable because too many countries would be confronted with losses in ecosystem protection compared with a Protocol without trade.
- Emission trading on the basis of cost-minimum exchange rates would not meet the least-cost solution and would save fewer costs (3 to 9 percent) than

single-zone trading. It also would not alter overall ecosystem protection. Small decreases in ecosystem protection might be expected in some countries without reducing their costs.
- Trading with constant-deposition rates would ensure that every country is at least as well off as under the Protocol, but cost savings would be small (0.5 to 1 percent).
- Emission trading subject to deposition constraints (trading rule) would save costs (0.5 to 3 percent). It is expected to reduce ecosystem protection in only one country; that country would save costs, however.

Regarding the conditions that affect the cost efficiency of emission trading, a number of findings have been made.

- Combining the national ceilings in the Protocol with existing emission and fuel standards considerably limits the cost savings of emission trading.
- Excluding the few non-signatories from trade or increasing transaction costs does not lead to significant losses in cost savings.
- If the bilateral trading sequence is random, the expected cost savings of constant-deposition rates trading are much lower. The expected cost savings of cost-minimum exchange rate trading remain largely unchanged.

If we compare these conclusions with the insights that emerged from the overview of economic theory, simulation models, and practice, the following observations emerge. First, although single-zone trading also leads to the highest cost savings, its application to sulfur emissions in Europe seems to be neither environmentally effective nor politically acceptable. If the potential cost savings of trading are not used to further reduce emissions, post-trade emission reductions are not expected to take place where needed to protect the environment and trading does not constitute a Pareto-dominant improvement compared with the Second Sulfur Protocol without trading. Second, the application of Europe-wide sulfur trading shows that regulatory constraints and, in some cases, bilateral trading with imperfect information can seriously limit the expected cost savings of trade. Market power and transaction costs are less relevant.

Finally this study has sought to prepare an institutional design for joint implementation under the Second Sulfur Protocol. Schemes using exchange rates to guide bilateral agreements would probably offer sufficient room for cost efficiency while keeping transaction costs low. Of the two different concepts of exchange rates, based on the least-cost solution or keeping countries' average depositions at least constant, the former rates would save more costs but the latter would probably be more effective in respecting third-party interests. The following was concluded.

Conclusions 311

- It might be necessary to augment the use of these exchange rates with an environmental impact assessment of each proposed agreement to obtain political approval of joint implementation agreements. However, the rules for acceptance should be clarified before trading starts in order to reduce transaction costs.
- The non-mandatory exchange rates could be used by parties to discover profitable trades.
- Joint implementation could be translated into domestic policies in any way a party desired, but changes in the form of adapting sectoral bubbles would be much easier to accomplish and would decrease transaction costs.
- International transactions would mainly take the form of bilateral case-by-case agreements between governments and could be supplemented by a clearinghouse if firms were to enter trading.
- Monitoring and enforcement would not have to be fundamentally changed with joint implementation, although the veracity of self-reported information on national emission levels could be checked more often. If individual firms become active joint implementation partners, international standards for domestic enforcement might have to be agreed on.

In closing, it appears that the potential cost savings from international trading of sulfur emission reduction commitments are constrained by the fact that trading must constitute a Pareto-dominant improvement for all signing parties involved and by the fact that trading is superimposed on existing regulations. This also implies that the only trading designs that would be politically acceptable are trading with constant deposition rates combined with rules on the allowable increases in sulfur deposition in particular grids, trading with rules on allowable deposition increases, or deposition permit trading. This confirms what was expected from the theoretical analysis of emission trading for non-uniformly dispersed pollutants in an international context. Before the international trading of sulfur emission reductions among nations is extended to trading among individual firms, it appears that it would be more useful to introduce national emission permit trading schemes. Their potential cost savings are large and their introduction could also form the first step toward international trade in SO_2 permits between firms.

Appendix

List of Symbols

Symbol	Description
a_{ji}	Transfer coefficient from source i to receptor j
B_i	Benefits to source i
C	Total costs of emission reductions
C_x, C_y, C_i	Costs of reducing emissions of sources x, y, and i, respectively
C'_x, C'_y, C'_i	Marginal costs of reducing emissions of sources x, y, and i, respectively
C'_i	Marginal costs in the cost minimum
$C'_{i,l}$	Marginal costs of region i at level l
D_j	Deposition at point j
D_j^0	Desired level of deposition at point j
E	Total level of emissions
E^0	Desired level of emissions
E^*	Emissions in the cost minimum
$\overline{E}_i$	Uncontrolled emissions of source i
E_i	Emissions after abatement
E_x^0, E_y^0, E_i^0	Pre-trade emissions of sources x, y, and i, respectively
E_x, E_y	Post-trade emissions of sources x and y, respectively
$E_i^{\min}, E_i^{\max}$	Minimum and maximum emission levels at source i, respectively
F	Emission charge or fee
F_e	Emission tax on emissions exported
F_j	Deposition charge for deposition at receptor j
G	Transaction costs
H	Lagrange function
I	Conventional production inputs
L	Total number of permits issued (licensed)

Appendix

Symbol	Description
L_i^0	Initial emission permits allocated to source i
L_i^*	Permits held by source i in equilibrium
L_j	Deposition permit for receptor j
L_{ij}	Permits held by source i to deposit at receptor j
L_{ij}^0	Initial permits held by source i to deposit at receptor j
M	Monetary side payments
N	Background emissions
N_j	Background deposition at receptor j
P	Permit price
P^*	Market equilibrium permit price
P_j	Deposition permit price
Q	Level of pollution
R_i	Emission reduction source i
R_i^*	Equilibrium emission reduction, source i
$\overline{R}_i$	Emission reductions under Nash equilibrium
$R_{i,l}$	Emission reduction region i at the lth level of emission reduction
$R_{i,l,\max}$	Maximum emission reduction in region i at the lth level of emission reduction
S_j	Surplus variable (extent to which actual deposition is below the desired deposition)
t	Time
T_x, T_y	Volume of emissions traded by sources x and y, respectively
w_{xy}	Exchange rate: the rate at which y has to decrease emissions if x increases emissions by one unit
U_i	Welfare level in country i
$\overline{U}_i$	Welfare levels of initial (Nash) agreement
X	Consumption as well as production
Z_i	Reduction in deposition in country i
$\overline{Z}_i$	Welfare levels of initial (Nash) agreement
λ	Shadow price of constraint on total level of emissions
λ_j	Shadow price of deposition constraint j
η	Price elasticity of demand
μ	Shadow price of exchange rate constraint on emissions
μ_i	Shadow price of constraint on welfare
τ	Interest rate

References

Adar, Z., and Griffin, J.M., 1976, Uncertainty and the choice of pollution control instruments, *Journal of Environmental Economics and Management*, **3**(3):178–188.

Alcamo, J., Hordijk, L., and Shaw, R., 1990, *The RAINS Model of Acidification: Science and Strategies in Europe*, Kluwer, Dordrecht, the Netherlands.

Altman, A., Amann, M., Klaassen, G., Ruszczyński, A., and Schöpp, W., 1994, Cost-effective Sulfur Emission Reduction Under Uncertainty, WP-94-119, International Institute for Applied Systems Analysis, Laxenburg, Austria.

Amann, M., 1992, Zur effizienten multinationalen Allokation von Emissionsminderungsmassnahmen zur Verringerung der sauren Deposition – Anwendungsbeispiel Österreich, PhD diss., Universität Fridericiana, Karlsruhe, Germany.

Amann, M., and Klaassen, G., 1995, Cost-effective strategies for reducing nitrogen deposition in Europe, *Journal of Environmental Management*, **43**:289–311.

Amann, M., Klaassen, G., and Schöpp, W., 1991, *UN/ECE Workshop on Exploring European Sulfur Abatement Strategies*, Status Report SR-91-03, International Institute for Applied Systems Analysis, Laxenburg, Austria.

Amann, M., Klaassen, G., and Schöpp, W., 1993, Closing the Gap between the 1990 Deposition and the Critical Sulfur Deposition Values, background paper prepared for the UN/ECE Task Force on Integrated Assessment Modelling, 7–8 June 1993, International Institute for Applied Systems Analysis, Laxenburg, Austria.

Amann, M., and Schöpp, W., 1993, Reducing Excess Sulfur Deposition in Europe by 60 Percent, background paper prepared for the UN/ECE Working Group on Abatement Strategies, 30 August–2 September, International Institute for Applied Systems Analysis, Laxenburg, Austria.

Anderson, R., Hofmann, L., and Rusin, M., 1990, The Use of Economic Incentive Mechanisms in Environmental Management, Research Paper #051, American Petroleum Institute, Washington, DC, USA.

Atkinson, S., 1983, Marketable pollution permits and acid rain externalities, *Canadian Journal of Economics*, **16**(4):704–722.

Atkinson, S., 1994, Tradable discharge permits: Restrictions on least-cost solutions, in G. Klaassen and F. Førsund, eds., *Economic Instruments for Air Pollution Control*, Kluwer, Dordrecht, the Netherlands: 3–21.

Atkinson, S., and Tietenberg, T.H., 1982, The empirical properties of two classes of design for transferable discharge permits markets, *Journal of Environmental Economics and Management*, **9**:101–121.

Atkinson, S., and Tietenberg, T.H., 1984, Approaches for reaching ambient standards in non-attainment areas, *Land Economics*, **60**(2):148–159.

Atkinson, S., and Tietenberg, T.H., 1987, Economic implications of emission trading rules for local and regional pollutants, *Canadian Journal of Economics*, **20**(2):370–386.

Atkinson, S., and Tietenberg, T.H., 1991, Market failure in incentive-based regulation: The case of emissions trading, *Journal of Environmental Economics and Management*, **21**:17–31.

Ausubel, J., and Victor, D., 1992, Verification of international environmental agreements, *Annual Review of Energy and Environment*, **17**:1–43.

Averch, H., and Johnson, L., 1962, Behavior of the firm under regulatory constraint, *The American Economic Review*, **52**:1052–1069.

Barakat, and Chamberlin, 1991, Study of Atmospheric Trading Programs in the United States. Final Report, Emission Trading Working Group, Canadian Council of Ministers of the Environment, Toronto, Canada.

Barde, J.-P., and Pearce, D.W., 1991, eds., *Valuing the Environment*, Earthscan, London, UK.

Barrett, S., 1990, The problem of global environmental protection, *Oxford Review of Economic Policy*, **6**(1):68–79.

Barrett, S., 1991, Economic instruments for climate change policy, in *Responding to Climate Change: Selected Economic Issues*, Organisation for Economic Cooperation and Development, Paris, France: 53–108.

Barrett, S., 1992, Convention on climate change, in *Economic Aspects of Negotiations*, Organisation for Economic Cooperation and Development, Paris, France.

Barrett, S., 1993, A Strategic Analysis of "Joint Implementation" Mechanisms in the Framework Convention on Climate Change, prepared for United Nations Conference on Trade and Development at Geneva, Switzerland, London Business School/Centre for Social and Economic Research on the Global Environment, London, UK.

Batterman, S.A., and Amann, M., 1991, Targeted acid rain strategies including uncertainty, *Journal of Environmental Management*, **32**:57–72.

Baumol, W., and Oates, W.E., 1971, The use of standards and prices for protection of the environment, *Swedish Journal of Economics*, **73**(1):42–54.

Baumol, W., and Oates, W.E., 1979, *Economics, Environmental Policy and the Quality of Life*, Prentice Hall, Englewood Cliffs, NJ, USA.

Baumol, W., and Oates, W.E., 1988, *The Theory of Environmental Policy*, Cambridge University Press, Cambridge, UK.

Becker, N., Baron, M., and Shechter, M., 1993, Economic instruments for emission abatement under appreciable technological indivisibilities, *Environmental and Resource Economics*, **3**:263–284.

Bennett, G., ed., 1991, *Air Pollution Control in the European Community*, Graham and Trotman, London, UK.

Bergman, H., Jörnsted, O., and Lövgren, K., 1993, *Five Economic Instruments in Swedish Environmental Protection Policy*, Swedish Environmental Protection Agency, Solna, Sweden.

Bohi, D., 1993, Utilities and State Regulators Fail to Take Advantage of the Benefits of Trading Emission Allowances, October 1993, Resources for the Future, Washington, DC, USA.

Bohi, D., Burtraw, D., Krupnick, A., and Stalon, C., 1990, Emission Trading in the Electric Utility Industry, Discussion Paper QE90-15, Resources for the Future, Washington, DC, USA.

Bohm, P., 1990, *Efficiency Issues and the Montreal Protocol on CFCs*, Environment Working Paper no. 40, The World Bank, Washington, DC, USA.

Bohm, P., 1994, Making carbon emissions quota agreements more efficient: Joint implementation versus quota tradability, in G. Klaassen and F. Førsund, eds., *Economic Instruments for Air Pollution Control*, Kluwer, Dordrecht, the Netherlands: 187–208.

Bohm, P., and Russell, C.S., 1985, Comparative analysis of alternative policy instruments, in A.V. Kneese and J.L. Sweeney, eds., *Handbook of Natural Resource and Energy Economics*, vol. 1, Elsevier, Amsterdam, the Netherlands: 395–460.

Boland, J., 1986, Economic Instruments for Environmental Protection in the United States, ENV/ECO/86/14, Organisation for Economic Cooperation and Development, Paris, France.

Borge, E., 1992, Letter of September 1992, State Pollution Control Authority (SFT), Oslo, Norway.

Borowski, A.R., and Ellis, H.E., 1987, Summary of the final federal emissions trading policy statement, *Journal of the Air Pollution Control Association*, **37**(7):798–800.

Bromley, D., 1989, Entitlements, missing markets, and environmental uncertainty, *Journal of Environmental Economics and Management*, **17**:181–194.

Buchanan, J., and Tullock, G., 1975, Polluter's profit and political response: Direct control versus taxes, *American Economic Review*, **65**:139–147.

Bressers, H., and Klok, P.J., 1988, Fundamentals for a theory of policy instruments, *International Journal of Social Economics*, **15**:22–41.

Burtraw, D., 1993, Interview, 16 September, Resources for the Future, Washington, DC, USA.

Burtraw, D., 1995, Personal communication, 3 March, Resources for the Future, Washington DC, USA.

Burtraw, D., Harrison, K., and Turner, P., 1994, Path Dependence in Bilateral Emission Trading. Paper prepared for the American Economics Association, 3–5 January at Boston, MA, USA, Resources for the Future, Washington DC, USA.

Carlin, A., 1992, The United States Experience with Economic Incentives to Control Environmental Pollution, EPA–230–R–92–001, US Environmental Protection Agency, Washington, DC, USA.

Cason, T., 1993, Seller incentive properties of EPA's emission trading auction, *Journal of Environmental Economics and Management*, **25**(2):177–195.

CDA, 1988, *Acid Rain and Photochemical Oxidants Control Policies in the European Community*, Cambridge Decision Analysts/Environmental Resources Limited, The Leopard Press, Woodfolds Oaksey, UK.

Cesar, H., and Klaassen, G., 1990, Costs, Sulfur Emissions and Deposition of the EC Directive on Large Combustion Plants, WP-90-006, International Institute for Applied Systems Analysis, Laxenburg, Austria.

Chiang, A., 1984, *Fundamental Methods of Mathematical Economics*, McGraw Hill, Singapore.

Chichilnisky, G., and Heal, G.M., 1993, Market Instruments for International Environmental Policy, 8 September, Columbia University, New York, NY, USA.

Christiansen, G.B., and Tietenberg, T.H., 1985, Distributional and macroeconomic impacts of environmental policy, in A.V. Kneese and J.L. Sweeney, eds., *Handbook of Natural Resource and Energy Economics*, vol. 1, Elsevier, Amsterdam, the Netherlands: 345–393.

Coase, R., 1960, The problem of social cost, *Journal of Law and Economics*, **3**:1–44.

Cofala, J., 1991, Personal communication, 12 March, International Institute for Applied Systems Analysis, Laxenburg, Austria.

Compliance Strategies Review, 1993, EATX Survey October 1993, Allowance prices fall from August survey, *Compliance Strategies Review*, 8 November, 1993.

Cowling, E.B., 1982, Acid precipitation in historical perspective, *Environmental Science and Technology*, **16**(2):110A–123A.

Crocker, T.D., 1966, The structuring of atmospheric pollution control systems, in H. Wolozin, ed., *The Economics of Air Pollution*, W.W. Norton, New York, NY, USA: 61–86.

Cropper, M.L., and Oates, W.E., 1990, *Environmental Economics: A Survey*, Resources for the Future, Washington, DC, USA.

Dales, J.H., 1968, *Pollution, Property and Prices*, University Press, Toronto, Canada.

Dasgupta, P., 1982, *The Control of Resources*, Basic Blackwell, Oxford, UK.

Dasgupta, P., and Heal, G.M., 1979, *Economic Theory and Exhaustible Resources*, Cambridge University Press, Cambridge, UK.

Dean, A., 1993, What do global models tell us about the carbon taxes required and the economic costs entailed in reducing CO_2 emissions? in Y. Kaya, N. Nakićenović, W.D. Nordhaus, and F.L. Toth, eds., *Costs, Impacts, and Benefits of CO_2 Mitigation*, CP-93-2, International Institute for Applied Systems Analysis, Laxenburg, Austria: 213–233.

De Clerq, M., 1983, *Economische aspecten van het vervuilingsbeleid*, Spruyt, Van Mantgem en de Does, Leiden, the Netherlands.

Dekkers, C., 1993, Telephone interview, 7 September, Ministry of Housing, Physical Planning and the Environment, The Hague, the Netherlands.

Demsetz, H., 1967, Toward a theory of property rights, *American Economic Review*, **LVII**(2):347–359.

Dewees, D., 1983, Instrument choice in environmental policy, *Economic Inquiry*, **21**:53–71.

Downing, P.B., 1984, *Environmental Economics and Policy*, Little, Brown and Company, Boston, MA, USA.

Downing, P.B., and White, L., 1986, Innovation in pollution control, *Journal of Environmental Economics and Management*, **13**:18–29.

Dudek, D., and Palmisano, J., 1988, Emission trading: Why is this thoroughbred hobbled? *Columbia Journal of Environmental Law*, **13**:217–256.

Dumas, R., 1995, Personal communication, The Netherlands Ministry for Public Housing, Physical Planning and Environmental Protection (VROM), Directorate Air, 20 January, The Hague, the Netherlands.

Dwyer, J., 1992, California's tradeable emissions policy and its application to the control of greenhouse gases, in *Climate Change, Designing a Tradeable Permit System*, Organisation for Economic Cooperation and Development, Paris, France: 41–77.

EC, 1991, Council Regulation (EEC) No. 594/91 of 4 March 1991 on substances that deplete the ozone layer, *Official Journal of the European Communities*, 14.3.91.

Edmonds, J., Barns, D., and Ton, M., 1993, The regional costs and benefits of participation in alternative hypothetical fossil fuel carbon emissions reductions protocols, in Y. Kaya, N. Nakićenović, W.D. Nordhaus and F.L. Toth, eds., *Costs, Impacts and Benefits of CO_2 Mitigation*, CP-93-2, International Institute for Applied Systems Analysis, Laxenburg, Austria: 291–314.

Elkraft, 1993, SO_2-NO_x-prognoseredegørelse 1990 for de danske elvaerker, 31 March, Elkraft, Ballerup, Denmark.

Elman, B., 1993, Interview, 13 September, US Environmental Protection Agency, Washington, DC, USA.

Elman, B., Tyler, T., and Doonan, M., 1992, Economic Incentives under the New Clean Air Act, paper presented at the 85th Annual Meeting of the Air and Waste Management Association, 21–26 June at Kansas City, Missouri, US Environmental Protection Agency, Washington, DC, USA.

Elsam/Elkraft, 1993, Indberetning i henhold til bekendtgorelse af 18.12.1991 om begraensing af udledning af svolvdioxid og kvaelstofoxider fra kraftvaerker, 11 May, Ballerup, Denmark.

Endres, A., 1986, Charges, permits and pollutant interactions, *Eastern Economic Journal*, **12**(3):327–336.

Endres, A., and Schwarze, R., 1993, Das Zertifikatsmodell vor der Bewährungsprobe? Eine ökonomische Analyse des "Acid Rain"-Programms des neuen US-Clean Air Acts, Diskussions beitrag Nr. 200, Februar 1993, Fernuniversität, Hagen, Germany.

Ermoliev, Y., Klaassen, G., and Nentjes, A., 1995, Adaptive cost-effective ambient charges under incomplete information, *Journal of Environmental Economics and Management*, (accepted for publication).

Federal Committee for the Environment, 1991, Regulation, Issued by the Federal Committee for the Environment, 1 October, 1991, Concerning the Law 309 of July 1991 on Air Pollution Control, Prague, Czech Republic.

Barrett, S., Bohm, P., Fisher, B.S., Kuroda, M., Mubazi, J., Shaj, A., and Stavins, R., 1994, An Assessment of Greenhouse Policy Instruments, Draft for IPCC Working Group III (September 1994), Intergovernmental Panel on Climate Change, Geneva, Switzerland.

Førsund, F., 1985, Input–output models, national economic models, and the environment, in A.V. Kneese and J.L. Sweeney, eds., *Handbook of Natural Resource and Energy Economics*, vol. 1, Elsevier, Amsterdam, the Netherlands: 325–341.

Førsund, F., 1992a, BAT and BATNEEC: An Analytical Interpretation, Memorandum no. 28, December, Department of Economics, University of Oslo, Oslo, Norway.

Førsund, F., 1992b, Emission Trading, paper presented at the first meeting of the Task Force on Economic Aspects of Abatement Strategies, 2–3 March, Geneva, Switzerland.

Førsund, F., 1993, Sulphur Emission Trading, paper prepared for the fourth meeting of the Task Force on Economic Aspects of Abatement Strategies, 10–11 June, London, UK.

Førsund, F., and Nævdal, E., 1993, Sulphur Emission Trading, paper prepared for the fifth meeting of the Task Force on Economic Aspects of Abatement Strategies, 9–10 December, Geneva, Switzerland.

Førsund, F., and Nævdal, E., 1994a, Trading sulphur emission quotas: Third party constraints, paper prepared for the sixth meeting of the Task Force on Economic Aspects of Abatement Strategies, 12–13 May, London, UK.

Førsund, F., and Nævdal, E., 1994b, Effect-based trade in sulphur emission quotas, paper prepared for the Task Force on Economic Aspects of Abatement Strategies, UN/ECE, 8–9 December, Geneva, Switzerland.

Førsund, F., and Nævdal, E., 1994c, Trading sulphur emission in Europe, in G. Klaassen and F. Førsund, eds., *Economic Instruments for Air Pollution Control*, Kluwer, Dordrecht, the Netherlands: 231–248.

Foster, V., and Hahn, R., 1994, ET in LA: Looking Back to the Future, ENRP Project 88/Round II, Project Report P-94-01, Kennedy School of Government, Harvard University, Cambridge, MA, USA.

Gaasbeek, P., 1993, Telephone interview, 11 November, Shell Nederland BV, Rotterdam, the Netherlands.

GAO, 1986, Vehicle Emissions, EPA Program to Assist Leaded-Gasoline Producers Needs Prompt Improvement, GAO/RCED-86-182, United States General Accounting Office, Washington, DC, USA.

Glatz, H., Krajasits, C., and Pohl, E., 1990, *Mehr Markt oder mehr Staat in der Umweltpolitik? Umweltzertifikate und Kontingente als Instrumente der Umweltpolitik*, Institut für Wirtschaft/Österreichischer Arbeiterkammertag, Vienna, Austria.

Goffman, J., 1993, Interview, 15 September, Environmental Defense Fund, Washington, DC, USA.

Gollop, F.M., and Roberts, M., 1985, Cost-minimizing regulation of sulfur emissions: Regional gains in electric power, *The Review of Economics and Statistics*, **67**(1):81–90.

Griffin, R.C., 1987, Environmental policy for spatial and persistent pollutants, *Journal of Environmental Economics and Management*, **14**:41–53.

Hahn, R.W., 1984, Market power and transferable property rights, *Quarterly Journal of Economics*, **99**(4):735–765.

Hahn, R.W., 1986, Trade-offs in designing markets with multiple objectives, *Journal of Environmental Economics and Management*, **13**:1–12.

Hahn, R.W., 1989, Economic prescriptions for environmental problems: How the patient followed the doctor's orders, *Journal of Economic Perspectives*, **3**(2):95–114.

Hahn, R.W., 1990, The political economy of environmental regulation: Towards a unifying framework, *Public Choice*, **65** (April): 21–45.

Hahn, R.W., and Hester, G.L., 1989a, Where did all the markets go? An analysis of EPA's Emissions Trading Program, *Yale Journal on Regulation*, **6**:109–153.

Hahn, R.W., and Hester, G.L., 1989b, Marketable permits: Lessons for theory and practice, *Ecology Law Quarterly*, **16**:361–406.

Hahn, R., and May, C., 1994, The behavior of the allowance market: Theory and evidence, *The Electricity Journal*, **7**(2):28–37.

Hahn, R.W., and McGartland, A., 1989, The political economy of instrument choice: An examination of the US role in implementing the Montreal Protocol, *Northwestern University Law Review*, **83**:592–611.

Hahn, R., and Noll, R., 1982, Designing a market for tradeable emission permits, in W. Magat, ed., *Reform of Environmental Regulation*, Ballinger, Cambridge, MA, USA: 119–146.

Hahn, R.W., and Stavins, R.N., 1992, Economic incentives for environmental protection: Integrating theory and practice, *American Economic Review*, **82**(2):464–468.

Haigh, N., 1989, New tools for European air pollution control, *International Environmental Affairs*, **1**(1):26–37.

Halkos, G., 1993, Sulphur abatement policy: Implications of cost differentials, *Energy Policy* (October): 1035–1043.

Hanisch, T., Selrod, R., Torvanger, A., and Aaheim, A., 1993, Study to Develop Practical Guidelines for "Joint Implementation" Under the UN Framework Convention on Climate Change, Report 1993: 2, Centre for International Climate and Energy Research (CICERO), University of Oslo, Oslo, Norway.

Harford, J., 1978, Firm behavior under imperfectly enforceable pollution taxes and standards, *Journal of Environmental Economics and Management*, **5**:26–43.

Hausker, K., 1992, The politics and economics of auction design in the market for sulfur dioxide pollution, *Journal of Policy Analysis and Management*, **11**(4):553–572.

Hettelingh, J.-P., Downing, R.J., and de Smet, P., 1991, Mapping Critical Loads for Europe, CCE Technical Report No. 1, RIVM Report No. 259101001, National Institute of Public Health and Environmental Protection, Bilthoven, the Netherlands.

Hoel, M., 1990, Emission Taxes in a Dynamic Game of CO_2 Emissions, memorandum from the Department of Economics, University of Oslo, Oslo, Norway.

Hoel, M., 1991, Efficient international agreements for reducing emissions of CO_2, *The Energy Journal*, **12**(2):93–107.

Hoel, M., 1992a, Emission taxes in a dynamic international game of CO_2 emissions, in R. Pethig, ed., *Conflicts and Cooperation in Managing Environmental Resources*, Springer-Verlag, Berlin, Germany: 39–68.

Hoel, M., 1992b, Carbon taxes: An international tax or harmonized domestic taxes? *European Economic Review*, **36**:400–406.

Hoel, M., 1993, Harmonization of carbon taxes in international climate agreements, *Environmental and Resource Economics*, **3**:221–231.

Hordijk, L., Shaw, R., and Alcamo, J., 1990, Background to acidification in Europe, in J. Alcamo, L. Hordijk, and R. Shaw, *The RAINS Model of Acidification: Science and Strategies in Europe*, IIASA/Kluwer Academic Publishers, Dordrecht, the Netherlands: 31–60.

Hourcade, J.C., and Baron, R., 1993, Tradeable permits, in *International Economic Instruments and Climate Change*, Organisation for Economic Cooperation and Development, Paris, France: 13–44.

ICF, 1989, Economic Analysis of Title V (Acid Rain Provisions) of the Administration's Proposed Clean Air Act Amendments, ICF Incorporated, September, US Environmental Protection Agency (Acid Rain Division), Washington, DC, USA.

ICF, 1991, Regulatory Impact Analysis of the Proposed Acid Rain Implementation Regulations, ICF Incorporated, 20 June, US Environmental Protection Agency (Acid Rain Division), Washington, DC, USA.

ICF, 1992, Regulatory Impact Analysis of the Final Acid Rain Implementation Regulations, ICF Incorporated, 19 October, US Environmental Protection Agency, Washington, DC, USA.

ICF, 1994, Economic Analysis of the Title V Requirements of the 1990 Clean Air Act Amendments of 1990, US Environmental Protection Agency, Washington, DC, USA.

Intriligator, M.D., 1971, *Mathematical Optimization and Economic Theory*, Prentice Hall, Englewood Cliffs, NJ, USA.

Johnson, S., and Corcelle, G., 1989, *The Environmental Policy of the Environmental Communities*, Graham and Trotman, London, UK.

Kambhu, J., 1990, Direct controls and incentive systems of regulation, *Journal of Environmental Economics and Management*, **18**:S-72–S-85.

Keeler, A., 1991, Noncompliant firms in transferable discharge permit markets: Some extensions, *Journal of Environmental Economics and Management*, **21**:180–189.

Kete, N., 1992, The US acid rain control allowance trading system, in *Climate Change, Designing a Tradeable Permit System*, Organisation for Economic Cooperation and Development, Paris, France: 78–108.

Klaassen, G., 1993a, Trade-offs in exchange rate trading for sulfur emissions in Europe, background paper prepared for the fourth meeting of the Task Force on Economic Aspects of Abatement Strategies, 10–11 June, London, UK.

Klaassen, G., 1993b, Sulfur Emission Trading and Regulation in Europe, WP-93-69, International Institute for Applied Systems Analysis, Laxenburg, Austria.

Klaassen, G., 1994, Joint Implementation in the Second Sulfur Protocol: A Tempest in a Teapot? Background paper prepared for the sixth meeting of the Task Force on Economic Aspects of Abatement Strategies, 12–13 May, London, UK.

Klaassen, G., 1995, Trade-offs in emission trading and regulation in Europe, *Environmental and Resource Economics*, **5**:191–219.

Klaassen, G., and Amann, M., 1992, *Trading of Emission Reduction Commitments for Sulfur Dioxide in Europe*, SR–92–03, International Institute for Applied Systems Analysis, Laxenburg, Austria.

Klaassen, G., Førsund, F., and Amann, M., 1994, Emission trading in Europe with an exchange rate, *Environmental and Resource Economics*, **3**:1–26.

Klaassen, G., and Nentjes, A., 1995, Emission Trading for Air Pollution Control in Practice, WP-95-21, International Institute for Applied Systems Analysis, Laxenburg, Austria.

Klaassen, G., and Opschoor, J.B., 1991, Economics of sustainability or the sustainability of economics: Different paradigms, *Ecological Economics*, **4**:93–115.

Kling, R.W., 1988, An institutionalist theory of regulation, *Journal of Economic Issues*, **22**:197–211.

Klink, J., Krozer, Y., and Nentjes, A., 1989, Technologische Ontwikkeling door Marktconform Milieubeleid, Nederlandse Organisatie voor Technologisch Aspecten Onderzoek, Groningen, the Netherlands.

Kneese, A.V., 1964, The Economics of Regional Water Quality Management, The Johns Hopkins Press for Resources for the Future, Baltimore, MD, USA.

Kneese, A.V., and Bower, B., 1968, *Managing Water Quality: Economics, Technology, Institutions*, The Johns Hopkins Press for Resources for the Future, Baltimore, MD, USA.

Kornai, J., 1986, The soft budget constraint, *Kyklos*, **39**(1):3–30.

Koutstaal, P., 1993, Verhandelbare CO_2 Emissierechten in Nederland en de EG, Beleidsstudies deel 1, Ministerie van Economische Zaken, the Hague, the Netherlands.

Kruitwagen, S., 1992, Tradeable Permits for SO_2 Emissions in Europe: A Pollution Offset Experiment, Wageningen Economic Papers, 1992–4, Wageningen Agricultural University, Wageningen, the Netherlands.

Krupnick, A.J., 1986, Costs of alternative policies for the control of nitrogen dioxide in Baltimore, *Journal of Environmental Economics and Management*, **13**:189–197.

Krupnick, A.J., Oates, W.E., and van de Verg, E., 1983, On marketable air pollution permits: The case for a system of pollution offsets, *Journal of Environmental Economics and Management*, **10**:233–247.

Kverndokk, S., 1992, Global CO_2 agreements: A cost-effective approach, *The Energy Journal*, **14**(2):91–112.

Leaf, D., 1993, Interview, 13 September, US Environmental Protection Agency, Washington, DC, USA.

Lipskey, M., 1993, Facsimile, Cantor Fitzgerald Environmental Brokerage Services, 29 November, New York, NY, USA.

Liroff, R., 1986, *Reforming Air Pollution Regulation: The Toil and Trouble of EPA's Bubble*, The Conservation Foundation, Washington, DC, USA.

Lobsenz, G., 1995, Allowance trading takes off, *Energy Daily*, February 15, 1995: 1–2.

Lövgren, K., 1994, Economic instruments for air pollution control in Sweden, in G. Klaassen and F. Førsund, eds., *Economic Instruments for Air Pollution Control*, Kluwer, Dordrecht, the Netherlands: 107–121.

Lubbers, R., 1993, Telephone interview, 2 September, Samenwerkende Electriciteits Producten, Arnhem, the Netherlands.

Makowski, M., and Sosnowski, S., 1988, User Guide to Mathematical Programming Package for Multi-Criteria Dynamic Linear Problems, WP-88-111, International Institute for Applied Systems Analysis, Laxenburg, Austria.

Malec, W., 1993, Emission allowances stall in marketplace, *Forum for Applied Research and Public Policy*, Summer: 45–48.

Mäler, K.-G., 1989, The acid rain game, in H. Folmer and E.C. van Ierland, eds., *Valuation Methods and Policy Making in Environmental Economics*, Elsevier, Amsterdam, the Netherlands: 231–252.

Mäler, K.-G., 1990, International environmental problems, *Oxford Review of Economic Policy*, **6**(1):80–108.

Mäler, K.-G., 1993, The Acid Rain Game II, Beyer Discussion Paper Series No. 32, Beijer International Institute of Ecological Economics, Stockholm, Sweden.

Mäler, K.-G., 1994, Acid rain in Europe: A dynamic perspective on the use of economic incentives, in E.C. van Ierland, ed., *International Environmental Economics: Theories, Models and Applications to Climate Change, International Trade and Acidification*, Elsevier, Amsterdam, the Netherlands: 351–372.

Malik, A., 1990, Markets for pollution control when firms are non-compliant, *Journal of Environmental Economics and Management*, **18**:97–106.

Maloney, M., and Yandle, B., 1984, Estimation of the cost of air pollution control regulation, *Journal of Environmental Economics and Management*, **11**:244–263.

Mason, R., and Swanson, T., 1994, Joint implementation of the Second Sulfur Protocol, Draft report, December 1994, Centre for Social and Economic Research on the Global Environment, University College London, London, UK.

Maxwell, J., and Weiner, S., 1993, Green consciousness or dollar diplomacy? The British response to the threat of ozone depletion, *International Environmental Affairs*, **5**(1):19–41.

McGartland, A.M., 1984, Marketable Permit Systems for Air Pollution Control: An Empirical Study, PhD diss., University of Maryland, MD, USA.

McGartland, A.M., 1988, A comparison of air pollution permits: The case for a system of pollution offsets, *Journal of Environmental Economics and Management*, **15**:35–44.

McGartland, A.M., and Oates, W.E., 1985, Marketable permits for the prevention of environmental deterioration, *Journal of Environmental Economics and Management*, **12**:207–228.

McLean, B., 1993, Interview, 13 September, US Environmental Protection Agency, Washington, DC, USA.

McMillan, J., 1986, *Game Theory in International Economics*, Harwood, New York, NY, USA.

Milliman, S., and Prince, R., 1989, Firm incentives to promote technological change in pollution control, *Journal of Environmental Economics and Management*, **17**:247–265.

Ministry of the Environment, 1991, Exploration of Economic Instruments for Implementation of Cost-effective Reductions of SO_2 in Europe Using the Critical Loads Approach, report from a designated expert meeting 29–30 May at Oslo, Norway, Ministry of Environment, Oslo, Norway.

Ministry of Housing, Physical Planning and the Environment, 1992, *A Reprint of the United Nations Framework Convention on Climate Change*, CCD/Paper 8, August 1992, Leidschendam, the Netherlands.

Misiolek, W., and Elder, H., 1989, Exclusionary manipulation of markets for pollution rights, *Journal of Environmental Economics and Management*, **16**:156–166.

Montgomery, W.D., 1972, Markets in licenses and efficient pollution control programs, *Journal of Economic Theory*, **5**:395–418.

MOSZNIL (Ministry of Environmental Protection, Natural Resources and Forestry), 1990, Ordinance of the MOSZNIL of 12th February 1990 on Protection on Air against Pollution, *Journal of Laws No. 15*, Warsaw, Poland [in Polish].

NAPAP, 1991, The U.S. National Acid Precipitation Assessment Program: 1990 Integrated Assessment Report, The NAPAP Office of the Director, Washington, DC, USA.

Nentjes, A., 1990a, Economische Instrumenten in het Milieubeleid: Financiering of Sturingsmiddel? in P. Nijkamp and H. Verbruggen, eds., *Het Nederlandse Milieu in de Europese Ruimte*, Preadviezen voor de Nederlandse Vereniging voor de Staathuishoudkunde, Leiden, the Netherlands: 145–166.

Nentjes, A., 1990b, An economic model of transfrontier pollution abatement, in *Public Finance, Trade and Development*, Proceedings of the International Institute of Public Finance, Istanbul, Turkey, 1988, Wayne State University Press, Detroit, MI, USA: 243–261.

Nentjes, A., 1991, Environmental Taxes as Instrument for Environmental Policy: When Are They Appropriate? Documentatieblad Ministerie van Financien, November–December, the Hague, the Netherlands: 265–282.

Nentjes, A., 1993, Groeidwang verplaatsen? in F. Biesboer, ed., *Greep op groei, het thema van de jaren negentig*, Aktie Strohalm/Uitgeverij Jan van Arkel, Utrecht, the Netherlands: 117–135.

Nentjes, A., 1994, Control of reciprocal transboundary pollution and joint implementation, in G. Klaassen and F. Førsund, eds., *Economic Instruments for Air Pollution Control*, Kluwer, Dordrecht, the Netherlands: 209–230.

Nentjes, A., and Dijkstra, B., 1993, The political economy of instrument choice in environmental policy, in M. Faure, J. Vervacle, and A. Waele, eds., *Environmental Standards in the European Union in an Interdisciplinary Framework*, Maklu/Nomos/Blackstone/Schulthess, Antwerp/Brussels, Belgium: 197–216.

Nentjes, A., and Wiersma, D., 1988, Innovation and pollution control, *International Journal of Social Economics*, **15**(3/4):51–70.

NERI, 1992, Emissions Trading Program for Stationary Sources of NO_x in Ontario, National Economic Research Associates, Cambridge (MA), prepared for the Advisory Group on Emission Trading/The Ontario Ministry of Energy, Canada.

Nicholson, W., 1989, *Microeconomic Theory: Basic Principles and Extensions*, Dryden Press, Chicago, IL, USA.

Nilsson, J., and Grennfelt, P., eds., 1988, Critical Loads for Sulphur and Nitrogen, Report from a workshop held at Skokloster, Stockholm, Sweden.

Nordberg, L., 1993, Combatting Air Pollution, Geneva, Switzerland (unpublished document).

Nowicki, M., 1993, *Environment in Poland: Issues and Solutions*, Kluwer, Dordrecht, the Netherlands.

Nussbaum, B., 1992, Phasing down lead in gasoline in the US: Mandates, incentives, trading and banking, in *Climate Change, Designing a Tradable Permit System*, United Nations Economic Commission for Europe: 25–40.

Oates, W., and McGartland, A., 1985, Marketable permits and acid rain externalities: A comment and some further evidence, *Canadian Journal of Economics*, **18**(3):668–675.

Oates, W., Portney, P., and McGartland, A., 1988, The Net Benefits of Incentive-based Regulation: The Case of Environmental Standard-setting in the Real World, discussion paper CRM-89-03, Resources for the Future, Washington, DC, USA.

OECD, 1976, *Economics of Transfrontier Pollution*, Organisation for Economic Co-operation and Development, Paris, France.

OECD, 1991, *Environmental Policy: How to Apply Economic Instruments*, Organisation for Economic Co-operation and Development, Paris, France.

OJ, 1988, Council Directive of November 1988 on the limitation of emissions of certain pollutants in the air from large combustion plants, *Official Journal of the European Communities*, L336, **31** (7 December): 1–13.

Ølsgaard, P., 1994, Interview, 20 January, Elkraft, Ballerup, Denmark.
Opschoor, J.B., 1994, Developments in the use of economic instruments for environmental policies in OECD member countries, in G. Klaassen and F. Førsund, eds., *Economic Instruments for Air Pollution Control*, Kluwer, Dordrecht, the Netherlands: 75–106.
Opschoor, J.B., de Savornin Lohman, A., and Vos, H., 1994, *Managing the Environment: The Role of Economic Instruments*, Organisation for Economic Cooperation and Development, Paris, France.
Opschoor, J.B., and Turner, R.K., eds., 1994, *Economic Incentives and Environmental Policies; Principles and Practice*, Kluwer, Dordrecht, the Netherlands.
Opschoor, J.B., and van der Straaten, J., 1992, Institutional aspects of sustainable development, in F.J. Dietz and J. van der Straaten, eds., *Sustainability and Environmental Policy*, Ed. Sigma, Berlin, Germany: 55–74.
Opschoor, J.B., and Vos, H., 1989, *Economic Instruments for Environmental Protection*, Organisation for Economic Cooperation and Development, Paris, France.
Palmer, A., Mooz, W., Quinn, T., and Wolf, K., 1980, Economic Implications of Regulating Chlorofluorocarbon Emissions from Nonaerosol Applications, EPA-560/12-80-001, EPA, Washington, DC, USA.
Palmisano, J., 1993, Interview, AER*X, 15 September, Washington, DC, USA.
Peaple, N., 1993, Interview, 15 December, EC DG Environment, Brussels, Belgium.
Pearce, D., 1989, Sustainable Development: An Economic Perspective, London Environmental Economics Centre, Gatekeeper series LEEC 89-01, London, UK.
Pearce, D., and Turner, R., 1990, *Economics of Natural Resources and the Environment*, The Johns Hopkins University Press, Baltimore, MD, USA.
Peeters, M., 1992, *Marktconform milieurecht? Een rechtsvergelijkende studie naar de verhandelbaarheid van vervuilingsrechten*, W.E.J. Tjeenk Willink, Zwolle, the Netherlands.
Persson, G., 1982, ed., What is acidification? *Acidification Today and Tomorrow*, Swedish Ministry for Agriculture, Environment '82 Committee, Risbergs Tryckeri AB, Uddevalla, Sweden: 30–49.
Peterson, F.M., and Fisher, A.C., 1977, The exploitation of extractive resources: A survey, *The Economic Journal*, **87**:681–721.
Portney, P., 1990, Air Pollution Policy, in P. Portney, ed., *Public Policies for Environmental Protection*, Resources for the Future, Washington, DC, USA: 27–96.
Pototschnig, A., 1994, Economic instruments for the control of acid rain in Britain, in G. Klaassen and F. Førsund, eds., *Economic Instruments for Air Pollution Control*, Kluwer, Dordrecht, the Netherlands: 22–45.
Quisthout, M., 1993, Telephone interview, 11 November, Ministry of Housing, Physical Planning and the Environment, The Hague, the Netherlands.
Rehbinder, R., and Sprenger, R.-U., 1985, The Emission Trading Policy of the United States of America: An Evaluation of its Advantages and Disadvantages and Analysis of its Applicability in the Federal Republic of Germany, EPA-230-07-85-012, US Environmental Protection Agency, Washington, DC, USA.
Rentz, O., Haasis, H.D., Morgenstern, T., Remmers, J., and Schons, G., 1990, Optimal Control Strategies for Reducing Emissions from Energy Conversion and Energy Use

in all Countries of the European Community, KfK–PEf72, Kernforschungszentrum Karlsruhe, Karlsruhe, Germany.

Rico, R., 1995, The US allowance trading system for sulfur dioxide: An update on market experience, *Environmental and Resource Economics*, **5**:115–129.

Roach, F., Kolstad, C., Kneese, A.V., and Williams, M., 1981, Alternative air quality policy options in the Four Corners region, *The Southwestern Review*, **1**(2):29–58.

Roland, K., 1992, From offsets to tradeable entitlements, in Combating Global Warming. Study on a Global System of Tradeable Carbon Emission Entitlements, UNCTAD/RDP/DFP/1. United Nations, New York, NY, USA: 23–25.

Rose, A., and Stevens, B., 1993, The efficiency and equity of marketable permits for CO_2 emissions, *Resource and Energy Economics*, **15**:117–146.

Rose-Ackermann, S., 1973, Effluent charges: A critique, *Canadian Journal of Economics*, **6**:512–528.

Russell, C.S., 1986, A note on the efficiency ranking of two second best policy instruments for pollution control, *Journal of Environmental Economics and Management*, **13**:13–17.

Russell, C.S., 1990, Monitoring and enforcement, in P. Portney, ed., *Public Policies for Environmental Protection*, Resources for the Future, Washington, DC, USA: 243–274.

Ruyssennaars, P., Sliggers, J., and Merkus, H., 1993, Joint Implementation in the Context of Acidification Abatement, Ministry of Housing, Physical Planning and Environment, Air Directorate, paper prepared for the fifth meeting of the Task Force on Economic Aspects of Abatement Strategies, 9–10 December, Geneva, Switzerland, the Hague, the Netherlands.

Sabogal, N., 1994, Letter, 11 January, UNEP/Ozone Secretariat, Nairobi, Kenya.

SEP (Samenwerkende Electriciteits Producenten), 1991, Plan van aanpak ter uitvoering van het convenant over de bestrijding van SO_2 and NO_x, Arnhem, the Netherlands.

Samuelson, P., 1954, The pure theory of public expenditure, *Review of Economics and Statistics*, **36**:387–389.

Sandnes, H., 1993, Calculated Budgets for Airborne Acidifying Components in Europe, 1985, 1987, 1988, 1989, 1990, 1991 and 1992, EMEP/MSC-W Report 1/93, Meteorological Synthesizing Center-West, The Norwegian Meteorological Institute, Oslo, Norway.

Sandor, R., Cole, J.B., and Kelly, M., 1993, Study on Model Rules and Regulations for a Global CO_2 Emissions Credit Market, Revised first draft as of 8 November, Centre Financial Products Limited, New York, NY, USA.

Schärer, B., 1979, Luftverschmutzung durch Schwefeldioxid: Ursachen, Wirkungen, Minderung, Umweltbundesamt, Berlin, Germany.

Schärer, B., 1981, Ergebnisse des SO_2-Berichtes des Umweltbundesamtes, *Umwelt*, 6/81:453–456.

Schärer, B., 1982, Saurer Regen – eine Herausforderung an die Luftreinhaltepolitik der achtziger Jahre, *Umwelt*, **88** (April): 7a–7b.

Schärer, B., 1993, Use of Economic Incentives in Germany in Air Pollution Control, paper prepared for the Conference on Economic Instruments for Air Pollution Control, 18–20 October, International Institute for Applied Systems Analysis, Laxenburg, Austria.

Schärer, B., 1995, Personal communication, 12 January, Umweltbundesamt, Berlin, Germany.

Seskin, E., Anderson, R., and Reid, R., 1983, An empirical analysis of economic strategies for controlling air pollution, *Journal of Environmental Economics and Management*, **10**:112–124.

Siebert, H., 1987, *Economics of the Environment*, Springer-Verlag, Berlin, Germany.

Sociaal Economische Raad (SER), 1990, Advies National Milieubeleidsplan Plus, Publicatie no. 17, The Hague, the Netherlands.

Solomon, B., and Rose, K., 1992, Privatization of pollution: Making the market for SO_2 emissions, *The Electricity Journal*, July.

Sørensen, E., 1993, Interview, 10 December, National Agency for Environmental Protection, Copenhagen, Denmark.

Spofford, W., and Paulsen, C., 1990, Efficiency Properties of Source Control Policies for Air Pollution Control: An Empirical Application to the Lower Delaware Valley, Discussion Paper QE88-13, Resources for the Future, Washington, DC, USA.

Sprenger, R.-U., 1989, Economic Incentives in Environmental Policies: The Case of Western Germany, paper prepared for the Prince Bertil Symposium on Economic Instruments in National and International Environmental Protection Policies, 12–14 June at Stockholm, Sweden.

Stavins, R., 1994, Transaction costs and tradeable permit markets, Submitted to the *Journal of Environmental Economics and Management*, Harvard University, Cambridge, MA, USA.

Stavins, R., and Hahn, R., 1993, Trading in Greenhouse Permits: A Critical Examination of Design and Implementation Issues, Faculty research working paper series R93-15, John F. Kennedy School of Government/Harvard University, Cambridge, MA, USA.

Streets, D.G., Hanson, D., and Carter, L.D., 1984, Targeted strategies for control of acid deposition, *Journal of the Air Pollution Control Association*, **34**(12):1187–1197.

Swaney, J., 1992, Market versus command and control environmental policies, *Journal of Economic Issues*, **26**(2):623–633.

Tietenberg, T.H., 1974, The design of property rights for air pollution control, *Public Policy*, **22**:275–292.

Tietenberg, T.H., 1978, Spatially differentiated air pollutant emission strategies: An economic and legal analysis, *Land Economics*, **54**(3):265–274.

Tietenberg, T.H., 1980, Transferable discharge permits and the control of stationary source air pollution: A survey and synthesis, *Land Economics*, **56**(4):391–416.

Tietenberg, T.H., 1985, Emission Trading: An Exercise in Reforming Pollution Policy, Resources for the Future, Washington, DC, USA.

Tietenberg, T.H., 1989, Designing Marketable Emission Permits: Lessons from the US Experience, paper prepared for the Prince Bertil Symposium on Economic Instruments in National and International Environmental Protection Policies, 12–14 June at Stockholm, Sweden.

Tietenberg, T.H., 1990, Economic instruments for environmental regulation, *Oxford Review of Economic Policy*, **6**(1):17–33.

Tietenberg, T.H., 1991, Reductions in Emissions: Command and Control or Market Based Mechanisms? Paper prepared for the Conference on Economy and Environment in the 1990s, 26–27 August at Neuchâtel, Switzerland.

Tietenberg, T.H., 1992a, *Environmental and Natural Resource Economics*, Harper Collins, New York, NY, USA.

Tietenberg, T.H., 1992b, Implementation issues: A general survey, in *Combating Global Warming. Study on a Global System of Tradeable Carbon Emission Entitlements*, UNCTAD/RDP/DFP/1, United Nations, New York, NY, USA: 127–150.

Tietenberg, T.H., 1995, Economic instruments for pollution control when emission location matters: What have we learned? *Environmental and Natural Resource Economics*, 5:95–113.

Tietenberg, T., and Victor, D., 1994, Administrative Structures and Procedures for Implementing a Tradeable Entitlement Approach to Controlling Global Warming, paper written for UNCTAD, Colby College, Waterville, ME, USA; International Institute for Applied Systems Analysis, Laxenburg, Austria.

Toman, M., Cofala, J., and Bates, R., 1994, Alternative standards and instruments for air pollution control in Poland, *Environmental and Resource Economics*, 4(5):401–418.

Tripp, J., and Dudek, D., 1989, Institutional guidelines for designing successful transferable rights programs, *Yale Journal on Regulation*, 6:369–391.

Tschirhart, J.T., 1984, Transferable discharge permits and profit-maximizing behavior, in T.D. Crocker, ed., *Economic Perspectives on Acid Deposition Control*, Butterworth, Boston, MA, USA: 157–171.

UN/ECE, 1987, National Strategies and Policies for Air Pollution Abatement, 1986 Review, ECE/EB.Air/14, United Nations Economic Commission for Europe, Geneva, Switzerland.

UN/ECE, 1991, Strategies and Policies for Air Pollution Abatement, 1990 Review, ECE/EB.Air/27, United Nations Economic Commission for Europe, Geneva, Switzerland.

UN/ECE, 1993, Strategies and Policies for Air Pollution Abatement, 1993 Review, Draft Report, EB.Air/R.76, United Nations Economic Commission for Europe Geneva, Switzerland (unpublished document).

UN/ECE, 1994a, Protocol to the 1979 Convention on Long-range Transboundary Air Pollution on Further Reduction of Sulphur Emissions, United Nations Economic Commission for Europe, Geneva, Switzerland.

UN/ECE, 1994b, Economic Aspects of Abatement Strategies, Progress Report by the Chairman of the Task Force, EB.AIR/WG.5/R.46, United Nations Economic Commission for Europe, Geneva, Switzerland (unpublished).

UN/ECE, 1994c, Joint Implementation Under the Oslo Protocol, Progress Report by the Chairman of the Task Force on Economic Aspects of Abatement Strategies, EB.AIR/WG.5/R.50, United Nations Economic Commission for Europe, Geneva, Switzerland (unpublished).

UNEP, 1987, Montreal Protocol on Substances That Deplete the Ozone Layer, United Nations Environmental Programme, New York, NY, USA.

References

UNEP, 1993a, Handbook for the Montreal Protocol on Substances That Deplete the Ozone Layer (third edition), Ozone Secretariat, United Nations Environmental Programme, Nairobi, Kenya.

UNEP, 1993b, The Reporting of Data by the Parties to the Montreal Protocol on Substances That Deplete the Ozone Layer, UNEP/OzL.Pro.5.5, 24 August 1993, United Nations Environmental Progamme, New York, NY, USA.

UNEP, 1993c, Transfer of Production Rights under Article 2 of the Montreal Protocol, UNEP/OzL. Pro. 5/8, 16 September 1993, Fifth meeting of the parties to the Montreal Protocol on Substances That Deplete the Ozone Layer in Bangkok, Thailand, United Nations Environmental Programme, New York, NY, USA.

UNIPEDE, 1993, Legal Status of Electricity Supply Undertakings in Europe, Ref08300, Ref9318, vols. I and II, International Union of Producers and Distributors of Electrical Energy, Paris, France.

US Congress, 1990, *Public Law 101–549 Nov. 15, 1990*, US Government Printing Office, Washington, DC, USA.

USEPA, 1979, *Bubble Policy*, 44 FR (Federal Register) 71779, 12/11/79, US Environmental Protection Agency, Washington, DC, USA.

USEPA, 1982, *Emissions Trading Policy Statement*, 47 FR 15076, 4/7/82, US Environmental Protection Agency, Washington, DC, USA.

USEPA, 1985a, *Regulation of Fuels and Fuel Additives; Banking of Lead Rights: Final Rule*, 40 CFR Part 80, 2/4/85, US Environmental Protection Agency, Washington, DC, USA.

USEPA, 1985b, *Costs and Benefits of Reducing Lead in Gasoline*, Report no. EPA-230-05-85-006 (February 1985), US Environmental Protection Agency, Washington, DC, USA.

USEPA, 1986, *Emissions Trading Policy Statement*, 47 FR 15076, 12/4/86, US Environmental Protection Agency, Washington, DC, USA.

USEPA, 1990, *Clean Air Act Amendments of 1990, Detailed Summary of Titles*, US Environmental Protection Agency, Washington, DC, USA.

USEPA, 1992a, *Protection of Stratospheric Ozone; Final Rule*, 40 CFR Part 82, 30/6/92, US Environmental Protection Agency, Washington, DC, USA.

USEPA, 1992b, *Guidance for the Stratospheric Ozone Protection Program*, US Environmental Protection Agency, Washington, DC, USA.

USEPA, 1992c, *Acid Rain Program*, Overview, US Environmental Protection Agency, Washington, DC, USA.

USEPA, 1992d, *Acid Rain Program, Allowance trading*, US Environmental Protection Agency, Washington, DC, USA.

USEPA, 1992e, *Acid Rain Program, Allowance Auctions and Direct Sales*, US Environmental Protection Agency, Washington, DC, USA.

USEPA, 1993, *Results of Auctions and Direct Sales*, US Environmental Protection Agency, Washington, DC, USA.

USEPA, 1994, *Results of Auctions and Direct Sales*, US Environmental Protection Agency, Washington, DC, USA.

USEPA, 1995, *1995 EPA Allowance Auction Results*, US Environmental Protection Agency, Washington, DC, USA.

Verbruggen, H., 1994, Environmental policy failures and environmental policy levels, in H. Opschoor and K. Turner, eds., *Economic Incentives and Environmental Policies*, Kluwer, Dordrecht, the Netherlands: 41–54.

Vernon, J., 1988, *Emission Standards for Coal-Fired Plants*, IEA Coal Research, London, UK.

Vivian, W., and Hall, W., 1981, *An Examination of US Market Trading in Pollution Offsets*, The University of Michigan, Ann Arbor, MI, USA.

Vlieg, F., 1993, Personal communication, The Netherlands Ministry for Public Housing, Physical Planning and Environmental Protection (VROM), Directorate Air, The Hague, the Netherlands.

Voight, P., 1993, Telephone interview, 24 November, US Environmental Protection Agency, Washington, DC, USA.

Voight, P., and Lee, D., 1993, Interview, 13 September, US Environmental Protection Agency, Washington, DC, USA.

Vos, J., Jansen, H., Oosterhuis, F., de Savornin Lohman, A., and Sterk, H., 1992, *Buitenlandse ervaringen met financiële instrumenten voor het milieubeleid*, DHV/Instituut voor Milieuvraagstukken, Vrije Universiteit, Amersfoort/Amsterdam, the Netherlands.

Weissman, A., 1993, Interview, 14 September, Clean Air Capital Markets, Washington, DC, USA.

Weitzman, M.L., 1974, Prices versus quantities, *Review of Economic Studies*, **XLI**(4):477–491.

Welsch, H., 1988, Cost-effective Control Strategies for Energy-related Transboundary Air Pollution: Results of a Simulation Model, Institute of Energy Economics/University of Cologne, Cologne, Germany (unpublished document).

Welsch, H., 1989, Kosten der SO_2–Minderung unter alternativen umweltpolitischen Strategien: Ergebnisse eines Simulationsmodells des westeuropäischen Kraftwerks sektors, *Zeitschrift für Energie und Wirtschaft*, 1/89:51–59.

Welsch, H., 1993, An equilibrium framework for global pollution problems, *Journal of Environmental Economics and Management*, 25:S64–S79.

Wenders, J., 1975, Methods of pollution control and the rate of change in pollution abatement technology, *Water Resources Research*, **11**(3):393–396.

Wiersma, D., 1989, *De Efficientie van een Marktconform Milieubeleid*, PhD diss., Rijksuniversiteit Groningen, Groningen, the Netherlands.

Wiersma, D., 1991, Static and dynamic efficiency of pollution control strategies, *Environmental and Resource Economics*, **1**:63–82.

Wüster, H., 1992, The Convention on Long-range Transboundary Air Pollution: Its achievements and potential, in T. Schneider, ed., *Acidification Research, Evaluation and Policy Applications*, Elsevier, Amsterdam, the Netherlands: 221–239.

Zylicz, T., 1994, Improving the environment through permit trading: The limits to a market approach, in E.C. van Ierland, ed., *International Environmental Economics: Theories, Models and Applications to Climate Change, International Trade and Acidification*, Elsevier, Amsterdam, the Netherlands: 283–306.

Index

accumulation, 186, 286
acid deposition, 1, 2, 111, 145, 155, 156, 193, 248
acid rain, 1, 7, 19, 145, 185–187, 190
acidification, 1, 2, 7, 44, 157, 161, 162, 185–188, 192, 208, 286
administrative costs, *see* costs, administrative
administrative practicability, 1, 4–6, 11, 12, 31, 41, 43, 44, 60, 63, 98, 99, 121, 127, 129, 130, 137, 141, 145, 155, 157, 160, 164, 170, 175, 179, 180, 270, 278, 303, 305, 307
agreements, 2, 3, 6, 64, 65, 70–72, 74, 75, 78–80, 91, 96, 97, 126, 162, 166, 167, 188, 273, 274, 284, 293, 294, 304
allocation, 12–14, 36, 37, 43, 46, 52, 59, 81, 84, 87, 99, 103, 107, 108, 113, 116, 117, 119, 121, 124, 126–129, 131, 146, 154, 158, 172, 179, 240, 273, 284, 290, 292, 305, 306
allowances, 142–144, 146, 147, 150–157, 168
ambient charges, 37, 47, 61–63, 119, 176, 180
ambient permits, 47–49, 60–63, 87, 95, 106–108, 110, 111, 124–126, 176, 177, 210, 258, 303, 304, 307
ambient standards, 49, 52–55, 57–60, 62, 63, 106–112, 114–117, 121, 127–129, 146, 156, 175–177, 286, 304, 305
atmospheric transport, 2, 8, 56, 60, 63, 180, 186, 241

auction, 17, 27, 40, 85, 92, 104, 121, 122, 124–126, 129, 142, 146–152, 181, 280, 306

banking of rights, 139, 141, 144, 151, 173, 283, 286
bargaining, 14, 15, 28, 29, 40, 41, 198, 293
benefits, 3–5, 8, 11, 12, 14–16, 35, 38, 40, 65, 66, 68, 69, 71, 73, 74, 77–86, 88–90, 92, 95, 96, 110, 111, 123, 126, 127, 181, 235, 260, 271, 274, 276, 288, 300, 309
bilateral trading, 6, 8, 93, 113–115, 128, 176, 178, 179, 210, 223, 225, 226, 229, 231, 237, 240, 242, 249, 250, 265, 280, 281, 293, 305, 307–310
bubbles, 100, 131, 135–138, 160, 163, 170–173, 179, 203, 207, 257, 290, 292, 296, 300, 306, 311

CAA, *see* Clean Air Act
CAC, *see* command-and-control
charges, 3–7, 11, 12, 16–18, 22–26, 31–44, 46, 47, 59, 61–66, 75, 76, 81, 96, 119, 121, 122, 126, 175, 180, 290
chlorofluorocarbons, 20, 21, 141
Clean Air Act, 121, 130, 131, 145, 292
clearinghouse, 293–295, 300, 311
command-and-control, 4, 99, 100, 102–108, 110, 111, 113–117, 119, 121–129, 176, 178–180, 186, 305, 306
common property, 6, 12, 14, 16, 41, 51, 94, 96

331

competition, 1, 4, 6, 14, 16, 27, 50, 55–57, 96, 99, 105, 146, 155, 157, 159, 161, 178, 265, 295
Convention on Long-range Transboundary Air Pollution, 2, 188, 208
cost efficiency, 1–6, 8, 11, 12, 22–25, 27, 29–31, 39, 41, 42, 44, 50, 54, 55, 57, 60, 61, 63–65, 70, 75, 76, 80, 88, 95–97, 99, 104–106, 108, 109, 113, 115, 116, 119, 127, 128, 130, 134, 137, 139, 144, 148, 153, 154, 161, 165, 168–173, 175–179, 210, 211, 223, 235, 239, 249, 267, 270, 271, 275, 276, 278, 282, 283, 286, 289, 290, 294, 296, 300, 302–310
cost minimization, 47, 71, 88, 90, 91, 97, 103, 119, 159, 275
cost savings
  estimated, 6, 34, 100, 104, 107, 108, 114, 124, 134, 135, 139, 144, 159, 163, 168, 210, 219, 264, 268, 269, 278, 279, 281, 309, 310
  expected, 30, 40, 54, 86, 102, 134, 148, 272, 293, 310
costs
  abatement, 14, 22, 29–31, 65, 69, 73, 91, 96, 102, 104–106, 116, 123–125, 151, 168, 238, 240, 270, 282, 304
  administrative, 4, 5, 11, 17, 32, 33, 43, 57, 61, 121, 122, 129, 138, 140–142, 148, 157, 173, 180, 278, 280, 307
  monitoring, 32, 121, 122, 148, 149, 157, 180
  pollution control, 2, 17, 21, 23, 25, 28, 30–33, 38, 47–49, 54, 58–60, 63, 70, 73, 75, 76, 78, 85, 86, 89, 98, 102–105, 114, 117, 118, 123, 124, 127–129, 134, 175, 180, 181, 193, 208, 218, 219, 225, 275, 298, 305–308
critical loads, 1, 2, 19, 192–194, 199, 208, 250, 274, 276, 277, 279

damage, 1, 3, 12, 14, 16–19, 21, 41, 64, 111, 166, 186, 187, 190, 192, 193, 202, 277
Denmark, 131, 157, 158, 172, 173, 190, 197, 198, 205, 207, 208, 261, 267, 268
deposition, 66
  constraints, 7, 59, 72, 73, 77, 88, 89, 91, 93, 95, 115, 210, 212, 214, 221, 223, 224, 226, 229, 235, 237, 239, 250, 256, 258, 259, 261, 262, 270, 271, 279–284, 287, 296, 308–310
  permit, 37, 48–52, 54–58, 60–62, 80, 87–95, 97, 125, 180, 258, 279, 303, 304, 311
  standards, 48, 52, 54, 56–58, 60, 71, 73, 93, 95, 96, 98, 115, 119, 128, 210, 262, 305
  target, 2, 45–49, 52, 54, 56–60, 70, 71, 73, 74, 77, 80, 88, 89, 93, 96, 181, 194, 200, 208, 211, 214, 216, 217, 219, 221, 223, 231, 232, 234–238, 240, 276, 277, 279, 280, 308, 309, *see also* acid deposition
directives, 26, 200, 201, 202, 208, 308
distribution
  of costs and benefits, 4, 5, 8, 11, 38, 40, 79, 123, 127, 181, 260, 271, 309
  of permits, 29, 30, 37, 40, 48–50, 104, 105, 123–126, 133, 146, 163, 169, 264
distributional impacts, 79, 123, 124, 239
dynamic efficiency, 6, 12, 33, 41, 43, 44, 130, 175

economic incentives, 4, 7, 12, 26, 33, 34, 38–40, 64, 65, 123, 130, 180
economic instruments, 1, 3–7, 12, 32, 33, 36, 37, 40, 43, 64, 66, 75, 97, 98, 123, 130, 175, 181, 185, 207, 274, 302
emission ceilings, 3, 6, 26, 74, 102, 103, 118, 121, 158, 179, 197, 198, 200, 203, 204, 208, 239, 250, 274, 283, 286, 290, 292, 302, 308

Index

emission permits, 12, 27, 28, 30, 31, 35, 42, 51, 54, 55, 58, 59, 65, 71, 75, 80, 86, 94, 96, 99, 101, 123, 125, 126, 129, 154, 163, 174, 175, 178, 216, 237, 242, 244, 246, 252, 264, 283, 290, 303, 304

emission standards, 2, 3, 23, 26, 27, 32, 35, 39, 100, 102, 107, 108, 118, 119, 131, 133, 135–137, 153, 158, 160–162, 164, 165, 171–175, 179, 180, 192, 198, 199, 201–209, 248, 257, 261, 283, 290, 297, 306, 308

emission trading rules, 57, 62, 63

enforcement, 3, 4, 8, 11, 17, 25, 27–32, 38, 40, 42, 43, 86, 92, 131, 141, 143, 148, 159, 161, 167, 174, 176, 178–180, 200, 275, 283, 284, 296–301, 303, 307, 311

environmental effectiveness, 4–8, 11, 23, 25–27, 29, 31, 38, 43, 50, 55, 63, 92, 98, 99, 103, 110, 130, 137, 144, 153, 156, 160, 162, 175, 179, 210, 211, 239, 278, 282, 303

environmental standards, 3, 19, 41, 179, 303

EPA, 53, 57, 122, 131, 132, 134, 135, 137–143, 146–150, 153, 155, 156, 170–173, 176, 178–180, 245, 258, 280, 285, 290–292, 306

equilibrium, 11, 27–30, 33, 36, 48, 50, 54, 57, 74–77, 82, 85, 86, 97, 98, 104, 105, 114, 115, 117, 124, 126, 129, 147, 223, 248, 250, 264, 266, 267, 281, 304

European Community (EC), 202, 208, 308

exchange rate, 6, 8, 185, 211–216, 218, 219, 221, 223–229, 231–235, 238–240, 242–245, 249, 250, 252, 256–259, 261, 262, 264, 267–272, 277, 280–282, 287, 288, 290–294, 300, 308–311

externalities, 1, 6, 12–14, 37, 41, 61, 64, 84, 86, 97, 111, 186, 304

fees, 40, 136, 155

financial burden, 123–127, 129, 181

firms, 11, 14, 25, 27, 29–31, 36, 39, 40, 43, 61, 76, 77, 81, 83–87, 91, 94–97, 112, 114, 115, 126, 131, 133, 135, 137, 144, 151, 158, 161, 163, 165, 172, 174, 179, 180, 207, 270, 285, 291, 292, 295, 296, 298–301, 304, 306, 311

flexibility, 95, 100, 103, 104, 106, 108, 123, 131, 132, 157, 163, 173, 180, 257, 278, 286, 292

free riding, 1, 6, 51, 55–58, 63, 78, 94–97

fuel standards, 104, 123, 160, 198–200, 204, 206, 208, 209, 240, 248, 256, 257, 264, 271, 272, 278, 281, 285, 290–292, 308–310

full cooperative solution, 5, 66, 68–70, 74, 79

gas oil, 198, 199, 201, 206–208, 247

Germany, 159, 164, 171, 172, 187, 190, 197, 198, 202, 205–208, 256, 261, 264, 265, 268, 290, 306

grandfathering, 27, 104, 124–127, 129, 133, 142

growth, 23, 25–27, 29, 32, 35, 37, 43, 166, 275, 303

industry, 29, 36, 37, 39–41, 43, 47, 141, 153, 181, 303

inflation, 23, 25, 29, 32, 35, 43, 148, 303

information requirements, 5, 11, 19, 31, 41, 43, 56–58, 130, 141, 157, 278, 303

innovation, 4, 11, 33–36, 43, 61, 99, 104, 122, 123, 126–130, 137, 138, 145, 155, 157, 160, 164, 165, 170, 173, 180, 276, 303, 305, 307

international tax, 75–78, 80, 181

joint implementation, 3, 8, 66, 70, 71, 73, 74, 77, 82, 185, 200, 209, 239, 242, 245, 249, 250, 252, 256, 259–261, 264, 270–274, 276–278, 280, 282–288, 290–302, 308–311

large combustion plants, 158, 159, 162, 198, 201–205, 208, 209
LCPs, *see* large combustion plants
lead, 131, 134, 138–141, 170, 172, 173, 177–180, 210, 286, 306
least-cost solution, 28, 55, 56, 59, 62, 99, 106, 110–112, 114, 115, 119, 128, 129, 175, 176, 210, 214, 215, 217–220, 223, 227, 228, 233, 303–305, 307–310
legislation, 2, 110, 153, 157, 160, 162, 164, 170, 186, 199, 200, 202, 204, 208, 209, 248, 256, 290, 299, 308
location
  receptor, 87, 106, 304
  source, 12, 16, 20, 33, 37, 44–46, 58, 61, 80, 83, 85, 97, 99, 106, 107, 109, 112, 117, 125, 162, 180, 259, 274, 277, 284, 303, 304, 308

marginal benefits, 14, 77–79, 82, 85, 87, 88
market equilibrium, 48, 49, 81–86, 147, 265, 267
market failure, 1, 6, 12, 153
market imperfections, 4, 5, 23, 24, 29, 41, 85, 104, 105, 129, 179
market power, 29, 31, 42, 51, 63, 85, 86, 92, 96, 104, 105, 126, 128, 144, 169, 178, 179, 266, 295, 303, 305, 307, 310
modified offset rule, *see* offset rule, modified
monitoring, 8, 11, 26–28, 31, 32, 38, 40, 43, 86, 121, 122, 131, 138, 140, 141, 148, 149, 157, 159, 161, 164, 167, 173, 174, 189, 200, 201, 275, 283, 284, 288, 296–300
Montreal Protocol, 64, 141–143, 166–168, 170, 171, 179, 273, 285, 306
multiple-zone trading, 48, 61–63, 176, 177, 179, 304

Nash equilibrium, 66–72, 74, 75, 77, 79, 80, 89, 90, 93, 96, 178, 192, 194, 198, 235, 275, 304
negotiations, 2, 8, 64, 70, 76, 157, 188, 193, 194, 197, 200, 250, 273, 299, 302, 308
net benefits, 3, 5, 11, 12, 14, 15, 18, 19, 67–71, 75, 76, 79, 81, 82, 85, 90, 91, 97, 276
Netherlands, 124, 131, 160, 162, 168, 172, 173, 178–180, 190, 197, 198, 202, 204–208, 260, 261, 264, 268, 276, 290, 291, 296, 306
netting, 133–136, 138, 173
non-degradation offset rule, *see* offset rule, non-degradation

offset rule, 52, 57, 93–97, 111, 112, 176, 178, 205, 209, 210, 304
  modified, 57, 93–95, 97, 111–113, 176, 178, 304
  non-degradation, 52, 53, 93, 95, 97, 112, 114, 176, 210, 211, 304
  pollution, 52, 93, 95–97, 112, 304
optimal policies, 18
optimization, 7, 8, 113, 118, 156, 240, 242, 244, 247, 250
ozone, 22, 64, 131, 141, 143, 144, 166, 168, 192, 193, 201, 273

Pareto, 13, 14, 66–71, 73–75, 77, 79, 80, 82, 84, 85, 87, 89–92, 95–98, 129, 178, 181, 210, 219, 225, 235, 237, 238, 276, 278, 279, 282–284, 288, 304, 309–311
permit market, 28, 29, 31, 34, 36, 48, 49, 55, 81–83, 86–88, 91, 96, 99, 104–106, 117, 125, 126, 129, 148, 153, 154, 163, 172, 176, 248, 258, 272, 292, 296, 303, 309
permit price, 17, 28–31, 33, 35, 49, 50, 61, 81, 86, 88–92, 116, 117, 128, 136, 142, 144, 169, 178, 248, 264, 265, 298, 305

Index 335

permits, 3, 4, 12, 17, 27–43, 48–53, 55, 58–62, 64, 65, 78, 81, 82, 85, 87, 89–91, 93–95, 97, 104–106, 110, 131, 179–181, 211, 237, 242, 258, 265–267, 292, 296, 302–307, 311

Pigovian tax, 3, 16, 17, 41

policy, 1, 3, 6–8, 12, 17, 20, 27, 29, 31, 33, 36–41, 43, 44, 48, 63, 65, 67, 99, 103, 108, 117, 121, 131, 132, 134, 136–138, 143, 144, 154, 155, 159, 170, 171, 173, 175, 176, 179, 180, 188, 194, 200, 241, 290–292, 299, 303, 306

political acceptability, 1, 6, 12, 19, 28, 41, 43, 44, 76, 86, 97, 126, 178–181, 202, 260, 271, 276, 279, 282, 283, 286, 294, 296, 303, 308–311

pollutants
  non-uniformly dispersed, 5, 6, 12, 20, 22, 41, 44, 46, 47, 54, 65, 66, 74, 76, 96, 99, 106, 124, 125, 175, 176, 181, 274, 286, 302–304, 307, 311
  uniformly dispersed, 12, 20–22, 31, 41, 60, 61, 99, 175

pollution offset rule, *see* offset rule, pollution

power plants, 12, 101–103, 108, 157–161, 163, 170–172, 179, 203, 205, 207, 209, 290, 291

practicability, *see* administrative practicability

price, 11, 14, 17, 21, 27–30, 32–39, 43, 45, 46, 49–51, 55, 61, 72, 77, 78, 80, 81, 87, 91, 92, 94, 97, 104, 105, 113, 124, 126, 128, 136, 139, 141, 146–148, 150–152, 154, 155, 159, 169, 173, 178, 180, 214–216, 236, 247, 248, 250, 262, 265–267, 275, 292–294, 296, 298, 300, 303, 305

property rights, 5, 6, 13–16, 28, 29, 37, 38, 40, 50, 51, 95, 131, 135, 137, 145, 153, 163, 171–173, 178–180, 273, 299, 301, 303, 306

protocol, 2, 8, 141, 166, 167, 185, 189, 190, 192, 194, 249, 273, *see also* sulfur, Second Sulfur Protocol

public good, 1, 4–6, 12–14, 16

quota, 39, 64, 78, 80, 86, 102, 103, 126, 142, 144, 158–161, 166–168, 170, 197, 198, 209, 308

receptor, 4, 7, 44–51, 54–57, 59–61, 63, 71, 77, 87, 91, 106, 110–112, 117–119, 175, 176, 179, 193, 194, 214–217, 220, 221, 223, 224, 226, 227, 229, 231–233, 235–238, 240, 244, 248, 249, 262, 271, 277, 283, 303, 304, 308, 309

refineries, 103, 138, 140, 141, 162, 163, 170, 172, 173, 207

regulation, 4, 6–8, 26, 27, 31–37, 39–44, 47, 59, 61–64, 86, 102, 103, 106, 116, 117, 119, 121, 122, 128, 131, 138, 140, 144, 153, 157, 159, 167, 169, 170, 176, 177, 180, 185, 186, 201, 204, 206, 208–210, 248, 256, 257, 259–261, 267, 270, 272, 278, 298, 302–307, 311

regulatory constraints, 171, 173, 178, 179, 240, 256, 271

regulatory instruments, 3, 5, 12, 32, 47, 178, 302

revenues, 25, 28, 32, 37, 40, 43, 51, 76–82, 85, 87, 89, 97, 126, 127, 142, 144, 148, 153, 155, 172, 179–181

sector covenants, 160

sequence, 8, 114, 223, 229, 231, 232, 240, 242, 246, 252, 256, 258, 268, 269, 271, 272, 309, 310

simulation models, 5, 6, 98, 99, 105, 123, 124, 127, 128, 130, 175, 177–180, 185, 210, 211, 278, 302, 305, 307, 309, 310

single-zone trading, 48, 58, 59, 63, 82, 115, 117, 118, 125, 176–179, 210, 304–307, 310

SO_2, 6, 7, 12, 20, 44, 65, 78, 99, 104, 108, 110, 121, 123, 127, 131, 134, 145–147, 153, 156, 157, 160, 162, 164, 175, 178, 185, 186, 190, 197, 198, 201–204, 206–209, 227, 244, 248, 252, 265, 266, 281, 284, 286, 291, 294, 295, 299, 301–303, 311
strategic behavior, 29, 51, 265, 266
sulfur
  content, 78, 164, 198, 199, 201, 207, 208, 247
  Second Sulfur Protocol, 2, 5, 8, 249, 250, 274, 277–279
  tax, 79, 205, 207, 209
  trading, 8, 101–105, 115, 121, 130, 131, 148, 149, 157, 170–174, 177, 179–181, 185, 210, 246, 260, 286, 292–295, 306, 310
sulfur dioxide, *see* SO_2

taxes, 17, 32, 46, 64, 65, 71, 75, 76, 78–80, 86, 181, 205, 207, 209, 290
tradable emission permits, 3, 4, 12, 30, 31, 35, 42, 65, 75, 82, 86, 96, 99, 163, 174, 175, 178, 283, 290, 296, 303, 304
tradable permits, 3, 17, 31–34, 36, 37, 39–43, 61, 64, 78, 104, 106, 122, 124, 126, 127, 130–132, 141, 142, 171, 248, 296, 298, 302, 303, 306, 307
trade, *see* single-zone trading, *see* multiple-zone trading
trading rules, 48, 52, 57, 61–63, 93, 95, 97, 132, 134, 138, 140, 176, 179, 181, 284, 304–307
transaction costs, 1, 4, 6, 8, 14–16, 28–31, 38, 42, 50–52, 54–56, 58, 59, 62, 63, 78, 85, 86, 91, 95–97, 99, 104–106, 110, 115, 117, 121, 129, 135–137, 144, 149, 155, 168, 170–173, 176, 178–181, 210, 240, 242, 245, 246, 256–258, 271, 272, 278, 280–285, 287, 289, 290, 293, 294, 296, 300, 303, 304, 306, 307, 309–311

transboundary pollution, 5, 12, 64, 187, 188, 193
transfers, 16, 124, 131, 143, 148, 166–168, 170, 296

uncertainty, 3, 6, 8, 11, 17, 19, 29, 41, 47, 55, 59, 99, 137, 142, 153–155, 168, 170, 172, 178, 179, 281, 289, 306, 307
uniform cutbacks, 2, 47, 118
United Nations, 2, 186–188
US Environmental Protection Agency, *see* EPA
utility, 12–14, 65, 66, 73, 89, 145, 146, 151, 154, 156

welfare, 3, 12–15, 17–19, 41, 66–69, 71, 73, 74, 77, 78, 82, 83, 85–90, 95–97, 127, 192, 288, 300